AGRICULTURAL TERMS

AGRICULTURAL TERMS

as used in the
Bibliography of Agriculture
from data provided by the
National Agricultural Library
U.S. Department of Agriculture.

2nd Edition 1978

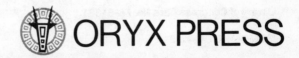 ORYX PRESS

Operation Oryx, started more than 15 years ago at the Phoenix Zoo to save the rare white antelope—believed to have inspired the unicorn of mythology—has apparently succeeded. The operation was launched in 1962 when it became evident that the animals were facing extinction in their native habitat of the Arabian peninsula.

An original herd of nine, put together through *Operation Oryx* by five world organizations now number 47 in Phoenix with another 38 at the San Diego Wild Game Farm, and four others which have recently been sent to live in their natural habitat in Jordan.

Also, in what has come to be known as "The Second Law of Return," rare biblical animals are being collected from many countries to roam freely at the Hai Bar Biblical Wildlife Nature Reserve in the Negev, in Israel, the most recent addition being a breeding herd of eight Arabian Oryx. With the addition of these Oryx, their collection of rare biblical animals is complete.

PREFACE

This edition of **Agricultural Terms** represents an extensive revision and expansion of the vocabulary first published in 1976 with the title **Thesaurus of Agricultural Terms.** The vocabulary is the source for the headings appearing in the subject index of the **Bibliography of Agriculture** and contains approximately 37,000 terms.

The list of terms is organized so that word variants (singular or plural versions of the same word, or different spellings of the same word) are grouped together. These synonymous terms — almost 12,000 of them — appear in parentheses, in italics, following the key term. There are 2,000 cross-reference terms which appear on the line below the key term, with an X-Ref indication. In addition, these cross-reference terms appear in their appropriate alphabetic place within the vocabulary.

The objective has been to include as many significant terms as possible in the authority file — over 23,000 — for a better machine-produced subject index. At the same time, it has not been possible to build a completely consistent structure for the vocabulary. Users will find, for example, that plural or singular terms as well as noun, verb, or adjective forms are used as main headings.

Both natural language and scientific terminology are used, including general (Botany), specific (Ethnobotany, Mycology, Paleobotany), popular (Leafhoppers) and scientific (Cicadellidae) terms.

Scientific names of genera and higher taxa in the animal or plant kingdoms are used as main terms, but not names of lower taxonomic units as species, varieties, etc. Names of certain species, however, are included as main terms because sometimes such names also designate a genus, e.g., Norway spruce, Picea abies and Balsam fir, Abies balsamea.

An attempt was made to list livestock breeds, soil types and cheese types directly. Some of these names consist of two or three words that, although meaningful when used as phrases (e.g., Black soils, Swiss cheese, Red Danish cattle), cannot be included as key words in the single-term vocabulary. Therefore, words like black, red and those having a geographic connotation were omitted. Users should look for citations under soils, cheeses and names of livestock.

Whenever necessary to enhance the retrieval of citations, specific terms have been listed separately as main headings. For example, Ectomycorrhizae, Endomycorrhizae, Ectendomycorrhizae are treated as main terms even though they represent specific types of the general term Mycorrhizae.

General terms that acquire specific meaning when combined with other terms are not listed separately (e.g., composition, analysis, processing, etc.). Users interested in milk composition or soil composition will find pertinent citations under milk and soils.

This subject authority file also controls 35,000 "stop" words, which have been specifically identified as terms which should *not* be included in the vocabulary.

The publishers have used their best efforts to include important key terms in this edition of the vocabulary, but have no legal responsibility for accidental omissions or errors in these listings. We will be grateful for any comments or suggestions, whether favorable or critical, for use in the preparation of future editions.

AGRICULTURAL TERMS

Abaca
Abacarus
Abalone
Abamagenin
Abate
Abattoirs see Slaughter
Abdomen (Abdomens,
    Abdominal)
    X-Ref: Celiac, Coeliac,
    Intraabdominal
Abedus
Abelmoschus see Hibiscus
Aberdeen
Aberration (Aberrant,
    Aberrations)
Abgrallaspis
Abies (Fir)
    X-Ref: Firs
Abietaceae
Abietic
Abiogenesis (Abiogenic)
Abiotic
Abiotrophy
Ablactation see Weaning
Ablation
Ablautus
Abnormality (Abnormal,
    Abnormalities,
    Abnormally, Abnormity,
    Anomalies, Anomalous,
    Anormal, Teratism)
    X-Ref: Anomaly,
    Teratisms
Abomasum (Abomasal,
    Abomasally)
Aborigines (Aboriginal,
    Aboriginals, Aborigine)
Abortion (Aborted,
    Abortifacient, Aborting,
    Abortions, Abortive,
    Miscarriage)
    X-Ref: Miscarriages
Abrasion (Abraded,
    Abrasions, Abrasive)
Abraxas
Abrin
Abroma
Abronia
Abrotanella
Abruptex
Abrus
Abscesses (Abscess)

Abscisate
Abscisic
Abscisin
    X-Ref: Dormin
Abscission (Abscising,
    Abscissing, Abscissions)
Absconditella
Absidia
Absidiomycosis
Abstracting
Abuta
Abutilon (Abutilons)
Acacia (Acacias)
Acacipetalin
Acaena
Acalypha
Acalypta
Acalypterate
Acanthaceae
Acanthamoeba
Acanthaster
Acanthiophilus
Acanthocalycium
Acanthocephala
    (Acanthocephalan)
Acanthocephaliasis
Acanthocinini
    (Acanthocinine)
Acanthocinus
Acanthocreagris
Acanthodrilidae
Acantholepis
Acantholimon
Acantholyda
Acanthomia
Acanthopanax
Acanthophthirius
Acanthoscelides
Acanthosomatidae
Acanthospermum
Acanthosyris
Acanthotermes
Acarapis
Acardia
Acariasis
Acaricides (Acaracide,
    Acaracides, Acaricidal,
    Acaricide)
Acaridae (Acarid,
    Acariform, Acariformes,
    Tyroglyph, Tyroglyphid)
    X-Ref: Tyroglyphidae

Acarina (Acari, Acarian,
    Acarians, Acarida,
    Acarofauna)
Acarine
Acaroceras
Acaroidea (Acaroid)
Acarology
Acaropsis
Acarus
    X-Ref: Tyroglyphus
Acasis
Acaudaleyrodes
Acaulon
Accedin (Accedine)
Accidents (Accident)
Accipiter
Acclimation (Acclimated,
    Acclimatisation,
    Acclimatization,
    Acclimatized,
    Acclimatizing)
Accothion see Fenitrothion
Accounting (Account,
    Accounts, Bookkeeper,
    Bookkeepers)
    X-Ref: Bookkeeping
Acellular see Cells
Acenaphthene
Acentria
Acephate
Acepromazine
Acer (Maple)
    X-Ref: Maples
Aceraceae
Aceratagallia
Acerella
Aceria
Acetabularia
Acetabulum
Acetal (Acetals)
Acetaldehyde
    (Acetaldehydes)
Acetamide (Acetamides)
Acetamidobiphenyls
Acetanilide (Acetanilides)
Acetates (Acetate)
    X-Ref: Monoacetate
Acetazolamide
Acetic
Acetoacetates (Acetoacetate)
Acetobacter
Acetohydroxy

Acetoin
Acetolactate
Acetolysis
Acetomonas
Acetone
Acetonemia (Acetonaemia)
Acetonide
Acetonitrile
Acetonuria see Ketonuria
Acetophenide
Acetophenone
Acetoxy
Acetoxylation
Acetoxymethyl
Acetoxyprogesterone
Acetyl
Acetylacetone
Acetylaminofluorene
Acetylaminomannuronic
Acetylase (Acetylases)
Acetylation (Acetylated)
Acetylcholine (Acetylcholin)
Acetylcholinesterases
    (Acetylcholinesterase)
Acetylcolletotrichin
Acetylene (Acetylenes,
    Acetylenic)
Acetylgalactosamine
Acetylglucosamine
Acetylglutamic
Acetylhexosaminidase
Acetylhomoserine
Acetylindolyl
Acetyllasiocarpine
Acetylneuraminic
Acetylnornantenine
Acetylornithine
Acetylpectulinarin
Acetylpromazine
Acetylricinoleate
Acetylsalicylic
Acetylserine
Acetyltransferase
    (Acetyltransferases)
Achaea
Achaearanea
Achalasia
Achatina
Achenes (Achene, Achenial,
    Achenium, Achenocarp,
    Akenium)
    X-Ref: Akene

Acherontia
Acheta
Achilidae
Achillea
Achimenes
Achlya
Achnanthes
Achoea
Acholeplasma
Achondroplasia
  (Chondrodystrophy)
  X-Ref:
    Chondrodystrophia
Achras see Sapodilla
Achroia
Achromobacter
  (Achromobacters)
Achromobacteraceae
Achyla
Achyranthes
Achyrophorus
Acidemia (Acidaemia)
Acidity (Acidities)
Acidogenesis
Acidolysis
Acidophilic (Acidophil,
  Acidophilous, Aciduric)
Acidophilin
Acidosis (Acidoses, Acidotic)
Acidulation (Acidification,
  Acidified, Acidulant)
  X-Ref: Disacidification
Aciduria
Acilius
Acinetobacter
Acinogaster
Acinus (Acinar)
Aciptilia
Aclerda
Acleris
Aclitus
Acmaea
Acmaeodera
Acne (Acnes)
Acnida
Acnistus
Acokanthera
Acomys
Aconidial see Conidium
Aconitase
Aconitate (Aconitates)
Aconitic
Aconitum (Aconite,
  Aconites)
  X-Ref: Monkshood
Aconurella
Acorenone
Acorns (Acorn)
Acorus
Acosta
Acostaea
Acoustics (Acoustic,
  Acoustical)
Acraea
Acraeidae
Acrasiales

Acrasin
Acreage (Acre, Acreages,
  Acres)
Acremonium
Acrida
Acridae
Acrididae (Acridid,
  Acridids, Acridinae)
Acridine
Acridoidea (Acridoid)
Acridology
Acridomorpha
Acridone (Acridones)
Acriflavine (Acriflavin)
Acripeza
Acrobasis
Acrocarpous
Acrocercops
Acroceridae
Acrochaetium
Acrocylindrium
Acrodermatitis
Acrolein (Acroleins)
Acrolepia
Acrolepiidae
Acrolepiopsis
Acromyrmex
Acroneuria
Acronine
Acronychia
Acronycta
Acronyctoides
Acropetal (Acropetally)
Acroptilon
Acroricnus
Acrosin (Acrosins)
Acrosomes (Acrosomal,
  Acrosome)
  X-Ref: Postacrosomal
Acrostichum
Acrotrichis
Acrotylus
Acrylamide
Acrylate (Acrylates)
Acrylic (Acrylics)
Acrylonitrile
Actaea
Actellic
Actenophthalmus
ACTH
Actia
Actias
Actidinia
Actidione see
  Cycloheximide
Actin
Actinedida
Actinidia
Actinin
Actiniopteris
Actinobacillosis
Actinobacillus
Actinocephalus
Actinodaphne
Actinometry (Actinometer)
Actinomucor

Actinomyces
Actinomycetaceae
Actinomycetales
  (Actinomycete,
  Actinomycetes)
Actinomycin
Actinomycosis
  (Actinomycoses)
Actinophage
Actinoplanaceae
Actinoplanes
Actinopodidae
Actinospora
Actinostachys
Actinostemon
Actinotropism
Actomyosin
Actornithophilus
Aculeata (Aculeate)
Aculops
Aculus
Acupuncture (Acupunctural)
Acyclic
Acyl
Acylamidase
Acylases (Acylase)
Acylation (Acylated,
  Acylating, Acylations)
Acylcarbamates
Acylglucoses see Glucoses
Acylglycerols
Acylhydrolases
Acylomus
Acyrthosiphon
Acytoplasmic see Cytoplasm
Acytostelium
Adactyly
Adaina
Adalia
Adamantane
Adansonia (Adansonian)
Adaptation (Adapt,
  Adaptability,
  Adaptational, Adaptations,
  Adaptive, Adapted,
  Adapting, Adaption,
  Adaptions, Adaptive,
  Adaptiveness, Adapts)
Additives (Additive,
  Additivity)
Adducts (Adduct)
Adelanthus
Adelesta
Adelges
Adelgidae
Adelginae
Adelina
Adelocephalidae
Adelomycetes (Adelomycete)
Adelphocoris
Adenanthera
Adenia
Adenine (Adenin,
  Adenineless, Adenines)
Adenocalymma
Adenocarcinoma

(Adenocarcinomas)
Adenocarpus
Adenocaulon
Adenohypophysectomy
  (Adenohypophysectomized)
Adenohypophysis
  (Adenohypophyseal,
  Adenohypophyses,
  Adenohypophysial)
  X-Ref:
    Intraadenohypophysial
Adenoids (Adenoid)
Adenolobus
Adenoma (Adenomas,
  Adenomatoid,
  Adenomatosis,
  Adenomatous)
Adenophorus
Adenosine (Adenosines)
Adenosinemonophosphate
Adenosinetriphosphatase
  (Adenosinetriphosphatases)
Adenostoma
Adenostyles
Adenosyl
Adenosylcobalamin
Adenosylethionine
Adenosylmethionine
Adenoviruses (Adenoviral,
  Adenovirus)
Adenyl (Adenylate,
  Adenylates, Adenylic)
Adenylyl
Adenylyltransferase
Adephaga
Aderidae
  X-Ref: Hylophilidae
Adesmia
Adhatoda
Adhesion (Adhered,
  Adherence, Adhesions,
  Adhesiveness)
Adhesives (Adhesive)
Adiantaceae
Adiantum
Adina
Adipic (Adipate)
Adipinate
Adipocyte see Trophocytes
Adipokinetic
Adipose (Adiposes, Adiposis,
  Adiposity)
Adistemia
Adjuvants (Adjuvant)
Adnexa (Adnexal)
Adolescence (Adolescent,
  Adolescents, Teen,
  Teenage, Teenagers)
  X-Ref: Preadolescent,
  Teens
Adonis
Adoretus
Adorrhinyptia
Adorrhinyptiini
Adoxa
Adoxinia

Adoxomyia
Adoxophyes
ADP
Adranes
Adrenal (Adrenals)
Adrenalectomy
 (Adrenalectomised,
 Adrenalectomized)
Adrenalin (Adrenaline)
Adrenergic
 X-Ref: Antiadrenergic
Adrenoceptor
Adrenocholinergic
Adrenocortical see Cortex
Adrenocorticoid see
 Corticoids
Adrenocorticosteroid see
 Corticoids
Adrenocorticotropic see
 Corticotrophic
Adrenodoxin
Adriamycin
Adrisa
Adromischus
Adsuki see Beans
Adulteration (Adulterant,
 Adulterants, Adulterated,
 Adulterations)
Adulticides (Adulticide)
Adults (Adult, Adulthood)
Adventitious
Advertising
Advising (Advise,
 Advisement, Advises,
 Advisor, Advisors)
Aechmanthera
Aechmea
Aecidium
Aeciospores (Aeciospore)
Aedeagus
Aedes
Aedimorphus
Aedine
Aegelenine
Aegeriidae
Aegialia
Aegialiini
Aegialina
Aegialomyia
Aegiceras
Aegilops
Aeginetia
Aegiphila
Aegle
Aegyptianella
Aelia
Aellenia
Aelurostrongylus
Aeneolamia
Aenictus
Aeolanthus
Aeolian
Aeolisaccus
Aeolothripidae
Aeonium
Aepophilus

Aequorin
Aerangis
Aeranthes
Aeration (Aerate, Aerated,
 Aerating, Aerator)
 ı X-Ref: Reaeration
Aerial (Aerially)
Aerides
Aerobacter
Aerobes (Aerobe, Aerobic,
 Aerobically, Aerobiosis)
Aerobiology
Aerochemistry
 (Aerochemical)
Aerococcus (Aerococci)
Aerodynamics
 (Aerodynamic,
 Aerodynamical,
 Aerodynamically)
Aerogel
Aerology (Aerological)
Aeromechanics
 (Aeromechanical)
Aeromicrobiology
 (Aeromicrobiological)
Aeromonas (Aeromonads)
Aeromycology
 (Aeromycologic,
 Aeromycological)
Aeropalynology see
 Palynology
Aeropedellus
Aerophagia (Aerophagy)
Aerophobia
Aerophotography
Aerophytes
Aeroplane see Aircraft
Aerosol (Aerosal,
 Aerosolized, Aerosols)
Aerotherapeutics
 (Aerotherapy)
Aerotropism
Aesalinae
Aesalus
Aeschynomene
Aescin
Aesculus
Aeshna (Aeschna)
Aeshnidae (Aeschnidae)
Aeshrion
Aesthetics (Aesthetic)
Aestivation see Estivation
Aethalion
Aethalionidae
Aetheroleum
Aethiolaspis
Aethus
Aethusia
Aetiology see Etiology
Afalon see Linuron
Afference (Afferents)
Affinine
Afforestation (Afforest,
 Afforestations, Afforested)
 X-Ref: Forestation
Aflagellate see Flagellate

Aflatoxins (Aflatoxicosis,
 Aflatoxigenic,
 Aflatoxigenous, Aflatoxin)
Aframomum
Africander
Africaspis
Afrogigagnathus
Afrosteles
Afterbirth see Placenta
Aftercrops see Crops
Afzelin
Agabus
Agalactia (Agalactosis)
Agallia
Agalliopsis
Agallis
Agallol
Agalmyla
Agamermis
Agamic see Sex
Agammaglobulinemia
 (Agammaglobulinaemia,
 Agammaglobulinaemic)
Agamospermy
Agaon
Agaonidae
Agapa
Agapanthia
Agapanthus
Agapostemon
Agar
Agaricaceae (Agaric,
 Agarics)
Agaricales
Agaricus
Agaristidae (Agaristid)
Agaroid
Agarose
Agasicles
Agatharchus
Agathelpis
Agathidium
Agathis
Agathomyia
Agathosma
Agation
Agavaceae
Agave (Agaves)
AGDEX
Agdistis
Agelaius
Agelena
Agelenidae
Agelenopsis
Agelon
Ageneotettix
Ageniaspis
Agenius
Ageratina
Ageratriol
Ageratum
Agglutination
 (Agglutinability,
 Agglutinating,
 Agglutinogen,
 Agglutinogenic)

X-Ref:
 Microagglutination
Agglutinins (Agglutinin)
 X-Ref: Heteroagglutinins
Aggression (Aggressive,
 Aggressiveness,
 Aggressivity)
Agigenin
Aging (Age, Aged, Ageing,
 Ages)
Agis
Agitation (Agitator,
 Agitators)
Aglaia
Aglais
Aglaonema
Aglia
Aglossa
Aglycones (Aglycon,
 Aglycone)
Aglycyderes
Aglycyderidae
Agmatine
Agmenellum
Agnus
Agonini
Agonistic
Agonoderus
Agonum
Agoristenidae
Agouti
 X-Ref: Aguti
Agramma
Agraptocorixa
Agrarianism (Agrarian)
Agrias
Agribusiness
 (Agribusinesses)
Agricol
Agriculturalists
 (Agriculturalist,
 Agriculturist,
 Agriculturists)
Agriinae
Agrilus
Agrimod
Agrimonia
Agrimonolide
Agriocnemis
Agrionidae
Agriostomum
Agriotes
Agriphila
AGRIS
Agrius
Agrobacterium
 (Agrobacteria)
Agrobiology (Agrobiological)
Agrobotany (Agrobotanic,
 Agrobotanical,
 Agrobotanically)
Agrocenosis see Cenosis
Agrochemicals
Agrochemistry
 (Agrochemical)
Agroclavine

Agroclimatology *see*
   Climate
Agrocybe
Agrodiaetus
Agroecology *see* Ecology
Agroeconomy *see*
   Economics
Agroecosystems *see* Ecology
Agroecotypes *see* Ecotypes
Agrohydrology
   (Agrohydrologic,
   Agrohydrological)
Agroindustry
   (Agroindustrial,
   Agroindustries)
Agrologists
Agrology *see* Pedology
Agromelioration *see*
   Melioration
Agrometeorology
   (Agrometeorological)
Agromicrobiology
Agromyza
Agromyzidae *(Agromyzid)*
Agronal
Agronomists *(Agronomist)*
Agronomy *(Agronomic,*
   Agronomical,
   Agronomically,
   Agrophytotechnical,
   Agrotechnical,
   Agrotechnics,
   Agrotechnique,
   Agrotechniques)
   X-Ref:
      Agrophytotechnics,
      Agrotechnology
Agroperlit
Agrophysics *(Agrophysical)*
Agrophytocoenosis *see*
   Cenosis
Agrophytotechnics *see*
   Agronomy
Agropyron *(Agropyrum)*
   X-Ref: Zeia
Agroreclamation *see*
   Reclamation
Agrostemma
Agrostis *(Agrestis)*
Agrotechnology *see*
   Agronomy
Agrotidae
Agrotinae
Agrotis
Agrumenia
Agrypnia
Agrypon
Agtron
Agulla
Aguti *see* Agouti
Ahlesia
Ahnfeltia
Aides
Ailanthus
Aiolopus
Air *(Aired, Airflow, Airing,*

Airless)
Airblast
Airborne *(Airborn)*
Airbrush
Aircooling
Aircraft *(Aeroplanes,*
   Aircrafts, Airplane,
   Airplanes, Airtanker)
   X-Ref: Aeroplane
Airline
Airphotos *see*
   Photogrammetry
Airports *(Airport)*
Airsacculitis
Aizoaceae
Ajania
Ajuga
Ajugasterone
Akabane
Akaganeite
Akaniaceae
Akebia
Akene *see* Achenes
Akinesia *(Akinesis)*
Akinetes *(Akinete)*
Aklomide
Alachlor
   X-Ref: Lasso
Alamethicin *(Alamethicins)*
Alangiaceae
Alangium
Alanine
Alanosine
Alanyl
Alaotypus
Alaptus
Alar *(Aminozid)*
   X-Ref: Aminozide,
      Daminozide, SADH
Alaria
Alate
Alatol
Albatrellus
Albedo
Albertine
Albiflorin *see* Paeoniflorin
Albinism *(Albino, Albinotic)*
Albizzia *(Albizia)*
Albugo
Albumins *(Albumen,*
   Albumin, Albuminate,
   Albuminous, Prealbumin)
   X-Ref: Prealbumins
Albuminuria
Alcaligenes
Alcea
Alces
Alchemilla
Alchemy *(Alchemist)*
Alchimilla
Alcian
Alcidion
Alcohol *(Alcoholic,*
   Alcoholism,
   Alcoholization, Alcohols)
   X-Ref: Nonalcoholic

Alcoholates
Alcoholimeter
Alcoholysis
Aldehydes *(Aldehyd,*
   Aldehyde)
Alderfly
Alders *see* Alnus
Aldicarb
   X-Ref: Temik
Aldimine
Alditols *(Alditol)*
Aldolases *(Aldolase)*
Aldonic
Aldose *(Aldoses)*
   X-Ref: Anhydroaldose
Aldosterone
Aldosteronism
Aldotriouronic
Aldrichina
Aldrin
Alectoria
Alectoris
Alectorobius
Alectra
Alectryon
Aleochara
Alethopteris
Aleuria
Aleuriaxanthin
Aleurites
Aleurocanthus
Aleurodidae *(Aleurodid,*
   Aleurodids)
Aleurodothrips
Aleurolobus
Aleurone *(Aleuronic)*
Aleurothrixus
Alewifes
Alexgeorgea
Aleyrodes
Aleyrodidae
Aleyrodoidea
Alfalfa *(Alfalfas, Lucern,*
   Lucerne)
   X-Ref: Lucernes
Algae *(Alga, Algal,*
   Algologic, Algological,
   Microalga, Microalgal,
   Phycological)
   X-Ref: Algology,
      Microalgae, Phycology,
      Unialgal
Algebra *(Algebraic)*
Algicides *(Algaecide,*
   Algaecides, Algicidal,
   Algicide)
Algin *(Alginate, Alginates,*
   Alginic)
Alginase
Algologists
   X-Ref: Phycologists
Algology *see* Algae
Algotoxins
Alicyclic
Aliesterases *(Aliesterase)*
Alimentation *(Alimentary)*

X-Ref:
   Hyperalimentation,
   Realimentation
Aliphatic
Alipur
Alisma
Alismataceae
Alismatales
Alisols *(Alisol)*
Alizarin
Alkali *(Alkalies, Alkaline,*
   Alkalinity, Alkalinization,
   Alkalinized, Alkalinizing,
   Alkalinous, Alkalis,
   Alkalization, Alkalized)
Alkaloids *(Alkaloid,*
   Alkaloidal)
Alkalosis *(Alkaloidosis,*
   Alkaloses)
Alkan
Alkanals
Alkane *(Alkanes)*
Alkanedicarboxylic
Alkanesulfonamides
Alkanesulfonates
   (Alkanesulfonate,
   Alkanesulphonates)
Alkanoic
Alkenals
Alkenes
Alkenyl
Alkoxy
Alkoxybenzimidazoles
Alkoxyisoxazoles
Alkoxylation
Alkoxylipides *see* Lipids
Alkyds *(Alkyd)*
Alkyl
Alkylamino
Alkylammonium
Alkylation *(Alkylated,*
   Alkylating)
Alkylbenzene
Alkylbenzenesulfonamides
Alkylcoumarins
Alkyldinitrophenol
Alkylene
Alkylguanidines
Alkylisothiouronium
Alkylphenols *(Alkylphenol)*
Alkylphenyl
Alkylphosphorylation *see*
   Phosphorylation
Alkylpyrazine
   (Alkylpyrazines)
Alkylthiophenes
Alkylthiophenylureas
Alkylureas
Allamanda
Allamandeae
Allancastria
Allantochorion
Allantoic
Allantoin
Allantoinase
Allantois *(Allantoides)*

Allatectomy
(Allatectomized)
Alleculidae (Alleculid)
Allele (Alleles, Allelic,
Allelism, Allelocytic,
Allelomorph,
Allelomorphism)
X-Ref: Interallelic,
Isoallelic, Nonallelic
Allelochemics
Allelopathy (Allelopathic)
Allene (Allenes, Allenic)
Allergy (Allergen,
Allergenic, Allergenicity,
Allergens, Allergic,
Allergies, Autoallergic)
X-Ref: Autoallergy,
Parallergic
Allescheria
Allethrin
Allethrolone
Alliaceae
Alliaria
Alliin
Allionia
Allium
Alloantigens see Antigen
Alloantisera
Allochernes
Allocyathin
Allodapula
Alloeochaete
Alloeorhynchus
Allogamy (Allogamic,
Allogamous)
Allogenic (Allogeneic)
Allognosta
Allograft (Allografts)
Allograpta
Allometry (Allometric,
Allometrical)
Allomones
Allomorphism (Allomorphic)
Allomyces
Allomyrina
Allopatry (Allopatric)
Alloperla
Allophane (Allophanic)
Allophycocyanin
Allophyes
Alloplasm (Alloplasmic)
Alloplastic
Alloploids (Alloploid,
Alloploidy)
Allopolyploids
(Allopolyploid,
Allopolyploidization,
Allopolyploidy)
Alloptidae
Allopurinol
Allose
Allosteric
Allotetraploids
(Allotetraploid)
Allothrips
Allothrombium

Allotransplantation see
Transplants
Allotriploids (Allotriploid)
Allotropa
Allotype (Allotypes,
Allotypic)
Alloxan
Alloxysta
Alloys (Alloy)
Allozymes (Allozyme,
Allozymic)
Allspice
Alluaudomyia
Alluvium (Alluvial,
Alluviums)
Allyl (Allylic)
Allylisothiocyanates
(Allylisothiocyanate)
Allylphenols (Allylphenol)
Almeria
Almonds (Almond,
Almondized)
Alnetum
Alnus (Alder)
X-Ref: Alders
Alocasia
Aloe (Aloes)
Aloeides
Aloin
Aloineae
Alopecia
X-Ref: Baldness
Alopecuridine
Alopecurus
Alopex
Aloysia
Alpacas (Alpaca)
Alpechin
Alphafarnesene see
Farnesene
Alphaviruses
Alpheus
Alphitobius
Alphonsea
Alpine
X-Ref: Prealpine
Alpinetin
Alpinia
Alsophila
Alstonia
Alstroemeria
Alternanthera
Alternaria (Alternarioses,
Alternariosis)
Alternariol
Althaea
Althenia
Altica
Alticidae (Alticid, Alticinae)
Altimetry (Altimetric)
Altitude (Altitudes,
Altitudinal, Altitudinally)
Altosid
Altozar
Altrose
Altruism

Alum
Aluminosilicates
(Aluminosilicate)
Aluminum (Aluminium)
Alveolus (Alveolar, Alveoli)
Alycomesis
Alydidae
Alysicarpus
Alysiinae
Alysiini
Alyssum
Alytosporium
Amalo
Amanita
Amanitins (Amanitin,
Amanitine)
Amantadine
Amara
Amaranthaceae
Amaranthin
Amaranthus (Amaranth,
Amaranths)
Amaryllidaceae (Amaryllid,
Amaryllids)
Amaryllis
Amathes
Amatidae
Amauromyza
Amauronematus
Amaurosis
Amaurosoma
Amber
Amberlite
Ambilhar see Niridazole
Ambivina
Amblycerus
Amblycoryphenes
Amblyiulus
Amblyomma
Amblyopia
Amblypelta
Amblypygi
Amblyscirtes
Amblyseiini
Amblyseius
Amblystegium
Amblyteles
Ambrosia
Ambrosiaceae
Ambrosiella
Ambulance
Ambulatory
Amelanchier
Amelioration see
Melioration
Ameloblastoma
Ameloblasts (Ameloblastic)
Amenia
Ameniinae
Amerhinus
Americhernes
Americium
Amerodiscusiella
Ameronothrus
Ameroseius
Ametryne

Amia
Amiben see Chloramben
Amic
Amidases (Amidase)
Amides (Amide)
Amidination
Amidines (Amidine)
Amidinotransferase
Amidolyase
Amidon
Amidophos
Amidopyrine
Amidostomum
Amidotransferases
(Amidotransferase)
Amines (Amination, Amine)
X-Ref: Superamine
Aminimides
Amino (Aminoacid,
Aminoacids)
Aminoacidemia
Aminoacyl (Aminoacylation)
Aminoacylases
(Aminoacylase)
Aminoadipic
Aminoalkyl
Aminobenzoic
Aminobutane
Aminobutenolide
Aminobutyrate
Aminobutyric
Aminocyclopropanol
Aminoelaidic
Aminoethanol
Aminoethyl
Aminoethylphosphonic
Aminoglutethimide
Aminohydrolases
(Aminohydrolase)
Aminoimidazole
Aminoisobutyric
Aminolevulinate
Aminolevulinic
Aminomethane
Aminomethyl
Aminomethylphenyl
Aminonicotinamide
Aminopeptidases
(Aminopeptidase)
Aminophenol
Aminophenyl
Aminopherase
Aminopolysaccharides
(Aminopolysaccharide)
Aminopropionic
Aminopropionitrile
Aminoproteinases
Aminopterin
Aminopurines (Aminopurin,
Aminopurine)
Aminopyridine
Aminopyrine
Aminosidine
Aminosuccinamic
Aminosugar
Aminotransferases

(Aminotransferase)
Aminouracil
Aminozide *see* **Alar**
Amiodarone
Amiphos
Amitermes
Amitermitinae
Amitosis
Amitrole *(Aminotriazole,*
Amitrol)
    X-Ref: Weedazol
Amitus
Ammi
Ammobates
Ammoecius
Ammonia *(Ammoniac,*
Ammoniacal, Ammonical)
Ammoniate *(Ammonation)*
Ammonification
(Ammoniated,
Ammoniation,
Ammonified,
Ammonisation,
Ammonization,
Ammonized)
Ammonium
Ammonolysis
Ammophila
Ammophos
Ammotragus
Amnesia
Amnion *(Amniotic)*
Amniota *(Amniote,*
Amniotes)
Amodiaquine
Amoeba *(Amoebae,*
Amoebic)
Amoebe *(Amoebal)*
Amoebiasis
Amoebidae
Amoebotaenia
Amoora
Amorpha
Amorphacarus
Amorphigenin
Amorphococcus
Amorphoidea
Amorphol
Amorphophallus
Amorphotheca
Amortization
Amoxycillin
AMP
Ampelamus
Ampeloglypter
Ampelomyces
Ampelopteris
Ampere
Amperometry
(Amperometric)
    X-Ref: Biamperometric
Amphetamines
(Amphetamine)
Amphibian *(Amphibians,*
Amphibious)
Amphibolus

Amphicarpaea
Amphidinium
Amphidiploids
(Amphidiploid,
Amphidiploidization)
Amphientomidae
Amphigymnia
Amphimallon
Amphimallus
Amphimerus
Amphinemura
Amphipathic
Amphiploids *(Amphiploid,*
Amphiploidy,
Amphipolyploid)
Amphipoda
Amphipoea
Amphipyra
Amphipyrinae
Amphisphaeria
Amphistomes
Amphistomiasis
Amphitecna
Amphitornus
Amphora
Amphoridium
Amphorogyneae
Amphorophora
Amphotericin
Ampicillin
    X-Ref: Binotal
Amplifiers *(Amplifier)*
Amplinus
Amprol
Amprolium
Ampullae
Amputation *(Amputations)*
Amrasca
Amsacta
Amsinckia
Amsonia
Amycelial *see* **Mycelia**
Amygdalar
Amygdalin
Amygdalus
Amyl
Amylascus
Amylases *(Amylase)*
Amylemia
Amylodextrin
Amyloglucosidase
Amylogram
Amylography *(Amylograph)*
Amyloidosis *(Amylosis)*
Amyloids *(Amyloid)*
Amylolysis *(Amylolytic)*
Amylopectin
Amyloplast
Amylose
Amynothrips
Amyrin
Anabaena
Anabasine *(Anabasin)*
Anabasis
Anabenopsis *(Anabaenopsis)*
Anabiosis

Anabolism *(Anabolic,*
Anabolics)
Anabrus
Anacampsis
Anacamptis
Anacardiaceae
Anacardium
Anacassis
Anacharis
Anacridium
Anactinotrichida
Anacystis
Anadevidia
Anaea
Anaerobes *(Anaerobe,*
Anaerobic, Anaerobically,
Anaerobiosis, Anerobic)
Anaeroplasma
Anagallis
Anagasta
Anagotus
Anagyrine
Anagyrus
Analcime
Analeptics
Analges
Analgesics *(Analgesia,*
Analgesic)
Analgidae
Analgoidea
Analyzers *(Analyser,*
Analyzer)
    X-Ref: Microanalyser
Anamnesis *(Anamnestic)*
Ananas
Anaphaeis
Anaphase
Anaphes
Anaphothrips
Anaphrodisia
Anaphylaxis *(Anaphylactic)*
Anaplasma
Anaplasmosis
Anapsectra
Anaptychia
Anarsia
Anarthrophyllum
Anas
Anaspis
Anastatus
Anastomoses *(Anastomose,*
Anastomosed,
Anastomosis)
    X-Ref: Enteroanastomosis
Anastrepha
Anatidae *(Anatid)*
Anatis
Anatomists *(Anatomist)*
Anatomy *(Anatomic,*
Anatomical, Anatomico,
Anatomicopathological,
Anatomo)
    X-Ref: Microanatomy
Anatoxin
Anaulaceae
Anaulacomera

Anax
Ancestral *see* **Progeny**
Anchoviella
Anchovies *(Anchovy)*
Anchusa
Ancistrocerus
Ancistrocladidine
Ancistrocladine
Ancistrocladisine
Ancistrocladus
Ancistropsylla
Anconeal *see* **Elbows**
Ancylis
Ancylol *see* **Disophenol**
Ancylopus
Ancylostoma
Ancylostomatidae
Ancylostomiasis
    X-Ref: Ankylostomiasis
Ancylostomidae
Ancymidol
Ancyrophora
Ancyrophorus
Andagistemus
Andamino
Andesite *(Andesitic)*
Ando
Andosol *(Andosols)*
Andrachne
Andranodorus
Andrena
Andrenidae
Andrenosoma
Andricus
Androctonus
Androcymbium
Androecium *(Androecious)*
Androgen *(Androgenesis,*
Androgenic,
Androgenization,
Androgenized,
Androgenous, Androgens)
    X-Ref: Antiandrogens
Androgenetics *see* **Genetics**
Andrographis
Androlaelaps
Andrology *(Andrological)*
Andropogon
Andropogoneae
Androsace
Androst
    X-Ref: Dhomoandrost
Androstane *(Androstan,*
Androstanes)
Androstene
Androstenedione
Androstenone
Androsterility *see* **Sterile**
Anechura
Aneilema
Anemia *(Anaemia,*
Anaemias, Anaemic,
Anemias, Anemic)
Anemomenotaxis
(Anemomenotactic)
Anemometry *(Anemometer)*

Anemone *(Anemones)*
Anemonin
Anemonopsis
Anemopsis
Anemotaxis
Anergy *(Anergics)*
Anesthesia *(Anaesthesia,*
  Anaesthesiology,
  Anaesthetic, Anaesthetics,
  Anaesthetised,
  Anaesthetization,
  Anaesthetized,
  Anesthesiology, Anesthetic,
  Anesthetics,
  Anesthetization,
  Anesthetized,
  Anesthetizing)
  X-Ref: Unanesthetized
Anestrus *see* Estrus
Anethole
Anethum
Aneudihaploids
Aneuhaploids
Aneuploids *(Aneuploid,*
  Aneuploidy)
Aneuraceae
Aneurin *see* Thiamine
Aneurophytales
Aneurus
Aneurysm *(Aneurysmal,*
  Aneurysms)
Aneusomaty
Angelica
Angina
Angiocardiography
Angiography *(Angiogram)*
Angiology
Angioma *(Angiomatic)*
Angiometopa
Angioneurosis
  (Angioneurotic)
  X-Ref: Vasoneurosis
Angiopteris
Angiosarcoma
Angiosorus
Angiospermae *(Angiosperm,*
  Angiospermatophyta,
  Angiospermatophyte,
  Angiospermic,
  Angiospermous,
  Angiosperms)
Angiostrongyliasis
Angiostrongylus
Angiotensin
Angophora
Angora
Angraecum
Anguillulidae
Anguina
Angulitermes
Angus
Angustalius
Anhalonidine
Anhydrase *(Anhydrases)*
Anhydrides *(Anhydride)*
Anhydrite

Anhydroaldose *see* Aldose
Anhydroerythritol *see*
  Erythritol
Anhydropentoses *see*
  Pentoses
Anhydroperforine *see*
  Perforine
Anhydrosugars *see* Sugar
Anhydrous
Aniba
Anigozanthos
Anilides *(Anilide)*
Anilines *(Aniline)*
Anilinonaphthalene
Animals *(Animal)*
  X-Ref: Beast
Animomyia
Anionites
Anisacanthus
Anisakiasis
Anisakis
Anise
Aniseia
Anisocentra
Anisodontea
Anisole
Anisomeles
Anisomitochondrial *see*
  Mitochondria
Anisomycin
Anisoploids *(Anisoploid,*
  Anisoploidy)
Anisops
Anisoptera
Anisopteromalus
Anisostephus
Anisota
Anisotome
Anisotomidae
Anisotropy *(Anisotropic,*
  Anisotropically)
Anit
Anitra
Aniulus
Ankistrodesmus
Ankle *see* Tarsus
Ankylosis *(Ankylosing)*
Ankylostomiasis *see*
  Ancylostomiasis
Annatto
Annelidae *(Annelid,*
  Annelids)
Annona
Annonaceae
Annuals
Annularia
Annulus
Anobiidae *(Anobiid,*
  Anobiids)
Anobium
Anocentor
Anode *(Anodal, Anodic)*
Anodendron
Anodonta
Anodynon *see* Chlorethyl
Anoestrus *see* Estrus

Anoetidae
Anogramma
Anoidiella
Anolina
Anomala
Anomalagrion
Anomalinae
Anomaly *see* Abnormality
Anomer *(Anomeric)*
Anomotaenia
Anopheles *(Anopheline,*
  Anophelines, Anophelism)
Anophelinae
Anophthalmia
  (Anophthalmic)
Anoplius
Anoplocephala
Anoplocephalidae
Anoplocnemis
Anoplognathus
Anoplonyx
Anoplura
Anopterus
Anorexia *(Anorectic,*
  Anorexigenic)
Anosmia
Anotylus
Anoxia *see* Hypoxia
Anser
Anserine
Antagonism *(Antagonisms,*
  Antagonist, Antagonistic,
  Antagonists)
Antechinus
Antelopes *(Antelope)*
Antemortem *see* Death
Antenatal *see* Birth
Antenna *(Antennae,*
  Antennal)
Antennapedia
Antennaria
Antennectomy
Antestia
Antestiopsis
Anthaxia
Anthaxiini
Anthelmintics
  (Anthelminthic,
  Anthelmintic)
Anthemideae
Anthemis
Anthephora
Antheraea
Antheridiogens
Antheridiol
Antheridium *(Antheridia,*
  Antheridial)
Anthers *(Anther)*
Anthesis *see* Flowering
Anthia
Anthicidae
Anthicus
Anthidium
Anthocaris
Anthocephalus
Anthocercis

Anthoceros
Anthocerotae
Anthocharis
Anthocleista
Anthocoptes
Anthocoridae *(Anthocorid)*
Anthocoris
Anthocyanidins
  (Anthocyanidin,
  Proanthocyanidin)
  X-Ref: Proanthocyanidins
Anthocyanins *(Anthocyan,*
  Anthocyanin,
  Anthocyaninless,
  Anthocyanized,
  Anthocyans)
Anthocyanogens
  (Anthocyanogen)
Anthodium
Anthomyiidae
  (Anthomyidae,
  Anthomyiid)
Anthomyzidae
Anthonominae
  (Anthonomine)
Anthonomus
Anthophila
Anthophily
Anthophora
Anthophoridae
Anthospermum
Anthostrongylus
Anthoxanthum
Anthracene
Anthracite
Anthracnose
Anthracobia
Anthranilate
Anthranilic
Anthraquinones
  (Anthraquinone,
  Anthraquinonic)
Anthraquinonoids
Anthrenus
Anthribidae *(Anthribid)*
Anthriscus
Anthrones *(Anthrone)*
Anthropogenesis
  (Anthropogenic,
  Anthropogenous)
Anthropologists
  (Anthropoligist)
Anthropology
  (Anthropological)
Anthropometry
  (Anthropometric)
Anthurium *(Anthuriums)*
Anthyllis
Antiadrenergic *see*
  Adrenergic
Antiandrogens *see* Androgen
Antiaphrodisiac *see*
  Aphrodisiac
Antiarrhythmic *see*
  Arrhythmia
Antibacterial *see*

Bactericides
Antibiogramy
  (Antibiograms)
Antibiosis
Antibiotherapy
Antibiotic (Antibiotically,
  Antibiotics, Antimicrobial,
  Antimicrobic,
  Microbiocidal)
  X-Ref: Antimicrobials,
  Microbicidal
Antibodies (Antibody,
  Isoantibody)
  X-Ref: Autoantibodies,
  Isoantibodies
Antibrachium see Limbs
Antibrucellar see
  Brucellosis
Anticancer see Cancer
Anticarsia
Anticellular see Cells
Antichlor
Anticholinesterase see
  Cholinesterases
Anticoagulants see
  Coagulants
Anticoagulation see
  Coagulation
Anticoccidials see
  Coccidiostat
Anticodon see Codon
Anticonvulsant see
  Convulsions
Anticorrosives see
  Corrosion
Anticrustants see Crusts
Anticytokinins see
  Cytokinins
Antidesiccants see
  Desiccants
Antidiphtheria see
  Diphtheria
Antidiuretic see Diuretics
Antidotes (Antidote)
Antidromy (Antidromic)
Antierosion see Erosion
Antiesterase see Esterases
Antiestrogens see Estrogen
Antiethylene see Ethylene
Antiexudative see Exudates
Antifeedants see Repellents
Antifertility see
  Contraceptives
Antifoaming see Foam
Antifolate
Antifreeze
Antifriction see Friction
Antifungal see Fungicides
Antigastra
Antigen (Alloantigenic,
  Antigenic, Antigenically,
  Antigenicities,
  Antigenicity, Antigens,
  Isoantigen)
  X-Ref: Alloantigens,
  Isoantigens

Antigibberellins see
  Gibberellins
Antiglobulins see Globulins
Antigonadotropin see
  Gonadotropins
Antihemagglutinins see
  Hemagglutinins
Antihematopoietic see
  Hematopoiesis
Antiherbivore see
  Herbivores
Antihistamines see
  Histamine
Antihormones see
  Hormones
Antihypercholesterol see
  Cholesterol
Antihypertensive see
  Hypertension
Antiinflammatory see
  Inflammation
Antiketogenic see
  Ketogenesis
Antiknock
Antileptospira see
  Leptospirosis
Antileukemic see Leukemia
Antilymphocyte see
  Lymphocytes
Antimacrophage see
  Macrophages
Antimalarial see Malaria
Antimetabolic see
  Metabolism
Antimetabolites see
  Metabolites
Antimicrobials see
  Antibiotic
Antimicrotubule see
  Tubules
Antimitotic see Mitosis
Antimony
Antimoulting see Molt
Antimutagenic see
  Mutagenesis
Antimycin
Antimycobacterial see
  Mycobacteriaceae
Antimycotic see Mycosis
Antineoplastic see
  Neoplasm
Antinutrients see Nutrients
Antinutritional see
  Nutrition
Antio see Formothion
Antiotensin
Antioxidation (Antioxidant,
  Antioxidants,
  Antioxidative,
  Antioxidizing)
Antiparasitic see Parasites
Antiperistaltic see
  Peristalsis
Antiphytopathogenic see
  Phytopathogens
Antiplatelet see Platelets

Antipleuropneumonia see
  Pleuropneumonia
Antipneumococcal see
  Diplococcus
Antipodes (Antipodal)
Antipollutants see
  Pollutants
Antiproteolysis see
  Proteolysis
Antiprotozoal
Antipyretic see Fever
Antirabies see Rabies
Antirachitic see Rickets
Antirinderpest see
  Rinderpest
Antirrhinum
Antisaccharine see
  Saccharine
Antisclerotic see Sclerosis
Antiscorbutic see Scurvy
Antisecretory see Secretion
Antiseepage see Seepage
Antiseptic (Antisepsis,
  Antiseptics)
Antiserum (Antisera,
  Antiserums)
Antispasmodic see Spasms
Antisporulant see
  Sporulation
Antistress see Stress
Antisuppressor see
  Suppression
Antiteuchus
Antithamnion
Antithiamine see Thiamine
Antithrombin see Thrombus
Antithymocyte see
  Thymocyte
Antithyroid see Thyroid
Antitoxic see Toxicity
Antitoxicants see Toxins
Antitoxicum
Antitoxoplasmic see
  Toxoplasmosis
Antitranspirants see
  Transpirants
Antitranspiration see
  Transpiration
Antitrust
Antitrypsin see Trypsin
Antituberculous see
  Tuberculosis
Antitumor see Tumor
Antitussive see Cough
Antityphus see Typhus
Antiulcerogenic see Ulcers
Antiurease see Urease
Antivaccina see Cowpox
Antivasopressin see
  Vasopressin
Antivermal see Vermin
Antiviral see Viruses
Antivirulent see Virulence
Antler
Antonina
Antrum (Antral)

Antrycide
Ants (Ant)
Anucleate see Nucleus
Anuraphis
Anuria
Anus (Anal)
  X-Ref: Circumanal,
  Perianal
Anystidae
Anystis
Anzia
Aonidiella
Aorta (Aortal, Aortas,
  Aortic)
  X-Ref: Subaortic
Aorticopulmonary
Aortography
Aortostenosis
Aotes (Aotus)
Aoudads (Aoudad)
Apacheiulus
Apamea
Apanteles
Apatania
Apathogenic see
  Pathogenesis
Apatite (Apatites)
Apatolestes
Apatrobus
Apatura
Apechthis
Apera
Aperture (Apertures)
Apes (Ape)
Apethymus
Apex (Apical, Apices)
Aphaenogaster
Aphaereta
Aphagia (Aphagic)
Aphananthy
Aphaniptera see
  Siphonaptera
Aphanizomenon
Aphanoascus
Aphanocapsa
Aphanocephalus
Aphanogmus
Aphanomyces
Aphanomycopsis
Aphanothece
Aphantopus
Aphasmatylenchus
Aphasmidia
Aphelandra
Aphelenchoides
Aphelenchoididae
Aphelenchus
Aphelidium
Aphelinidae
Aphelinus
Aphelocheirus
Aphelochirus
Aphicides (Aphicide)
Aphididae (Aphidae,
  Aphidiidae, Aphidiinae)
Aphidina

Aphidius
Aphidoletes
Aphidophages
 (Aphidophagous)
Aphids (Aphid, Aphidoidea)
Aphileus
Aphis
Aphodiidae
Aphodiinae
Aphodius
Apholate
Aphonopelma
Aphotic
Aphrodes
Aphrodina
Aphrodisiac
 X-Ref: Antiaphrodisiac
Aphrophora
Aphthosin
Aphylla
Aphyllophorales
 (Aphyllophorous)
Aphytis
Apiaceae
Apiaries (Apiarian, Apiary,
 Beehouse)
 X-Ref: Beehouses
Apiarists
Apicomplexa
Apiculture (Apicultural)
Apiculturists (Apiculturist)
Apiculus
Apidae
Apigenin
Apiocera
Apioceridae
Apioideae
Apiomerus
Apion
Apionidae (Apioninae)
Apioporthe
Apiose
Apiosphaeria
Apis
Apitong
Apium
Aplasia (Aplastic)
Apluda
Apnea
Apocarbonic see Carbons
Apocarpy see Carpel
Apocheima
Apochthonius
Apocroce
Apocynaceae
 (Apocynaceous)
Apocynum
Apodachlya
Apodemus
Apoenzyme see Enzymes
Apoferritin see Ferritin
Apogamia see
 Parthenogenesis
Apoidea
Apolipoproteins see
 Lipoproteins

Apollophanes
Apoloniinæ
Apolysis
Apomixis see
 Parthenogenesis
Apomorphine see Morphine
Aponogeton
Aponogetonaceae
Apoperoxidase see
 Peroxidases
Apoplast
Apoplastocyanin see
 Plastocyanin
Apoplexy
Apoproteins see Enzymes
Aporis
Aporodesmus
Aporphine (Aporphines)
Aposematism (Aposematic)
Apospory
Apostasiaceae
Aposymbiotic see Symbiosis
Apote
Apothecium (Apothecia,
 Apothecial)
Apovitellenin
Apparel see Clothing
Appendages (Appendage)
 X-Ref: Microappendages
Appetite (Appetites)
Appetizers (Appetizer)
Apples (Apple)
Applesauce
Appliance (Appliances)
Applicators (Applicator)
 X-Ref: Microapplicator
Appraisal (Appraisals,
 Appraise, Appraised,
 Appraisers, Appraising)
Apprentice (Apprentices,
 Apprenticeship,
 Apprenticeships)
Appressorium (Appressoria)
Appropriations
 (Appropriation)
Apricots (Apricot)
Apriona
Aprionus
Aprotic
Apterae
Apterin
Apterous see Wing
Apterygota (Apterygote)
Aptosimum
Apyrases (Apyrase)
Aquaculture see Aquiculture
Aquarium
Aquaspirillum
Aquatic (Aquatics)
 X-Ref: Semiaquatic,
 Subaquatic
Aqueducts (Aqueduct)
Aqueous
 X-Ref: Nonaqueous
Aquiculture (Aquacultural,
 Aquicultural)

X-Ref: Aquaculture
Aquifers (Aquifer)
Aquifoliaceae
Aquilaria
Aquilegia
Arabidopsis
Arabinan
Arabino
Arabinofuranose
Arabinofuranosidase
Arabinofuranosides
 (Arabinofuranoside)
Arabinogalactan
Arabinoglucuronoxylan
Arabinose (Arabinosic)
Arabinoside
Arabis
Arabitol
Arabogalactan
Arabonic
Araceae (Aroids)
Arachidene
Arachidonate
Arachidonic
Arachidonoyl
Arachin
Arachis
Arachnida (Arachnid,
 Arachnids, Arachnoidea)
Arachnidism
Arachniodes
Arachnomelia
Arachnomyces
Aradidae
Aradus
Araeothrips
Aragonite
Araldite
Aralia
Araliaceae
Araliopsis
Aramite
Aranda
Aranea
Aranei
Araneida (Araneae,
 Araneid)
Araneidae
Araneus
Araucaria
Araucariaceae
Araucarioxylon
Araujia
Arazate
Arboreal
Arboretum (Arboreta,
 Arboretums)
Arboricides see Silvicides
Arboriculture
 (Arboricultural)
Arboridia
Arborifera
Arborine
Arbors
Arboviruses (Arboviral,
 Arbovirus)

Arbuscule (Arbuscular)
Arbutin
Arbutus
Arceuthobium
Archaeopteridales
Archaeopyles
Archangiopteris
Archeology (Archaeological,
 Archaeology,
 Archeological)
Archibaccharis
Archichauliodes
Archiearis
Archips
Architectonics
Architects (Architect)
Architecture
Archives (Archive)
Arctagrostis
Arctia
Arctiidae
Arctium
Arctocorisa (Arctocorixa)
Arctopoa
Arctostaphylos
Arctotheca
Ardeicola
Ardisia
Ardostachys
Areca
Arecaceae
Arecoline
Arecophaga
Areflexia (Areflexive)
Arenaria
Arenarin
Arenaviruses
Arenga
Arenophthalide
Aresin see Monolinuron
Aretan
Arethusana
Aretit
Argas
Argasidae (Argasid)
Argemone
Argenine
Argentation
Argidae
Argillaceous see Clay
Arginase (Arginaseless)
Arginine (Arginin)
Argininosuccinate
Arginyl
Argiope
Argiopidae
Argon
Argynnis
Argyope
Argyranthemum
Argyresthia
Argyresthiidae see
 Yponomeutidae
Argyrodendron
Argyrodes
Argyrogramma

Argyroneta
Argyroploce
Argyrotaenia
Argyroxiphium
Arhar see Pigeonpeas
Arica
Arichanna
Aricia
Arid (Aridity, Aridness)
    X-Ref: Semiarid, Subarid
Aridelus
Aridius
Arils (Aril)
Ariocarpus
Arion
Arisarum
Aristapedia
Aristeromycin
Aristida
Aristocracy
Aristolochia
Aristolochic
Arkhar
Armadillidium
Armadillo
Armagnac see Brandies
Armature
Armeria
Armigeres
Armillaria
Armoracia
Army
Armyworms (Armyworm)
Arnica
Aroclor (Aroclors)
Aroga
Aroma (Aromas, Aromatic,
    Aromatics, Aromatizable,
    Aromatization,
    Aromatizers)
Aronia
Arracacia
Arrenurus
Arrest
Arrhenatherum
Arrhopalites
Arrhythmia
    (Antiarrhythmics,
    Arrhythmias, Arrhythmic,
    Arrhythmically,
    Arrhythmogenic)
    X-Ref: Antiarrhythmic
Arrowhead see Sagittaria
Arrowleaf see Sagittaria
Arsanilate
Arsanilic
Arsenic (Arsenate,
    Arsenates, Arsenical,
    Arsenicals, Arsenite)
Arsonic
Artabotrys
Artemia
Artemidiol
Artemisia
Artemisietalia
Arteries (Arteria, Arterial,

Artery)
Arteriography
Arterioles (Arteriole)
Arteriosclerosis
    (Arteriosclerotic)
Arteriovenous
Arteritis
    X-Ref: Periarteritis,
    Polyarteritis
Artesian
Arthritis (Arthritic,
    Arthritides)
    X-Ref: Polyarthritis
Arthrobacter
    (Arthrobacters)
Arthrobotrys
Arthrochilus
Arthroconidia
Arthroderma
Arthrodesis
Arthrodesmus
Arthrodia (Arthrodial)
Arthrogenesis
Arthrography
Arthrogryposis
Arthropathy (Arthropathia,
    Arthropathies)
Arthroplasty
Arthropoda (Arthropod,
    Arthropods,
    Microarthropod)
    X-Ref: Microarthropods
Arthroschista
Arthroscopy
Arthrosis (Arthrosic)
Arthrospores (Arthrospore,
    Arthrosporic,
    Arthrosporous)
Articerodites
Artichokes (Artichoke)
Articulata
Articulation (Articular,
    Articulated)
Articulus
    X-Ref: Intraarticular
Artifacts (Artefact,
    Artefacts, Artifact)
Artocarpus
Artyfechinostomum
Aruba
Arum
Arumalon
Arundina
Arundinaceae
Arundinaria
Arundinoideae
Arundo
Arvicola
Arxiella
Aryl
Arylacetic
Arylamidases (Arylamidase)
Arylbenzofuran
Arylcarbamate
Arylene
Arylesterases (Arylesterase)

Aryloxyprostaglandin
Arylsulfinyl
Arylsulfonyl
Arylsulphatases
    (Arylsulphatase)
Arylterpenoid
Arylthiourea
Arylurea
Arzama
Asaphes
Asarkina
Asarone
Asarum
Asbestos
Ascalaphus
Ascaricides (Ascaricide)
Ascaridae (Ascarid,
    Ascarids)
Ascaridia
Ascaridiasis (Ascariasis)
Ascaridoidea
Ascaris
Ascarops
Aschersonia
Aschiza
Asci (Ascitic, Ascus)
Ascia
Ascidae
Ascidiacea (Ascidians)
Ascites (Ascite)
Asclepiadaceae
Asclepias
Asclepin
Ascobolaceae
Ascobolus
Ascocarps (Ascocarp)
Ascocenda
Ascochyta
Ascofuranone
Ascomycetes (Ascomycete,
    Ascomycetous)
Ascophanus
Ascophyllum
Ascorbate
Ascorbic
Ascorbyl
Ascosphaera
Ascospores (Ascospore,
    Ascosporic,
    Ascosporogenesis)
Ascotis
Ascozonus
Aselgeoides
Asellus
Asemum
Asepsis (Aseptic,
    Aseptically)
Aserica
Asexual see Sex
Ashes (Ash, Ashing)
Asiago
Asias
Asilidae (Asilid)
Asilus
Asimina
Askanian

Asobara
Asolcus
Aspalathus
Asparagaceae
Asparagate
Asparaginase
Asparagine (Asparaginic)
Asparaginyl
Asparagopsis
Asparagus
Asparagusate
Asparate
Aspartase
Aspartic
Aspartyl
Aspasia
Aspens (Aspen)
Asperenone
Aspergillin
Aspergilloma
Aspergillomarasmine
    (Aspergillomarasmin)
Aspergillopeptidase
Aspergillosis
Aspergillus (Aspergilli,
    Aspergillic)
Asperisporium
Aspermy
Asperula
Asperumine
Asphalt (Asphaltic)
Asphodelus
Asphondylia
Asphyxia (Asphyxiation,
    Suffocation)
    X-Ref: Suffocating
Aspiculuris
Aspidiaceae (Aspidiaceous)
Aspidiotini
Aspidiotus
Aspidolea
Aspidosperma
Aspidotis
Aspilota
Aspiration (Aspirated,
    Aspirates, Aspirations)
Aspirators (Aspirator)
Aspirin
Aspleniaceae
Asplenium
Asporogenetic see
    Sporogenesis
Assay (Assayed, Assaying,
    Assays, Bioassays,
    Biotests)
    X-Ref: Bioassay, Biotest,
    Microbioassay
Asses (Ass)
Assets (Asset)
Astasia
Astata
Astatinae
Astaxanthin
Asteraceae
Astereae
Asterella

Asterionella
Asterocampa
Asterolecaniidae
Asteromella
Asteronema
Asters *(Aster)*
Asthenia
Asthma
Astigmata *(Astigmates)*
Astilbe *(Astilbes)*
Astomum
Astraea
Astraeus
Astragalus
Astrebla
Astriazine
Astrocaryum
Astrocytes *(Astrocyte)*
Astroloba
Astronia
Astronomy
Astrophytum
Asulam *(Asulox)*
Asymbiotic *see* Symbiosis
Asymmetrasca
Asymptomatic *see*
 Symptoms
Asynapsis *see* Synapsis
Asynchronous *see*
 Synchronization
Asystasia
Ataenius
Atalantia
Atavism
Ataxia
Atelespoda
Atethmia
Athalamia
Athanasia
Atherigona
Atherix
Atherogenesis *(Atherogenic)*
Atherosclerosis
 (Atherosclerotic)
Atheta
Athiorhodaceae
Athletes *(Athletic)*
Athous
Athripsodes
Athymia *(Athymic)*
Athyreosis *(Athyreotic)*
Athyriaceae
Athyrioideae
Athyrium
Atimia
Atisane
Atlantoaxial
Atlases *(Atlas)*
Atmometers *(Atmometer)*
Atmosphere *(Atmospheres,*
 Atmospheric)
Atomaria
Atomization *(Atomize,*
 Atomizer, Atomizers)
Atoms *(Atom, Atomic)*
Atopomelidae

Atopotrophos
Atopy *(Atopic)*
ATP
ATPases *(ATPase)*
Atrachya
Atractomorpha
Atractotomus
Atractyloside
Atranorin
Atraphaxis
Atraton *(Atratone)*
Atrazine *(Atrazin)*
 X-Ref: Gesaprim
Atresia
Atrichonotus
Atrinal
Atrioventricular
Atriplex
 X-Ref: Saltbush
Atrium *(Atria, Atrial)*
Atropa
Atropellis
Atrophy *(Atrophic,*
 Atrophied)
Atropine
Atta
Attacidae
Attacus
Attagenus
Attaphila
Attapulgite
Attelabidae
Atteva
Attheya
Attini
Attractants *(Attractant,*
 Attracting)
 X-Ref: Coattractants
Aturidae
Atylotus
Atyphloceras
Atypia *(Atypism)*
Atypidae
Auchenorrhyncha
Auchmeromyia
Aucoumea
Auctions *(Auction,*
 Auctioneer)
Aucuba
Aucubin
Audiospectrography
 (Audiospectrographic)
Audiovisual
Audit *(Auditing, Auditor,*
 Audits)
Auditory *see* Hearing
Augasma
Augers *(Auger)*
Augochlora
Augochlorini
Augosoma
Aujeszky
Aulacaspis
Aulacodiscus
Aulacomnium
Aulacophora

Aulacorthum
Aulax
Aulocara
Aulolaimoides
Aulolaimoididae
Aulonia
Aulonocneminae
Aulosira
Auplopus
Aural *see* Ears
Aurantioideae
Auraptene
Aureobasidium
Aureofungin
Aureomycin *see*
 Chlortetracycline
Auricle *(Auricles, Auricular)*
Auriculariales
Aurintricarboxylic
Aurones
Aurosomes
Aurothioglucose
Auscultation
Ausimi
 X-Ref: Ossimi
Ausobskya
Austdiol
Australorp
Australotettix
Austrobaileya
Austrobaileyaceae
Austrocephalocereus
Austrolestes
Austroperlidae
Austrosimulium
Austrothrips
Autecology *(Autecological)*
Autoactivation
Autoallergy *see* Allergy
Autoanalyzer *(Autoanalyser)*
Autoantibodies *see*
 Antibodies
Autobasidia *(Autobasidium,*
 Holobasidium)
 X-Ref: Holobasidia
Autobiography *see*
 Biography
Autocatalytic *see* Catalysis
Autoclave *(Autoclavable,*
 Autoclaved, Autoclaving)
Autodegradation *see*
 Degradation
Autoecism *(Autoecious)*
Autofertility *see* Fertility
Autofertilization *see*
 Fertilization
Autoflocculation *see*
 Flocculation
Autofluorescence *see*
 Fluorescence
Autogamy *(Autogamous)*
Autogenesis *(Autogenetic,*
 Autogenity, Autogenous,
 Autogeny)
Autographa
Autohexaploid *see*

Hexaploids
Autohistoradiography *see*
 Autoradiography
Autohydrolysis *see*
 Hydrolysis
Autoimmunity *see*
 Immunity
Autoinfection *see* Infection
Autoinhibition *see*
 Inhibition
Autointoxication *see*
 Intoxication
Autolysis *(Autolysate,*
 Autolysates, Autolytic,
 Autolyzed, Autolyzing)
Automatic *(Automatical,*
 Automatically,
 Automaticity,
 Automatism)
 X-Ref: Semiautomatic
Automation *(Automated,*
 Automates, Automating,
 Automatization,
 Automatized,
 Automatizing,
 Semiautomated)
Automixis
Automobiles *see* Cars
Automotive
Automutagens *see*
 Mutagenesis
Autonomy *(Autonomic,*
 Autonomous,
 Autonomously)
Autooxidation *see*
 Oxidation
Autoparasitism *see*
 Parasites
Autophagous
Autopodium *(Autopod)*
Autopolyploids
 (Autopolyploidy)
Autopsies *(Autopsy)*
 X-Ref: Postmortem
Autoradiography
 (Autohistoradiographic,
 Autoradiographic,
 Autoradiographical,
 Autoradiographs,
 Radioautographic)
 X-Ref:
  Autohistoradiography,
  Microautoradiography,
  Radioautography
Autoregulation
Autorhythm *see* Rhythm
Autosomes *(Autosomal,*
 Autosome)
Autospores *(Autospore)*
Autotetraploids
 (Autotetraploid,
 Autotetraploidy)
Autotomy
Autotransplantation *see*
 Transplants
Autotriploids *(Autotriploid)*

Autotripping
Autotrophism *(Autotrophic,*
  Autotrophical,
  Autotrophically)
Autovinificators *see*
  Vinification
Autoxidation *see* Oxidation
Autumn *(Autumnal)*
Auturus
Auxanography
  (Auxanographic)
Auxanometers
  (Auxanometer)
Auxinoxidases
  (Auxinoxidase,
  Auxinoxydase)
Auxins *(Auxin, Auxinic)*
Auxis
Auxometry *(Auxometric)*
Auxospores *(Auxospore)*
Auxotrophism *(Auxotroph,*
  Auxotrophic, Auxotrophs,
  Auxotrophy)
Avadex *see* Diallate
Avalanche *(Avalanches)*
Avena
Avenaciolide
Avenacosides
Aveneae
Avenochloa
Aversion
Aves *see* Birds
Avianization *(Avianized)*
Aviary
Aviation *(Aviators)*
Avicelase
Avicennia
Avicenniaceae
Avicennol
Avicides *(Avicide)*
Avicularin
Aviculture
Avidin
Avifauna *see* Birds
Avirulence *see* Virulence
Avitaminosis
Avitellina
Avitellinosis
Avocados *(Avocado,*
  Avocadoes)
Awamorin
Awards *(Award, Awarded,*
  Awardee, Awarding)
Awn *(Awned, Awnless,*
  Awns)
Axenic *(Axenical,*
  Axenically)
Axes *(Axe)*
Axils
Axis *(Axial)*
  X-Ref: Coaxial
Axles *(Axle)*
Axonemes *(Axoneme)*
Axonopsinae
Axonopsis
Axonopus

Axons *(Axon, Axonal)*
  X-Ref: Extraaxonal,
  Intraaxonal
Axostyles *(Axostyle)*
Axyris
Aylostera
Ayurvedic
Azadirachta
Azadirachtin
Azaguanine
Azaleas *(Azalea)*
Azanza
Azaperone *(Azaperon)*
Azaphilones
Azaplant
Azaserine
Azasteroids
Azathioprine *(Azathioprin)*
Azauracil
Azauridine
Azelaaldehydic
Azelia
Azeotropic *(Azeotrope)*
Azetidine
Azide
Azima
Azinphos *see*
  Azinphosmethyl
Azinphosmethyl
  X-Ref: Azinphos,
  Gusathion, Guthion
Aziridine
Aziridinecarboxamide
Aziridinyl
Azo
Azobenzene
Azodrin
Azoesters *(Azoester)*
Azoles *(Azole)*
Azolla
Azomethine
Azometrin
Azoreductases
Azotobacter
Azotobacterin
Azoturia
Azoxymethane
Azteca
Aztekium
Azulene
Azure
Azureocereus
Azygophleps
Babesia *(Babesial,*
  Babesiella)
Babesiasis *see*
  Piroplasmosis
Babesiidae
Baboons *(Baboon)*
Baccharis
Bacidia
Bacillaceae
Bacillariophyceae
Bacillariophyta
Bacillidae
Bacillus *(Bacillary, Bacilli,*

  Bacilliform)
Bacitracin
Backcross *see*
  Crossbreeding
Backfat *(Backfats)*
Backfires *see* Fire
Backhoes *(Backhoe)*
Backscattering
  (Backscatter)
Backwater
Baclofen
Bacogenins
Bacon *(Bacons)*
Baconers
Bacopa
Bacosides
Bacteremia
Bacteria *(Bacterial,*
  Bacterially, Bacterias,
  Bacterioflora,
  Bacteriologic,
  Bacteriological,
  Bacterium, Bacteroid,
  Bacteroids)
Bacteriastrum
Bactericides *(Bactericidal,*
  Bactericide, Bacteriocide)
  X-Ref: Antibacterial
Bacterins *(Bacterin)*
Bacteriochlorophyll
Bacteriocins *(Bacteriocin)*
Bacteriodes
Bacteriologists
  (Bacteriologist)
Bacteriology
Bacteriolysis *(Bacteriolytic)*
Bacteriophages *see* Phages
Bacterioscopy
Bacteriosis
Bacteriostasis
  (Bacteriostatic)
Bacteriotherapy
Bacteriuria
Bacterization
Bacteroidaceae
Bacteroides
Bactofugation
Bactra
Bactris
Bactrodesmium
Baculoviruses *(Baculovirus)*
Badhamia
Badrakemin
Badula
Baeomyces
Baetidae
Baetis
Bagasse *(Bagasses)*
Bagels
Bagging *(Bag, Bagged,*
  Bags)
  X-Ref: Sacks
Bagoini
Bagrada
Bagworms *see* Psychidae
Bahiagrass

Baileya
Baits *(Bait, Baited, Baiting)*
  X-Ref: Unbaited
Bajra *see* Pearlmillet
Bakanae
Bakeridesia
Bakerocoptes
Bakhtiari
Baking *(Bake, Baked,*
  Bakeries, Bakery)
Bakuchiol
Balan *see* Benefin
Balanites
Balanitis
Balanogastris
Balanophora
Balanophoraceae
Balanoposthitis
Balansia
Balantidiasis *(Balantidiosis)*
Balantidium
Balaustium
Balbas
Baldellia
Baldness *see* Alopecia
Balenine
Bales *(Bale, Baled, Baler,*
  Balers, Baling)
Baliochila
Baljusa
Balloons *(Balloon)*
Ballota
Ballus
Balm
Balneology *(Balneological)*
Balsa
Balsam *(Balsams)*
Balsaminaceae
Balyana
Bambermycins
Bamboo *(Bamboos)*
Bambusa
Bambusoideae
Bananas *(Banana)*
Banchinae
Banchus
Bandeiraea
Bandicoot
Bandicota
Bandsaws *see* Sawing
Bangia
Bangiaceae
Bangiales
Bangiophyceae
Bangiophycidae
Banisteriopsis
Bankesia
Bankia
Banks *(Bank, Banker,*
  Bankers, Banking)
  X-Ref: Multibank
Banksia
Banksinella
Banminth
Bannur
Bantam

Banti
Banvel *see* Dicamba
Baobab
Baoule
Baptisia
Barathra
Barbacenioideae
Barban
Barbarea
Barbary
Barbecuing *(Barbecue)*
Barberries *(Barberry)*
Barbilophozia
Barbital
Barbiturates *(Barbiturate)*
Barbutia
Barbutiidae
Bareogonalos
Barex
Bargaining *(Bargain)*
Barging
Barite
Barium
Bark *(Barked, Barking,*
  Barkless, Barks)
  X-Ref: Unbarked
Barki
Barleria
Barley *(Barleys)*
Barneoudia
Barns *(Barn, Cowhouse)*
  X-Ref: Cowhouses
Barnyard *(Barnyards)*
Barnyardgrass
Barometers *(Barometer,*
  Barometric, Barometry)
Baronia
Baroniinae
Baroreceptor
  (Baroreceptors)
Barrels *(Barrel)*
Barringtogenol
Barringtonia
Barrows *(Barrow)*
Bartonellosis
Bartschella
Barypithes
Basagran *see* Bentazon
Basalts *(Basalt, Basaltic)*
Basamid *see* Dazomet
Basella
Basellaceae
Basenji
Basfapon
Basidiobolus
Basidiocarp *(Basidiocarps)*
Basidiodendron
Basidiomycetes
  (Basidiomycete)
Basidiospores *(Basidiospore)*
Basidium *(Basidia, Basidial)*
Basil
Basilia
Basins *(Basin)*
  X-Ref: Interbasin
Basipetal

Baskets *(Basket)*
Basophilia
Basophils *(Basophilic)*
Bass
Bassaris
Bassia
Basswood *see* Tilia
Bast
Bastardia
Batatasins *(Batatasin)*
Batesian
Batholith
Bathophenanthroline
Bathurin
Bathycoelia
Bathyplectes
Bathysciinae
Batillipes
Batocera
Batophora
Batrachedra
Batrachospermum
Batrisini
Batrisomellus
Bats *(Bat)*
Batteries *(Battery)*
Batting
Bauhinia
Bauxite *(Bauxit)*
Bavistin *see* Carbendazim
Baygon *see* Propoxur
Baytex *see* Fenthion
Bazzania
Bdellovibrio
Bdelyropsis
Beaches
Beachgrass
Beagles *(Beagle)*
Beaks *(Beak)*
Beams *(Beam)*
Beans *(Adzuki, Bean)*
  X-Ref: Adsuki
Bearberries *(Bearberry)*
Beast *see* Animals
Beautification *(Beautify)*
Beauveria
Beauvericin
Beavers *(Beaver)*
Beccarianthus
Beckerella
Beckmannia
Bedbugs *(Bedbug)*
Bedfordia
Bedrock
Beds *(Bed, Bedded,*
  Bedding)
Bedsonia *see* Chlamydia
Bedsoniasis *see*
  Chlamydiosis
Beeches *see* Fagus
Beechnuts *(Beechnut)*
Beechwood *(Beechwoods)*
Beef *(Beefiness)*
Beefsteak *see* Steaks
Beehives *(Beehive)*
Beehouses *see* Apiaries

Beekeeping *(Beekeeper,*
  Beekeepers)
Beers *(Beer)*
Bees *(Bee, Bumblebee)*
  X-Ref: Bumblebees
Beeswax
Beetal
Beetles *(Beetle)*
Beets *(Beet, Beetroot,*
  Mangel)
  X-Ref: Mangels
Beggarweed
Begonia *(Begonias)*
Begoniaceae
Beidellite
Beijerinckia
Beilschmiedia
Belamcanda
Belang
Belba
Belching *see* Eructation
Belgrandiella
Beliceodendron
Bellardia
Bellatocheles
Bellevalia
Bellflowers *(Bellflower)*
Bellicositermes
Bellis
Bellona
Bellura
Belonolaimidae
Belonolaimus
Belonopelta
Beloperone
Belostoma
Belostomatidae
  (Belostomatid,
  Belostomidae)
Belts *(Belt, Belted)*
Bembicini *(Bembicine)*
Bembidiini
Bembidinae
Bembidion
Bembix
Bemisia
Benches *(Bench)*
Bending *(Bend, Bent)*
Benedyne
Benefin
  X-Ref: Balan, Benfluralin
Benfluralin *see* Benefin
Benlate *see* Benomyl
Bennettiaceae
Bennetticarpus
Benomyl
  X-Ref: Benlate
Bensulide
Bentazon *(Bentazone)*
  X-Ref: Basagran
Bentgrass *(Bentgrasses)*
Benthos *(Benthic,*
  Benthonic)
  X-Ref: Epibenthos,
  Macrobenthos
Bentonite *(Bentonites)*

Benzaldehyde
Benzamide
Benzamidine
Benzathine
Benzazepines
Benzene
Benzenesulfonic
Benzidine
Benzimidazolecarbamate
Benzimidazoles
  (Benzimidazol,
  Benzimidazole,
  Benzimidazolic)
Benzimidazolyl
Benzine
Benzisothiazole
Benzoate *(Hydroxybenzoate)*
Benzobisbenzofurans
Benzochinon
Benzodiazepines
  (Benzodiazepine)
Benzodioxan
Benzodioxaphosphorin
Benzodioxoles
Benzofuran
Benzofuranyl
Benzohydroxamic
Benzohydroximate
Benzoic
Benzoin
Benzonitrile *(Benzonitriles)*
Benzophenanthridine
Benzophenone
Benzophosphate
  X-Ref: Fozalon
Benzopyran
Benzopyrene
Benzopyrones
Benzoquinones
  (Benzoquinone)
Benzosemiquinone
Benzothiadiazoles
Benzothiazole
Benzothiazolyl
Benzothienyl
Benzoxazine *(Benzoxazin)*
Benzoxazolinone
  (Benzoxazolinones)
Benzoxepin
Benzoyl
Benzoylcholine
Benzoylpaeoniflorin *see*
  Paeoniflorin
Benzoylurea
Benzpyrene
Benzyl
Benzyladenine
Benzylamine
  (Benzylamines)
Benzylaminopurine
Benzylidene
Benzylisoquinoline
  (Benzylisoquinolines)
Benzyloxycarbonyl
Benzylpenicillin
Berbamine

Berberentulus
Berberidaceae
Berberine *(Berberines)*
Berberis
Berbine
Berenice
Berenil
Bergamot
Bergapten
Bergenia
Bergenin
Beriberi
Beridinae
Bermudagrass
(Bermudagrasses)
Bernese
Berrichon
Berries *(Berry, Berryfruit,*
Berryfruits)
Berrya
Bersama
Berseem
Berteroa
Bertholletia
Bertya
Beryllium
Berytidae
Berytinidae
Beskaragai
Besnoitia
Bessa
Bessel
Betacyanins *(Betacyanin,*
Betacyanine)
Betaine *(Betain)*
Betalains *(Betalain)*
Betamethasone
Betanal
X-Ref: Phenmedipham
Betanin *(Betanine)*
Betaxanthins
Betel *(Betelvine)*
Bethylidae *(Bethylids)*
Bethyloidea
Betonica
Betony
Betula *(Birch)*
X-Ref: Birches
Betulaceae
Betulaphis
Beverages *(Beverage)*
Bezoar *(Bezoars)*
Bezzia
Bhadarwah
X-Ref: Gaddi
BHC *see*
Hexachlorocyclohexane
Bhima
Bhindi *see* Okra
Bhoosa *(Bhusa)*
Biamperometric *see*
Amperometry
Bianthraquinonyls
Bibenzyls
Bibio
Bibionidae

Bible *(Biblical)*
Bibliographies
(Bibliographic,
Bibliographical,
Bibliography)
Bibos
Bicarbonate *(Bicarbonates)*
Bicoumarins *see* Coumarins
Bicoumarinyl
Bicuculline
Biddulphiaceae
Bidens
Biebersteinia
Biennials *(Biennial)*
Bifenox
Bifidobacterium
(Bifidobacteria)
Biflavanone
Biflavones
Biflavonoids
Biflavonyl
Bifrenaria
Bifurcaria
Bifurcation
Biggerellins
Bignonia
Bignoniaceae
Bihormonal *see* Hormones
Bikaneri
X-Ref: Nali
Bilateral
Bilberries *(Bilberry)*
Bile *(Biliary)*
Bilevon
X-Ref: Menichlopholan,
Niclofolan
Bilharzia *see* Schistosoma
Bilharziasis *see*
Schistosomiasis
Biliproteins *(Biliprotein)*
Bilirubin
Biliverdin
Bill
Billbergia
Billbugs *(Billbug)*
Billia
Bilobectomy
Binapacryl
Binaphthaquinone
Bincoside
Bindweeds *(Bindweed)*
Binocular *see* Eyes
Binodoxys
Binotal *see* Ampicillin
Bins *(Bin)*
Binucleate *see* Nucleus
Bioacoustics
Bioactivation *(Bioactive,*
Bioactivity)
Bioallethrin
Bioantioxidants *see*
Oxidants
Bioassay *see* Assay
Bioautography
(Bioautograph,
Bioautographic)

Biocatalytic *see* Catalysis
Biocenology
Biocenosis *(Biocenose,*
Biocenoses, Biocenotic,
Biocoenological,
Biocoenoses, Biocoenosis,
Biocoenotic, Bioecogroups)
Biochemistry *(Biochemical,*
Biochemically)
X-Ref: Physiobiochemical
Biochemists *(Biochemist)*
Biochronometry
Biocides *(Biocidal)*
Bioclimate *see* Climate
Biocorrosion *see* Corrosion
Biocybernetic *see*
Cybernetics
Biodegradation
(Biodegradability,
Biodegradable,
Biodegradative,
Biodeteriogenic,
Biodeterioration)
Biodisc
Biodyl
Biodynamics *(Biodynamic,*
Biodynamical)
Bioecology *see* Ecology
Bioeconomy *(Bioeconomic)*
Bioelectric *(Bioelectrical)*
Bioelectrochemistry
(Bioelectrochemical)
Bioelements
Bioenergetics *(Bioenergetic)*
Bioengineering
Bioenvironment *see*
Environment
Bioethanomethrin
Biofilters *see* Filters
Bioflocculation *see*
Flocculation
Biogarde
Biogen
Biogenesis *(Biogenetic,*
Biogenetically, Biogenic,
Biogenous)
Biogeocenosis
(Biogeocenoses,
Biogeocenotic,
Biogeocoenoses,
Biogeocoenosis)
Biogeochemistry
(Biogeochemical)
Biogeography
(Biogeographer,
Biogeographic,
Biogeographical,
Chorological)
X-Ref: Chorology
Bioghurt
Biography *(Biographic,*
Biographical, Biographies,
Memoriam, Memories,
Obituaries)
X-Ref: Autobiography,
Memory, Obituary

Biohazards *(Biohazard)*
Biohydrogenation
Bioindicators *(Bioindicator)*
Biokinetics
Biologists *(Biologist)*
Biology *(Biologic,*
Biological, Biologically,
Biologics, Biologies)
X-Ref: Nonbiological
Bioluminescence
Biomagnetism
Biomass *(Biomasses)*
Biomathematics
(Biomathematic,
Biomathematical)
Biome
Biomechanics
(Biomechanical)
Biomedicine *(Biomedical)*
Biomembranes
Biometeorology
(Biometeorological)
Biometrics *(Biometric,*
Biometrical, Biometry)
Biomimetic *see* Mimicry
Biomin
Biomorphological *see*
Morphology
Biomycin *see*
Chlortetracycline
Bionics
Bionomics *(Bionomic,*
Bionomy)
Biooxidation *see* Oxidation
Biophenology *see* Phenology
Biophysics *(Biophysical)*
Biophytum
Bioplast
Biopolymers *see* Polymers
Biopreparations
(Biopreparation)
Bioproduction
(Bioproductive,
Bioproductivity)
Biopsy *(Biopsied, Biopsies)*
Biopterin
Bioreactors *(Bioreactor)*
Bioregeneration *see*
Regeneration
Bioresmethrin
Biorhiza
Biorhythms *see* Rhythm
Biosciences *(Bioscience)*
Biosociology
Biosphere *(Biospheric)*
Biostabilization
Biostatics *(Biostatic)*
Biostatistics *(Biostatistical)*
Biosteres
Biostimulation
(Biostimulant,
Biostimulants,
Biostimulating,
Biostimulator,
Biostimulators)
Biostratigraphy *see*

Stratification
Biosuper
Biosynthesis *(Biosyntheses,*
Biosynthetic)
Biosystematics *see*
Taxonomy
Biosystematists *see*
Taxonomists
Biota
Biotaxonomy *see* Taxonomy
Biotechnics *(Biotechnic,*
Biotechnical)
Biotechnology
(Biotechnological)
Biotelemetry *(Biotelemetric)*
Biotest *see* Assay
Biotherapy *see* Therapy
Biothermal
Biotic *(Biotically, Biotics)*
Biotin
Biotite
Biotopes *(Biotope)*
Biotoxins *see* Toxins
Biotransformation *see*
Transformation
Biotrol
Biotropism *see* Tropism
Biotypology *(Biotype,*
Biotypes)
Biovit
Biparental *see* Parent
Biphenylol
Biphenyls *(Biphenyl)*
X-Ref: Diphenyl
Biphenylyl
Biphosphate *see* Phosphates
Bipolaris
Bipyridyl
Bipyridylium
Birches *see* Betula
Birchwood
Birds *(Avian, Bird)*
X-Ref: Aves, Avifauna
Birdsfoot
Birefringence *see* Refraction
Bironella
Birth *(Births, Birthweight,*
Caesarian, Cesarian,
Childbirth, Natality,
Neonatally, Neonate,
Neonates, Newborns,
Parturient, Postnatally,
Postpartal, Postparturient,
Prenatally, Preparial,
Prepartum, Puerperal,
Stillbirth, Stillborn)
X-Ref: Antenatal,
Caesarean, Cesarean,
Childbearing,
Intranatal, Natal,
Neonatal, Newborn,
Paripartal, Partal,
Parturition, Perinatal,
Peripartal, Postnatal,
Postpartum, Prenatal,
Preparturient,

Puerperium, Stillbirths
Bisabolane
Bisabolene
Bisabolol
Bisbenzylisoquinoline
(Bisbenzylisoquinolin)
Biscarbamates *see*
Carbamates
Bischofia
Biscoclaurine
Biscuits *(Biscuit)*
Bisdiazoacetylbutane
Bisditerpene *see* Diterpenes
Bisdithiocarbamate
Bisection *(Bisected)*
Bisexual *see* Sex
Bisflavones *see* Flavones
Bisindole *see* Indole
Bismethanesulfonate
Bismuth
Bisnor
Bisolvon
Bison *(Bisons)*
Bisphosphatase *see*
Phosphatases
Bisporomyces
Bisulfate
Bisulfite *(Bisulphite)*
X-Ref: Metabisulfite
Bitches *see* Dogs
Bithionol
Bithorax
Biting *(Bite, Bites)*
Bittacidae
Bittacus
Bitter *(Bittering, Bitterness,*
Bitters)
Bitterbrush
Bittergourd
Bittersweet
Bitterweed
Bitto
Bitumens *(Bitumen,*
Bituminous)
Biuret
Biuron
Bivoltine *see* Voltinism
Bivulvarity *see* Vulva
Bixa
Bixaceae
Bizotia
Blabera
Blaberidae *(Blaberinae)*
Blaberus
Blacinae
Blackberries *(Blackberry)*
Blackbirds *(Blackbird)*
Blackbrush
Blackbutt
Blackcurrants *see* Currants
Blackface
Blackflies *(Blackfly)*
Blackleg
Blackstonia
Blackthorn
Blacus

Bladders *(Bladder)*
Bladderworms
(Bladderworm)
Bladderwort
Blades *(Blade)*
Bladex
Blaesoxipha
Blakea
Blakeslea
Blanching
Blaniulidae
Blaps
Blapstinus
Blasia
Blast *(Blasting)*
Blasticidin
Blasticotoma
Blastocladia
Blastocladiales
Blastocladiella
Blastocysts *(Blastocyst)*
Blastodacna
Blastoderm *(Blastoderms)*
Blastodisks *(Blastodisc,*
Blastodiscs)
Blastogenesis *(Blastogenic)*
Blastokinin
Blastomyces
Blastomycetes
Blastomycin
Blastomycosis
Blastophaga
Blastophagus
Blastopore
Blastula
Blatta
Blattaria
Blattariae
Blattella
Blattellinae
Blattidae *(Blattellidae)*
Blattodea
Blattoidea
Bleaching *(Bleach,*
Bleachability, Bleached)
X-Ref: Photobleaching,
Unbleached
Bleasdalea
Bleb
Blechnaceae
Blechnum
Bledius
Bleeding *(Bleeder,*
Bleedings)
Blending *(Blend, Blended,*
Blender, Blendings,
Blends)
Blepharipa
Blepharis
Blepharolepis
Blepharoplast
Bletilla
Blidingia
Blighia
Blights *(Blight,*
Blisterblight)

Blindness *(Blind, Blinding)*
Blissinae
Blissus
Blister *(Blistered, Blistering,*
Blisters)
Blizzards
Bloat *(Bloating)*
X-Ref: Nonbloating,
Tympanites
Blood *(Bloodflow, Bloodless,*
Bloods, Haematic,
Hematinic, Hemato,
Hematochemical)
X-Ref: Hematic
Bloodhounds *(Bloodhound)*
Bloodmeal
Bloodroot
Bloodsucking
Bloodwood
Bloomeria
Blossom *see* Flowers
Blosyris
Blotch
Blow *(Blower, Blowers,*
Blowing, Blown)
Blowflies *(Blowfly)*
Blueberries *(Blueberry)*
Bluebush
Bluecomb
Bluegills *(Bluegill)*
Bluegrasses *(Bluegrass)*
Bluejoint
Bluestem *(Bluestems)*
Bluetongue
Blumea
Blyttiomyces
Boards *(Board)*
Boarmia
Boars *(Boar)*
Boats *(Boat)*
Bobartia
Bobcats *(Bobcat)*
Bocchus
Bocconia
Bocconine
Bochartia
Boehmenan
Boehmeria
Boeng
Boenninghausenia
Boergesenia
Boerhavia *(Boerhaavia)*
Boettcherisca
Bogs *(Bog, Bogged,*
Bogland, Boglands)
Boilers *(Boiler)*
Boiling *(Boil, Boiled,*
Boilings)
Bolbitiaceae
Bolbocerinae
Bolbomyia
Boles *(Bole)*
Boletaceae *(Boletaceous,*
Bolete, Boletes)
Boletales
Boletol

Boletopsis
Boletus
Bolitophila
Bollea
Bollweevils *see* Weevils
Bollworms *(Bollworm)*
Boloria
Bolti
Bolting *(Bolt, Bolts)*
Bolyphantes
Bombacaceae
Bombaceae
Bombacopsis
Bombax
Bombesin
Bombidae *(Bombinae,
    Bombini)*
Bombus
Bombycidae
Bombykol
Bombyliidae
Bombylius
Bombyx
Bonafousia
Bonafousine
Bonding *(Bond, Bonded,
    Bonds)*
Bones *(Bone, Boned,
    Boneless, Boning, Bony)*
Bongkrekic
Bonitos *(Bonito)*
Bonnemaisonia
Bonnemaisoniaceae
Bonsai
Bookkeeping *see*
    Accounting
Boophilus
Boophthora
Boopidae
Booponus
Boraginaceae
Boran
Borates *(Borate)*
Borax
Bordetella
Boreal
Boreholes *see* Wells
Boreoiulus
Boreovespula
Borers *(Borer)*
Boric
Borodinia
Borohydride
Boron *(Boronic)*
Boronia *(Boronias)*
Borrelia
Borrelidin
Borreria
Borrichia
Borya
Borzicactinae
Borzicactus
Bos
Bosistoa
Bosistoin
Bostrichidae *(Bostrychidae)*

Bostrychia
Bostrychus
Boswellia
Boswellic
Botanists *(Botanist)*
    X-Ref: Geobotanists
Botany *(Botanic, Botanical,
    Geobotanic, Geobotanical)*
    X-Ref: Geobotany
Botflies *(Botfly)*
Bothriochloa
Bothriuridae
Bothroponera
Bothynoderes
Bothynus
Botrychium
Botryococcus
Botryodiplodia
Botryomycosis
Botryopteris
Botryosphaeria
Botryotinia
Botrytis
Bottlebrush
Bottling *(Bottle, Bottled,
    Bottles)*
Bottomland *(Bottom,
    Bottomlands, Bottoms)*
Botulism
Bougainvillea
    (Bougainvilleas)
Bourdotia
Bourletides
Bourletiella
Bouteloua
Bouvardia
Boverin
Bovicola
Bovidae *(Bovid, Bovids)*
Bovine *(Bovin, Bovines)*
Bovisynchron
Bowdichia
Bowels *(Bowel)*
Bowringia
Boxcars
Boxer
Boxes *(Box, Boxed)*
Boxsaws *see* Sawing
Boxthorn
Boxwood *see* Buxus
Boycotts *(Boycott)*
Boydaia
Boys *(Boy)*
Boysenberries
Brachiaria *(Brachiarias)*
Brachinus *(Brachinids)*
Brachium *(Brachial)*
Brachybasidiaceae
Brachycaudus
Brachycentrus
Brachycephalism
    *(Brachycephalia,
    Brachycephalic)*
Brachycera *(Brachycerous)*
Brachychaeta
Brachychthoniidae

Brachycome
Brachycoryphus
Brachycylix
Brachydactyly
    X-Ref: Brachyphalangy
Brachydeutera
Brachyelytrum
Brachyglottis
Brachygnathia
Brachygrammatella
Brachylabis
Brachylaena
Brachymeria
Brachypanorpa
Brachyphalangy *see*
    Brachydactyly
Brachyphylla
Brachyphyllum
Brachypoda
Brachypodium
Brachypogon
Brachyptera
Brachypterous
Brachyrhinus
    (Brachyrrhinus)
Brachystegia
Brachystelma
Brachystomella
Brachythemis
Brachytricha
    (Brachytrichia)
Bracken
Bracon
Braconidae *(Braconid,
    Braconids)*
Bradycardia
Bradykinin
Bradysia
Brahea
Brahman
Braiding *(Braid)*
Brain *(Brains, Brainstem,
    Cerebellar, Cerebral,
    Cerebrocortical,
    Cerebroid, Intercerebralis,
    Intracerebellum,
    Intracerebrally,
    Intracerebroventricular)*
    X-Ref: Cerebellum,
    Cerebrum,
    Extracerebral,
    Intercerebral,
    Intracerebral
Bran
Branches *(Branch,
    Branched, Branching,
    Branchlet, Branchy)*
Branchiopoda *(Branchiopod)*
Brandies *(Brandy, Cognacs)*
    X-Ref: Armagnac,
    Cognac
Branding *(Brand, Brands)*
Brangus
Brasenia
Brassaia
Brassavola

Brassia
Brassica *(Brassicas)*
Brassicaceae *(Brassicaceous)*
Brassicasterol
Brassidic
Brassins
Brassolidae
Bratwurst
Braula
Brazoria
Bread *(Breadmaking,
    Breads)*
Breadfruit
Breakfast *(Breakfasts)*
Bream
Breast *(Breasts)*
Breathing *see* Respiration
Bredinin
Breeding *(Bred, Breds,
    Breeder, Breeders,
    Breedings, Interbred,
    Interbreeding)*
    X-Ref: Interbreed,
    Outbred, Rebreeding,
    Straightbred
Breeds *(Breed)*
Brefeldin
Breinlia
Bremia
Brennandania
Brenthis
Brentidae
Brestan
Brettanomyces
Bretylium
Brevennia
Brevianamides
Brevibacterium
Brevicolline
Brevicomin
Brevicoryne
Brevifolin
Brevilabris
Brevilegnia
Brevipalpus
Brevoortia
Brewing *(Brew, Brewer,
    Breweries, Brewers,
    Brewery, Brewhouse)*
Breynia
Breynin *(Breynins)*
Breynolide
Brickellia
Brie
Brigalows *(Brigalow)*
Brimeura
Brining *(Brine, Brined,
    Brines)*
Brinjal *see* Eggplant
Briquettes *(Briquet,
    Briquets, Briquetted,
    Briquetting)*
Brisket
Bristles *(Bristle, Bristly)*
Bristletails *(Bristletail)*
Brix

Brixia
Briza
Broadbeans (Broadbean)
Broadcasting (Broadcast,
   Broadcaster, Broadcasters)
Broadleaved (Broadleaf)
Broccoli
Brochymena
Broilers (Broiler)
Brokerage
Bromacil
Bromates (Bromate)
Bromatology (Bromatologic)
Bromegrass (Brome,
   Bromes)
Bromelain see Bromelin
Bromelia
Bromeliaceae (Bromeliad,
   Bromeliads)
Bromelin (Bromelins)
   X-Ref: Bromelain
Bromethyl
Bromhexine (Bromhexin)
Bromides (Bromide)
Bromine (Brominated,
   Brominating,
   Bromination)
Bromoacetate
Bromoacetyl
Bromobenzene
   (Bromobenzenes)
Bromobenzoate
Bromocholest
Bromocresol
Bromocriptine
Bromodeoxyuridine
Bromolysergic
Bromomethyl
Bromonium
Bromophenol
   (Bromophenols)
Bromophos
   X-Ref: Nexton
Bromosuccinimide
Bromosulphalein
Bromosulphthalein
Bromothiazole
Bromothymol
Bromouracil
Bromoxynil
Bromus
Bronchi (Bronchial,
   Bronchus)
   X-Ref: Parabronchial
Bronchiectasis
Bronchitis
Bronchoconstriction
   (Bronchoconstrictors)
Bronchodilation
   (Bronchodilator)
Bronchogenic
Bronchopneumonia
Bronchopulmonary
Bronchoscopy
Bronco
Bronopol

Brontispa
Brood (Brooder, Brooders,
   Brooding, Broods)
Broomcorn
Broomrape
Broomweed
Brosimum
Broths (Broth)
Broussonetia
Browallia
Browntop
Browsing (Browse)
Brucea
Bruceantin
Brucein
Brucella (Brucellae,
   Brucellas)
Brucellaceae
Brucellosis
   X-Ref: Antibrucellar
Brucheiser
Bruchidae (Bruchid,
   Bruchids, Bruchinae)
Bruchophagus
Bruchus
Brucine
Brugella
Brugia
Brugmansia
Bruguiera
Bruising (Bruise, Bruises)
Brumptomyia
Brush (Brushes, Brushland,
   Brushlands)
Brushwood
Brusone
Bryaceae
Bryales
Bryobia
Bryology (Bryological)
Bryonia
Bryophyllum
Bryophyta (Bryoflora,
   Bryophyte, Bryophytes)
Bryopogon
Bryopsida
Bryopsidophyceae
Bryopsis
Bryothinusa
Bryozoa (Bryozoan)
Bryum
Bubalus see Buffalo
Buccal
   X-Ref: Peribuccal
Bucculatrix
Bucephalophora
Buchanania
Buchloe
Buchnera
Buckeye
Buckleya
Buckthorn
Buckwheat
Budding (Budbreak,
   Budded)
Buddleia

Buddleja
Buddlejaceae
Budgerigars
Budgets (Budget,
   Budgetary, Budgeting)
Buds (Bud)
Budwood
Budworms (Budworm)
Buellia
Buelliaceae
Buenoa
Buergenerula
Buergersiochloa
Bufadienolides
   (Bufadienolide)
Buffalo (Buffaloes)
   X-Ref: Bubalus
Buffet
Bufo see Toads
Buformin
Bugs see Insects
Building (Build, Builders,
   Buildings, Builds)
Bulaea
Bulbocastanum
Bulbochaete
Bulbocodium
Bulbophyllinae
Bulbophyllum
Bulbothrix
Bulbourethral
Bulbs (Bulb, Bulbar, Bulbil,
   Bulbils, Bulbing, Bulblets,
   Bulbous, Bulbus)
Bulbuls
Bulgur
Bulinus
Bulla
Bulldogs (Bulldog)
Bulldozers (Bulldozer,
   Bulldozing)
Bullfinches
Bulls (Bull, Bullock,
   Bullocks)
Bulrushes (Bulrush)
Bumacris
Bumblebees see Bees
Bumilleriopsis
Bunaeopsis
Bunamidine
   X-Ref: Scolaban
Bunchgrass
Bunium
Bunostomum
Buns
Bunt
Bupalus
Bupleurum
Buprestidae (Buprestid)
Buprestis
Buquinolate
Burdock
Burlap (Burlapped)
Burmannia
Burning (Burn, Burned,
   Burner, Burners, Burnout,

Burns)
   X-Ref: Unburned
Burros (Burro)
Burrowing (Burrow,
   Burrows)
Bursa (Bursal)
   X-Ref: Intrabursal
Bursaphelenchus
Bursectomy (Bursectomized)
Bursera
Burseraceae
Bursicon
Bursitis
Buscopan
Bushes (Bush)
Bushveld
Business (Businesses)
Busseola
Busulfan (Busulphan)
Butadiene
Butadienediepoxide
Butalidis
Butanediol
Butanedione
Butanol
Butanone
Butchering see Slaughter
Butchers (Butcher)
Butea
Butein
Butene
Butenyl
Buteo
Buthidae
Buthus
Butiphos
Butomus
Butonate
Butoxide
Butter (Buttermaking,
   Butters)
Butterbur
Buttercup
Butterfat (Butterfats)
Butterflies (Butterfly)
Butterhead
Buttermilk (Buttermilks)
Butternuts (Butternut)
Buturon
Butyl
Butylamines
Butylarterenol
Butylate (Butylated)
Butylcarbamoyl
Butylene
Butylhydroxyanisole
Butylphenyl
Butylurea
Butyrate
Butyric
Butyrivibrio
Butyrolactone
Butyrometer
Butyrophenone
Butyrospermum
Butyryl

Butyrylcholine
Buvinol
Buxaceae
Buxbaumia
Buxus *(Boxwoods)*
  X-Ref: Boxwood
Buyers *(Buyer)*
Byblidaceae
Byctiscus
Byproducts *(Byproduct)*
Byrrhidae
Byrsonima
Byrsotria
Byssinosis
Byssochlamys
Bytiscus
Byturidae
Byturus
Caaverine
Cabbages *(Cabbage)*
Cabbageworms
  (Cabbageworm)
Cabins *(Cabin)*
Cables *(Cable)*
Cabombaceae
Cabucala
Cacalia
Cacao
Caccinia
Caccoplectus
Cachexia
Cachonina
Cachrys
Cacodylate
Cacodylic
Cacoecia
Cacoecimorpha
Cacopaurus
Cactaceae
Cactoblastis
Cactus *(Cacti)*
Cadaba
Cadang
  X-Ref: Kadang
Cadastre *(Cadastral)*
Cadaverine
Cadia
Cadinol
Cadmium
Cadoxen
Cadra
Caeculidae
Caeculus
Caecum *see* Cecum
Caedicia
Caelifera
Caenidae
Caenis
Caenopsylla
Caenorhabditis
Caenoscelis
Caerulein
Caesalpinia
Caesalpiniaceae
Caesalpinioideae
Caesarean *see* Birth

Caesulia
Cafeterias *(Cafeteria)*
Caffeic
Caffeine *(Coffein)*
  X-Ref: Coffeine
Caffeoyl
Caffeyl
Cages *(Cage, Caged,*
  Caging)
CAIN
Cairina
Cajanus
Cakes *(Cake)*
Cakile
Calacarus
Caladium *(Caladiums)*
Calamagrostis
Calameuta
Calamintha
Calamondin
Calamovilfa
Calandra
Calanthe
Calathidium
Calatropis
Calaustralin
Calcareous *(Calcarious)*
  X-Ref: Noncalcareous
Calcemia *(Calcemic)*
Calceolaria
Calciferol *see* Ergocalciferol
Calcification *(Calcific,*
  Calcifications, Calcified)
Calcifuges *(Calcifuge)*
Calcination *(Calcined)*
Calcinosis
Calciphily *(Calcicolous)*
Calciphos
Calciphylaxis
  (Calciphylactic)
Calcite *(Calcites, Calcitic)*
Calcitonin
Calcium
Calcofluor
Calculators
Calculi *(Calculus)*
  X-Ref: Microcalculi
Calculosis *see* Lithiasis
Caldariomyces
Caldesia
Calea
Caledia
Calendula
Calendulosides
Calephelis
Calibration *(Calibrate,*
  Calibrated, Calibrating,
  Calibrator)
Caliciales
Caliciopsis
Caliciviruses *(Caliciviral,*
  Calicivirus)
Calico
Caligo
Caligonellidae
Calineuria

Calipers *(Caliper)*
Calixin
Calleida
Calliandra
Callicarpa
Callicera
Callichila
Callicorixa
Calliergon
Calligonum
Calligypona
Callimastix
Callimorpha
Callinectes
Calliphora
Calliphoridae *(Calliphorid)*
Calliphorini
Calliptamus
Callirhytis
Calliscia
Callistemon
Callistephus
Calliterpenone
Callithrix
Callitrichaceae
Callitriche
Callitris
Callitroga
Callomyia
Callophrys
Callophyllum
Callopsylla
Callosamia
Callose
Callosobruchus
Callotia
Calluna
Calluses *(Callus)*
Callyntrura
Calobryales
Calocalpe
Calocoris
Calocybe
Caloglyphus
Calometry *(Calometric)*
Calomicrus
Calomycterus
Caloncoba
Calonectria
Calonyction
Calophyllum
Calopterygidae
Calopteryx
Caloptilia
Calories *(Caloric, Calorie)*
  X-Ref: Isocaloric
Calorimetry *(Calorimetric,*
  Microcalorimeter)
  X-Ref: Microcalorimetry
Calosoma
Calotarsa
Calotheca
Calotropis
Caloxanthin
Calpicarpum
Calpodes

Calpurnia
Calsequestrin
Caltha
Calvatia
Calves *(Calf)*
Calving *(Calve, Calved,*
  Calvings)
Calycadenia
Calycanthine
Calycomorphum
Calycopteris
Calyculosphaeria
Calylophus
Calypogeia
Calypso
Calyptra *(Calyptrate)*
Calyptridium
Calyptus
Calystegia
Calyx
Camarosporium
Camarotis
Camassia
Cambalidae
Cambalomorpha
Cambalopsis
Cambendazole
Cambium *(Cambial)*
  X-Ref: Procambium
Cambrian
Camelidae
Camelliaceae
Camelliagenins
Camellias *(Camelia,*
  Camellia)
Camels *(Camel)*
Camelus
Camembert
Camera
Cameraria
Camerobiidae
Camille
Camisia
Camissonia
Camnula
Camoensia
Camomile *(Camomille,*
  Chamomile)
Camouflage
Campanella
Campanula
Campanulaceae
Campelia
Camphor
Camphorosma
Camphorsulfonate
Campine
Camping *(Camper,*
  Campers, Campground,
  Campgrounds, Camps,
  Campsite, Campsites)
Campion
Campnosperma
Campodea
Campoletis
Camponotus

Campsis
Campsosternus
Camptobrochis
Camptochironomus
Camptosorus
Camptotheca
Camptothecine
   (Camptothecin)
Camptothlipsis
Camptylonemopsis
Campylium
Campylobacter
Campylocentrum
Campylomma
Campylopus
Campylothrombium
Campylotropis
Cams (Cam)
Canace
Canaceidae
Canaceoides
Canadensolide
Canals (Canal)
Cananga
Canangium
Canaries
Canarigenin
Canarigenone
Canarium
Canarygrass
Canavalia
Canavanine
Canbra
Cancellidium
Cancer (Cancerigenesis,
   Cancerigenic,
   Cancerogenic,
   Cancerogenicity,
   Cancerous, Cancers)
   X-Ref: Anticancer,
   Noncancerous,
   Precancerous
Candicanin
Candicidin
Candida
   X-Ref: Syringospora
Candidiasis
Candies (Candy)
Candivac
Candles (Candle)
Candling
Cane (Canefields, Canes)
Caneberries (Caneberry)
Canestrini
Canidae (Canids)
Canines see Dogs
Canis
Cankers (Canker,
   Cankered)
Cankerworm
Canna (Cannas)
Cannabidaceae
   (Cannabaceae)
Cannabidiol
Cannabidiolic
Cannabigerol

Cannabinoids (Cannabinoid)
Cannabinols (Cannabinol)
Cannabis
Canneries (Cannery)
Cannibalism (Cannibalistic)
Canning (Canned, Canners,
   Cans)
   X-Ref: Tinned, Tins
Cannulation (Cannula,
   Cannulae, Cannulas,
   Cannulated, Cannulating)
Canopy (Canopies)
Canotia
Canscora
Canteens (Canteen)
Cantelopes (Cantalope,
   Cantalopes, Cantaloup,
   Cantaloupe, Cantaloupes,
   Cantaloups, Cantelope)
Cantharellaceae
   (Cantharelloid)
Cantharellula
Cantharellus
Cantharidae
Cantharidin
Cantharis
Cantharoidea (Cantharoid)
Canthaxanthin
Cantheconidea
Canthon
Canthus
Capacitance
Capeweed
Capillanol
Capillaria
Capillariasis
Capillary (Capillar,
   Capillaries, Capillarity)
   X-Ref: Intercapillary,
   Transcapillary
Capital
Capitalism (Capitalist,
   Capitalistic)
Capitophorus
Capitulum
Capniidae
Capnioneura
Capnobotes
Capnodiaceae
   (Capnodiaceous)
Capnodiastrum
Capnodiopsis
Capnodis
Capnokyma
Caponization
Capons (Capon)
Capparidaceae
Capparis
Capra (Caprine)
Capraria
Caprate
Capreolus
Caprifoliaceae
Caprini
Caprolactam
Caprylate

Caprylic
Caprylohydroxamic
Capsaicin
Capsella
Capsicidin
Capsicum
Capsidiol
Capsule (Capsular,
   Capsulation, Capsuled,
   Capsules)
   X-Ref: Subcapsular
Capsulorrhaphy
Capsus
Captafol
Captan
Captivity (Captive,
   Captures)
   X-Ref: Recapture
Capuchin
Carabaos (Carabao)
Carabidae (Carabid,
   Carabids)
Carabini
Carabodes
Carabodidae
   X-Ref: Xenillidae
Caraboidea
Carabus
   X-Ref: Dysmictocarabus
Caradrinidae
Caragana
Caragard
Caralluma
Caramel (Caramels)
Carapa
Carapace
Caraphractus
Carausius
Caraway
Carbachol
   X-Ref: Carbocholine
Carbadox see Mecadox
Carbamates (Carbamate,
   Carbamated)
   X-Ref: Biscarbamates
Carbamic
Carbamide see Urea
Carbamoyl
Carbamoylethyl
   (Carbamoylethylation)
Carbamoylhydrazine
Carbamoylmethyluridine
Carbamoyloxymethyl
Carbamoyltransferase
Carbamyl (Carbamylation)
Carbamylcholine
Carbamylphosphate
Carbanide
Carbanilates (Carbanilate)
Carbanilides
Carbanions (Carbanion)
Carbaryl
   X-Ref: Sevin
Carbathione (Carbathion)
   X-Ref: Vapam
Carbazole (Carbazol,

   Carbazoles)
Carbazones (Carbazone)
Carbendazim
   (Carbendazime)
   X-Ref: Bavistin, MBC
Carbenicillin
Carbetamide
Carbides (Carbide)
Carbinols (Carbinol)
Carboammophoska
Carboanhydrase
Carbocholine see Carbachol
Carbodox
Carbofuran
Carbohydrases
   (Carbohydrase)
Carbohydrates
   (Carbohydrate)
Carboligase see Ligases
Carboline
Carbonarius
Carbonates (Carbonate,
   Carbonated)
Carbonic
Carboniferous
Carbonization (Carbonized)
Carbonmonoxyhemoglobin
Carbons (Carbon)
   X-Ref: Apocarbonic,
   Radiocarbon
Carbonylamine
Carbonylic
Carbonyls (Carbonyl)
Carbophenothion
Carbophos see Malathion
Carbosieve
Carbowax
Carboxamide
Carboxanilide
Carboxin
   X-Ref: Vitavax
Carboxyatractylate
Carboxybutyl
Carboxycathepsin
Carboxyesterase
Carboxyethyl
Carboxykinases
   (Carboxykinase)
Carboxyl (Carboxyls)
Carboxylamides
Carboxylases (Carboxylase)
Carboxylates (Carboxylate)
Carboxylation
   (Carboxylated,
   Carboxylating,
   Carboxylative)
Carboxylesterases
   (Carboxylesterase)
Carboxylic
Carboxylyase
Carboxymethyl
Carboxymethylation
   (Carboxymethylated)
Carboxymuconolactone
Carboxypeptidases
   (Carboxypeptidase)

X-Ref:
  Procarboxypeptidase
Carburetors
Carburo
Carcass *(Carcase, Carcases,*
  Carcasses)
Carcelia
Carceliopsis
Carcharolaimus
Carcinogenesis
  (Carcinogenic,
  Carcinogenicities,
  Carcinogenicity,
  Carcinogenous)
  X-Ref: Procarcinogenic
Carcinogens *(Carcinogen,*
  Cocarcinogenic,
  Cocarcinogens,
  Noncarcinogenic)
  X-Ref: Noncarcinogen
Carcinoma *(Carcinomas)*
Carcinophoridae
  (Carcinophorinae)
Carcinosarcoma
Cardamine
Cardamom *(Cardamon)*
Cardamonin
Cardanol
Cardaria
Cardboard *see* Fiberboard
Cardenolides *(Cardenolide)*
Cardiac
Cardiaspina
Cardigan *(Cardigans)*
Carding *(Carded)*
Cardioacceleration
  (Cardioaccelerator,
  Cardioaccelerators)
Cardiochiles
Cardiocondyla
Cardiolipin
Cardiology
Cardiomegaly
Cardiomyopathy
  (Cardiomyopathic)
Cardiopathy *(Cardiopathies,*
  Cardiopathogenicity,
  Cardiopathological,
  Cardiopathology)
Cardiopulmonary
Cardiorespiratory
Cardiospermum
Cardiotonic
Cardiotoxicity *(Cardiotoxic)*
Cardiovascular
  (Cardivascular)
Cardoon *(Cardoons)*
Carduus
Careers *see* Occupations
Carex
Careya
Caria
Cariblatta
Caribou
Carica
Caricaceae

Caricoideae
Caridina
Carina
Carineta
Carissa
Carlina
Carlinosin
Carmenta
Carmine
Carminomycin
Carnations *(Carnation)*
Carnauba
Carnegiea
Carniolan
Carnitine
Carnivora *(Carnivore,*
  Carnivores, Carnivorous)
Carnosine
Carnus
Carobs *(Carob)*
Carolenalone
Caropodium
Carotene *(Carotenes,*
  Carotenogenesis,
  Carotenogenic, Carotin)
Carotenoidoplasts
Carotenoids *(Carotenoid,*
  Carotinoid, Carotinoids)
Carotenol
Carotenone
Carotid
  X-Ref: Intracarotid
Carp *(Carps)*
Carpel *(Carpels)*
  X-Ref: Apocarpy
Carpenteria
Carpesium
Carpesterol
Carpet *(Carpets, Rug)*
  X-Ref: Rugs
Carpinus
Carpobrotus
Carpocapsa
Carpoglyphus
Carpology *(Carpologic)*
Carpophagus
Carpophilus
Carpophores *(Carpophore)*
Carpospores *(Carpospore)*
Carpus *(Carpal)*
Carrageen
Carrageenan *(Carrageenans)*
Carrageenin
Carrion
Carrots *(Carrot)*
Cars *(Automobile, Car)*
  X-Ref: Automobiles
Carterocephalus
Carthamus
Cartilage *(Cartilages,*
  Cartilaginous)
Cartography *(Cartographic)*
Cartons *see* Containers
Carum
Caruncula *(Caruncles)*
Carunculitis

Carya
  X-Ref: Hicoria
Caryedes
Caryedon
Caryokinetic *see* Mitosis
Caryophyllaceae
Caryophyllales
Caryophyllata
Caryophyllene
Caryopses *(Caryopsis)*
Caryopteris
Caryota
Caryotype
Casbene
Cascara
Cascarilla
Casearia
Casehardening
Casein *(Caseins)*
Caseinate
Caseolytes
Cashews *(Cashew,*
  Cashewnut, Cashewnuts)
Cashmere
Casia
Casings *(Casing)*
Casks
Cassava
  X-Ref: Mandioca,
  Manioc
Casseroles
Cassette
Cassia *(Cassias)*
Cassida
Cassididae *(Cassidinae,*
  Cassids)
Cassine
Cassinia
Cassytha
Castanea
Castanopsis
Caste *(Castes)*
Castilleja
Castor
Castorbeans *(Castorbean)*
Castration *(Castrate,*
  Castrated, Castrates,
  Castrating)
  X-Ref: Emasculation,
  Uncastrated
Casuarina *(Casuarinas)*
Casuarinaceae
Catabolism *(Catabolic,*
  Catabolite, Catabolizing)
Cataglyphis
Catalases *(Catalase,*
  Catalasic)
Cataline
Catalogs *(Catalog,*
  Cataloging, Catalogue,
  Catalogued, Catalogues)
Catalpa
Catalponol
Catalysis *(Catalysed,*
  Catalyses, Catalysing,
  Catalyst, Catalysts,

Catalytic, Catalyzed,
  Catalyzer, Catalyzes,
  Catalyzing)
  X-Ref: Autocatalytic,
  Biocatalytic,
  Noncatalytic
Catanthera
Catantopinae
Catapodium
Cataracts *(Cataract)*
Catarrh *(Catarrhal)*
Catasetinae
Catasetum
Catatropis
Cataulacus
Cataxia
Catchflies *(Catchfly)*
Catchment *(Catchments)*
Catchweed
Catechins *(Catechin)*
Catecholaminergic
Catecholamines
  (Catecholamine)
Catechols *(Catechol)*
Catenaria
Catenas *(Catena)*
Catering *(Catered, Caterer)*
Caterpillars *(Caterpillar)*
Catfacing
Catfish
Catha
Catharanthine
Catharanthus
Catharinea
Cathepsins *(Cathepsin,*
  Cathepsines, Catheptic)
Catheterization *(Catheter,*
  Catheterised, Catheters)
Cathidine
Cathodoluminescence
Cationite
Catkins *(Catkin)*
Catnip
Catocala
Catopidae *(Catopinae)*
Catoplatus
Catops
Catopsis
Catostomus
Cats *(Cat, Kitten)*
  X-Ref: Kittens
Catsup *see* Ketchup
Cattails *(Cattail)*
Cattalo
Cattle
Cattlehide
Cattlemen *(Cattleman)*
Cattleya *(Cattleyas)*
Caucalis
Cauda *(Caudae, Caudal,*
  Caudate)
Caulerpa
Caulerpales
Cauliflowers *(Cauliflower)*
Caulobacter
Caulobacteraceae

Caulophyllum
Cauloside
Caustic
Cauterization *(Cauterized.*
Cautery)
X-Ref: Electrocautery,
Microcautery
Cavaraiellia
Cavariella
Caves *(Cave, Cavernicole,*
Cavernicoles,
Cavernicolous, Cavernous)
Cavia
Caviar
Cavitation
Cavities *(Cavity)*
Cavum *(Cava)*
Cayratia
Ceanothus
Cebocephaly
Cebrionidae
Cebus
Cebysa
Cecectomy
Cecidiology *(Cecidia,*
Cecidium, Cecidogenesis,
Cecidology)
Cecidomyia
Cecidomyiidae
(Cecidomyidae)
X-Ref: Itonididae
Cecidophyopsis
Cecidozoaire
Cecropia
Cecum *(Caeca, Caecal,*
Ceca, Cecal)
X-Ref: Caecum, Coecum
Cedars *(Cedar)*
Cediopsylla
Cedrela
Cedrol
Cedrus
Ceiba
Celaenopsidae
Celaenopsoidea
Celandines *(Celandine)*
Celastraceae
Celastrales
Celastrus
Celerio
Celery
Celiac *see* Abdomen
Celite
Cellars *(Cellar, Cellaring)*
Cellia
Cellobiase
Cellobiitol
Cellobiose
Cellobiuronic
Cellogel
Celloidin
Cellolignin
Cellon *see*
Tetrachloroethane
Cellophanes *(Cellophane)*
Cells *(Cell, Celled, Cellular,*

Cellularity)
X-Ref: Acellular,
Anticellular,
Endocellular,
Exocellular,
Extracellular,
Intercellular,
Intracellular,
Noncellular,
Subcellular
Cellulases *(Cellulase)*
Cellules
Cellulolysis *(Cellulolytic)*
Cellulomonas
Cellulosates *(Cellulosate)*
Cellulose *(Celluloses,*
Cellulosic)
Celmisia
Celnovocaine
Celosia
Celsia
Celtidaceae
Celtis
Celyphidae
Cement *(Cemented,*
Cementing, Cements)
Cementum
Cenchrus
Cenocephalus
Cenococcum
Cenocorixa
Cenomanian
Cenopalpus
Cenosis *(Agrocenoses,*
Cenoses, Cenotic,
Coenoses, Coenosis)
X-Ref: Agrocenosis,
Agrophytocoenosis,
Coenotic
Census
Centaurea
Centaureinae
Centdarol
Centella
Centenarians *see* Elderly
Centipedes *(Centipede)*
Centrales
Centranthus
Centrechinus
Centrifuge *(Centrifugal,*
Centrifugally,
Centrifugals, Centrifugate,
Centrifugation,
Centrifugations,
Centrifuged, Centrifuges,
Centrifuging,
Ultracentrifugal,
Ultracentrifugation)
X-Ref: Ultracentrifuge
Centrioles *(Centriole)*
Centripetal
Centris
Centrococcus
Centrolepidaceae
Centrolepis
Centromeres *(Centromere,*

Centromeric)
Centromerus
Centrophantes
Centropogon
Centrosema
Centrosomes
Centrospermae
(Centrosperm)
Centrospora
Centrotinae
Centrotus
Centruroides
Cepha
Cephalanthus
Cephalaria
Cephalenchus
Cephaleuros
Cephalic
Cephalobidae
Cephalobolus
Cephalodesmius
Cephalodia
Cephalonomia
Cephalophora
Cephalopina
Cephalopoda *(Cephalopods)*
Cephaloridine
Cephaloscymnus
Cephalosporin
(Cephalosporins)
Cephalosporium
Cephalotaxus
Cephalozia
Cephapirin
Cephenomyia
Cephidae *(Cephid)*
Cephonodes
Cephus
Ceraclea
Cerambycidae *(Cerambycid,*
Cerambycids,
Cerambycinae)
Cerambycini
Cerambyx
Ceramiaceae
Ceramiales
Ceramic
Ceramide *(Ceramides)*
Ceramidia
Ceramium
Ceraphronidae
Ceraria
Cerastipsocus
Cerastium
Ceratina
Ceratitis
Ceratixodes
Ceratocystis
Ceratodon
Ceratoides
Ceratonia
Ceratonyx
Ceratophyllaceae
Ceratophyllidae
Ceratophyllum
Ceratophyllus

Ceratophyus
Ceratopogon
Ceratopogonidae
(Ceratopogonid,
Ceratopogonids)
Ceratopogonini
Ceratopteris
Ceratosanthes
Ceratosporium
Ceratostigma
Ceratostomella
Ceratozamia
Ceratozetes
Ceratozetidae
Cercal *see* Tail
Cercaria *(Cercariae)*
Cercerini
Cerceris
Cerci
Cercidiphyllum
Cercidium
Cercion
Cercis
Cercocarpus
Cercomegistus
Cercopidae *(Cercopid)*
Cercopithecus
Cercospora *(Cercosporae)*
Cercosporella
Cercosporidium
Cercyon
Cercyonis
Cereals *(Cereal)*
Cerebellum *see* Brain
Cerebrosides *(Cerebroside)*
Cerebrospinal
Cerebrum *see* Brain
Cerenopus
Ceresan *see* Granosan
Cereus
Ceriagrion
Ceriomydas
Cerium *(Ceric, Cerous)*
Cerodontha
Ceroid
Ceropalidae
Ceropegia
Ceroplastes
Ceroptin
Cerotoma
Certification *(Certified)*
Cerulenin
Ceruloplasmin
(Ceruloplasmine)
Cerura
Cervicola
Cervidae *(Cervids)*
Cervix *(Cervical)*
Cervus
Cerylonidae
Cesalpinia
Cesarean *see* Birth
Cesium *(Caesium)*
Cestocides *(Cestocidal)*
Cestoda *(Cestode, Cestodes,*
Metacestodes)

X-Ref: Metacestode
Cestodiasis *(Cestodiases)*
Cestrum
Cetacea *(Cetacean,*
Cetaceans)
Cetonia
Cetoniidae
Cetoniinae
Cetraria
Cetyl
Cetyltrimethylammonium
Ceuthophilus
Ceuthorhynchidius
Ceuthorrhynchus
Ceuthospora
Chabertia
Chactidae
Chadefaudiellaceae
Chaenomeles
Chaerophyllum
Chaeta
Chaetadoretus
Chaetanaphothrips
Chaetocanace
Chaetoceros
Chaetocnema
Chaetocoelopa
Chaetomiaceae
Chaetomium
Chaetomorpha
Chaetopeltis
Chaetophoraceae
Chaetophorineae
Chaetopsylla
Chaetosargus
Chaetosiphon
Chaetosticta
Chaetotaxy
Chafers *(Chafer)*
Chaff
Chaffcutter
Chagas
Chagasia
Chainsaw
Chaitophorus
Chalastogastra
Chalaza
Chalazion
Chalcididae *(Chalcid,*
Chalcids)
Chalcidoidea *(Chalcidoid,*
Chalcidoids)
Chalcionellus
Chalcodermus
Chalcoela
Chalcone *(Chalcones)*
Chalcosmia
Chalcosoma
Chalicodoma
Chalkland
Chalone
Chalybion
Chamaecyparis
Chamaedaphne
Chamaedorea
Chamaemyia

Chamaemyiidae
Chamaenerion
(Chamaenerium)
Chamaesiphon
Chamazulene
Chambers *(Chamber)*
Chameleon
Chamigrene *(Chamigrenes)*
Chamise
Chamois
Chamonixia
Champagne
(Champagnization,
Champagnized)
Champia
Champignon *see*
Mushrooms
Channels *(Channel,*
Channeling,
Channelization,
Channelized,
Channelling)
X-Ref: Multichannel
Chanoclavine
Chaoboridae
Chaoborus
Chaparral *(Chapparal)*
Char
Chara
Characeae *(Characean)*
Characiaceae
Charaxes
Charaxinae
Charcoal *(Charcoals)*
Chards *(Chard)*
Charips
Charlock
Charolais *(Charollais)*
Charophyceae
(Charophycean)
Charophyta *(Charophyte,*
Charophytes)
Charts *see* Mapping
Chascolytrum
Chasmanine
Chasmodon
Chasmogamy
Chauliognathus
Chaulmoogra
Chavicol
Chaya
Chayote
Cheatgrass
Cheddar *(Cheddaring)*
Cheeseburgers *see*
Hamburgers
Cheesecake *(Cheesecakes)*
Cheeses *(Cheese,*
Cheesemaking, Cheesy)
Chefs *(Chef)*
Cheilanthes
Cheiloneurus
Cheilotrichia
Cheimatobia
Cheiranthus
Cheirodendron

Cheiroplatys
Chelacaropsis
Chelatases *(Chelatase)*
Chelates *(Chelate, Chelated,*
Chelating, Chelation,
Chelators)
X-Ref: Unchelated
Cheletomimus
Cheleutoptera
Chelicerae *(Cheliceral)*
Chelicerata
Chelidonium
Chelidurella
Cheliomyrmex
Chelis
Chelisochidae
Chelonarium
Chelone
Chelonethidea
(Chelonethida)
Chelonus
Chelostoma
Chemicalization
Chemiluminescence
(Chemiluminescent,
Chemoluminescence)
X-Ref:
Photochemiluminescence
Chemistry *(Chemical,*
Chemically, Chemicals,
Chemism, Chemistries)
Chemists *(Chemist)*
Chemoautotrophy
(Chemoautotrophic)
Chemochorial *see* Chorion
Chemodectoma
Chemodenervated *see*
Denervation
Chemogenetical *see*
Genetics
Chemomutants
Chemoprophylaxis
(Chemoprophylactic)
Chemoreception
(Chemoreceptive,
Chemoreceptor,
Chemoreceptorial,
Chemoreceptors,
Chemoresponsive)
Chemosensitivity
(Chemosensitive,
Chemosensory)
Chemostat *(Chemostats)*
Chemosterilants
(Chemosterilant,
Chemosterilization,
Chemosterilized,
Chemosterilizing)
Chemosynthesis
(Chemosynthetic)
Chemosystematics *see*
Chemotaxonomy
Chemotaxis *(Chemotactic,*
Chemotactics)
Chemotaxonomy
(Chemosystematic,

Chemotaxonomic)
X-Ref: Chemosystematics
Chemotherapy
(Chemotherapeutants,
Chemotherapeutic,
Chemotherapeutical,
Chemotherapeutically,
Chemotherapeuticals,
Chemotherapeutics)
Chemotropism
(Chemotropic)
Chemurgy *(Chemurgic)*
Chen
Chenilles
Chenodeoxycholic
Chenopodiaceae
(Chenopodiaceous)
Chenopodium
Cherimoya
Chermes
Chermidae *see* Psyllidae
Chernetidae
Chernozem *(Chernozemic,*
Chernozems,
Nonchernozemic)
X-Ref: Nonchernozem
Cherries *(Cherry)*
Chert
Chervil
Chest *see* Thorax
Chestnuts *(Chestnut)*
Cheviot
Chevon
Chewing *(Chew)*
Cheyletidae *(Cheyletid,*
Cheyletids)
Cheyletiella
Cheyletiellidae
Cheyletus
Chiana *(Chianina)*
Chiasma *(Chiasmata)*
Chickens *(Chicken)*
Chickpeas *(Chickpea)*
Chicks *(Chick)*
Chickweed
Chicory *(Chickory)*
Chiggers *see* Trombiculidae
Chikungunya
Chikuwa
Childbearing *see* Birth
Children *(Child, Childhood)*
X-Ref: Preschoolers,
Schoolchildren, Toddler
Chilis *(Chillies)*
X-Ref: Chilli
Chill *see* Cool
Chilli *see* Chilis
Chillproofing
Chilo
Chilocorini
Chilocorus
Chilophaga
Chilopoda *(Chilopod,*
Chilopods)
Chiloscyphus
Chilotraea

Chilus
Chimaphila
Chimera *(Chimaera,*
  Chimaeras, Chimaeric,
  Chimaerism, Chimeral,
  Chimeras, Chimeric,
  Chimerical, Chimerism)
Chimneys *(Chimney)*
Chimpanzees *(Chimpanzee)*
Chinchillas *(Chinchilla)*
Chinoin
Chinone
Chinosol
Chionachne
Chionanthus
Chionaspis
Chionea
Chionochloa
Chionodoxa
Chionoecetes
Chionographis
Chios
Chip *(Chipper, Chippers,*
  Chipping, Chips)
Chipboard *(Chipboards)*
  X-Ref: Flakeboard
Chipmunks *(Chipmunk)*
Chiracanthium
Chiranthodendron
Chiraplothrips
Chirodiscoides
Chiromyzinae
Chironomidae *(Chironomid,*
  Chironomids,
  Chironominae,
  Chironomini)
  X-Ref: Tendipedidae
Chironomus
  X-Ref: Tendipes
Chiroptera *(Chiropteran,*
  Chiropterans)
Chiropterophily
  (Chiropterophils)
Chirosia
Chirothripoides
Chirothrips
Chirping *(Chirp)*
Chiseling *(Chisel)*
Chitin *(Chitinophilic,*
  Chitinous)
Chitinase
Chitosan
Chitting *see* Sprout
Chives *(Chive)*
Chlaenaceae
Chlamisus
Chlamydia *(Chlamydiae,*
  Chlamydial)
  X-Ref: Bedsonia
Chlamydiaceae
Chlamydiosis
  X-Ref: Bedsoniasis
Chlamydobotrys
Chlamydocarya
Chlamydomonadaceae
Chlamydomonas

Chlamydospores
  (Chlamydospore)
Chlamydosporites
Chloral
Chloramben
  X-Ref: Amiben
Chloramine
Chloramphenicol
Chloraniformethan
  X-Ref: Imugan
Chloranil
Chloranthus
Chlorate
Chlorcamphene
Chlorcyclizine
Chlordane
Chlordecone
  X-Ref: Kepone
Chlordene
Chlordiazepoxide
Chlordimeform
Chlorella
Chlorellaceae
Chlorethyl
  X-Ref: Anodynon
Chlorethylphosphonic
Chlorfenvinphos
  (Chlorofenvinphos)
Chlorflurenol
Chlorhexidine
Chloric
Chloridea
Chloridella
Chlorides *(Chloride)*
Chloridolum
Chlorina
Chlorination *(Chlorinated,*
  Chlorinating, Chlorinator)
Chlorine
Chloriona
Chloris
Chlorite *(Chlorites)*
Chlormadinone
Chlormephos
Chlormequat
Chloroacetamides
  (Chloroacetamide)
Chloroacetyl
Chloroacyl
Chloroamphetamine
Chloroanilines
  (Chloroaniline)
Chloroanisoles
Chloroazobenzene
Chlorobenzamide
Chlorobenzilate
Chlorobenzoic
Chlorobenzoyl
Chlorobenzylammonium
Chlorobiaceae
Chlorobium
Chlorobotrys
Chlorocarbanilate
Chlorocholine
  (Chlorcholine)
Chlorochroa

Chloroclystis
Chlorocnemis
Chlorococcales
Chlorococcum
  (Chlorococcal)
Chlorocrotyl
Chlorodeoxycellulose
Chlorodibenzofurans
Chlorodimeform
Chlorodiphenyls
Chloroethane *(Chlorethane)*
Chloroethanephosphonic
Chloroethanol
Chloroethylarylamines
Chloroethylphosphonic
Chloroethylthiamine
Chlorofluoromethanes
Chloroform
Chlorofos *see* Trichlorfon
Chlorogenic
Chlorogloea
Chlorogonium
Chlorohydrate
Chlorohydrin
Chlorohydroxide
Chloroisocyanates
Chloromequat
Chloromerus
Chloromonas
Chloronaphthalenes
Chloroneb
Chloroorganic
Chloroperlidae
Chloroperoxidase
Chlorophacinone
  X-Ref: Drat
Chlorophanus
Chlorophenols
  (Chlorophenolic)
Chlorophenoxy
Chlorophenoxypropionic
Chlorophenyl
Chlorophenylacetate
Chlorophenylalanine
Chlorophenylhydrazone
Chlorophos *see* Trichlorfon
Chlorophyceae
  (Chlorophycean)
Chlorophyll *(Chlorophyl,*
  Chlorophyllic,
  Chlorophyllous,
  Chlorophylls)
  X-Ref: Pyrochlorophyll
Chlorophyllases
  (Chlorophyllase)
Chlorophyllides
  (Chlorophyllide)
Chlorophyllin
Chlorophyta *(Chlorophyte)*
Chlorophytum
Chloropicolinic
Chloropicrin
Chloropidae
Chloroplasts *(Chloroplast,*
  Chloroplastic,
  Chloroplastid)

X-Ref: Subchloroplast
Chloroplatinate
Chloroplatinic
Chloropromazine
Chloropropane
Chloropropham
  (Chlorpropham)
Chloropropylate
Chloroprothixene
Chlorops
Chloropulvinaria
Chloroquine
Chloroquinoline
Chlorosis *(Chlorotic)*
Chlorothalonil
Chlorothioformates
Chlorotriazine
Chlorotryptophane
Chlorouracil
Chlorous
Chlorowax
Chloroxuron
  X-Ref: Tenoran
Chloroxylon
Chlorphenamidine
Chlorpromazine
Chlorpyrifos
  X-Ref: Lorsban
Chlorquinaldol
Chlortetracycline
  X-Ref: Aureomycin,
    Biomycin
Chlorthiamid
  (Chlorothiamid)
Chlortoluron
Chlosyne
Choanephora
Choanephoraceae
Chocolate
Chodatella
Choetospila
Choisya
Chokeberries *(Chokeberry)*
Choking
Chokla
Cholam *see* Durra
Cholate
Cholecalciferol
Cholecyst *see* Gallbladder
Cholecystitis
Cholecystokinesis
Cholecystokinin
Cholelithiasis
Cholera
Choleragen
Choleresis *(Choleretic)*
Cholest
Cholestane *(Cholestan,*
  Cholestanes)
Cholestasis *(Cholestatic)*
Cholesteremia
  (Cholesterolemia)
Cholesterol
  (Antihypercholesterolemic,
  Cholesteric, Cholesterin,
  Cholesterols, Cholesteryl)

X-Ref:
  Antihypercholesterol
Cholestyramine
Choleva
Cholinae
Cholinergic
Cholines *(Choline)*
Cholinesterases
  (Anticholinesterasic,
  Cholinesterase,
  Cholinesterasic,
  Pseudocholinesterase)
X-Ref:
  Anticholinesterase,
  Pseudocholinesterases
Cholinomimetic
Cholla
Chomelia
Chondrilla
Chondrococcus
Chondrocytes *(Chondrocyte)*
Chondrodystrophia *see*
  Achondroplasia
Chondrogeneses
  (Chondrogenesis,
  Chondrogenetic,
  Chondrogenic,
  Chondrogenous)
Chondroitin
Chondroplea
Chondrosarcoma
Chondrostoma
Chondrus
Chopardia
Chopping *(Chop, Chopped,*
  Chopper, Choppers,
  Chops)
Chorchorus
Chorda
Chordeumida
Chordotonal
Choreocolaceae
Choreutis
Chorinaeus
Chorioallantois
  (Chorioallantoic)
Choriomeningitis
Chorion *(Chorionic)*
X-Ref: Chemochorial
Choriongonadotropin
Chorioptes *(Chorioptic)*
Chorioretina
  (Chorioretinitis)
Chorisia
Chorismate
Choristoneura
Chorizococcus
Choroid *(Chorioid)*
Chorology *see* Biogeography
Choroterpes
Chorthippus
Chortoicetes
Chortophaga
Chorusing
Chosenia
Chriodes

Christella
Chromaffin
Chroman
Chromanols
Chromate
Chromatic
Chromatids *(Chromatid)*
X-Ref: Subchromatid
Chromatin *(Chromatine,*
  Chromatins)
Chromatium
Chromatography
  (Chromatogram,
  Chromatograms,
  Chromatograph,
  Chromatographic,
  Chromatographical,
  Chromatographically,
  Chromatographs,
  Radiochromatographic)
X-Ref:
  Radiochromatography
Chromatomyia
Chromatophores
  (Chromatophore,
  Chromatophorotropic)
Chromazonarol
Chromelosporium
Chromen
Chromenoflavanones
Chromenones
Chromic
Chromite
Chromium *(Chrome,*
  Chromous)
Chromobacterium
Chromocenters
  (Chromocenter,
  Chromocentre,
  Chromocentres,
  Chromocentric)
Chromogens *(Chromogen,*
  Chromogenic)
Chromolaena
Chromomeres
  (Chromomeric)
Chromones *(Chromone)*
Chromophores
  (Chromophore)
Chromoplasts *(Chromoplast)*
Chromoproteins
  (Chromoprotein)
Chromosomes
  (Chromosomal,
  Chromosomally,
  Chromosome,
  Chromosomic)
X-Ref:
  Extrachromosomal,
  Interchromosomal,
  Intrachromosomal
Chromotrope *(Chromatrope)*
Chromotropism
  (Chromatotropic,
  Chromatotropism)
Chronobiology

Chronology *(Chronological,*
  Chronologies)
Chronometer
Chronosequence
Chronotropism
  (Chronotropic)
Chroococcaceae
Chroococcales
Chrotogonus
Chrysactinia
Chrysalises *(Chrysalid,*
  Chrysalids, Chrysalis)
Chrysanthemates
  (Chrysanthemate)
Chrysanthemic
Chrysantheminae
Chrysanthemoides
Chrysanthemums
  (Chrysanthemum)
Chrysauginae
Chrysididae
Chrysis
Chrysobalanaceae
Chrysobothris
Chrysocapsa
Chrysocarabus
Chrysocharis
Chrysochloa
Chrysochroma
Chrysococcus
Chrysocoma
Chrysocoris
Chrysodeixis
Chrysoeriol
Chrysogaster
Chrysogonum
Chrysolina
Chrysomela
Chrysomelidae
  (Chrysomelid)
Chrysomelinae
Chrysomeloidea
Chrysomonadina
  (Chrysomonad)
Chrysomphalus
Chrysomya *(Chrysomyia)*
Chrysomyxa
Chrysopa
Chrysophanus
Chrysophyceae
Chrysophyta *(Chrysophyte,*
  Chrysophytes)
Chrysopidae *(Chrysopid)*
Chrysopilus
Chrysopinae
Chrysopogon
Chrysops
Chrysopsis
Chrysospermum
Chrysosplenium
Chrysosporium
Chrysoteuchia
Chrysothamnus
Chrysotoxum
Chrysozona
Chthoniidae

Chub *(Chubs)*
Chukar
Chunroides
Churches
Churning *(Churn)*
Churro
Chutes *(Chute)*
Chutney *(Chutneys)*
Chyle
Chylocladia
Chylomicrons
  (Chylomicron)
Chylothorax
Chyme
Chymomyza
Chymopapain
Chymosin
Chymotrypsin
  (Chymotrypsinogen,
  Chymotrypsins,
  Chymotryptic)
Chytridiaceae
Chytridiales *(Chytrid,*
  Chytridiaceous, Chytrids)
Chytridiomycetes
  (Chytridiomycete)
Chytridiopsis
Chytridium
Chytriomyces
Ciboriopsis
Cibotium
Cicada *(Cicadas, Cicads)*
Cicadellidae *(Cicadellid,*
  Cicadellinae)
Cicadelloidea
Cicadetta
Cicadidae
Cicadina *see* Cicadoidea
Cicadinea
Cicadoidea
  X-Ref: Cicadina
Cicadulina
Cicatrix *(Cicatricial)*
Cicer
Cichlidae *(Cichlid)*
Cichoriaceae
Cichorieae
Cichorioideae
Cichorium
Cicindela
Cicindelidae *(Cicindelids,*
  Cicindelinae)
Cicinnobolus
Cicuta
Cidaria
Cidarus
Cider *(Ciders)*
Cidial
Cienfuegosia
Cigarettes *(Cigarette)*
Cigars *(Cigar)*
Ciguatoxin
Ciidae *see* Cisidae
Cilia *(Ciliary, Ciliate,*
  Ciliated, Ciliogenesis)
  X-Ref: Unciliated

Ciliatine
Cimbicidae
Cimetidine
Cimex
Cimicidae
Cimicifuga
Cinara *(Cinaran)*
Cinaropsis
Cinchona
Cinclidotus
Cineangiocardiography
Cinefluoroscopy
  (Cinefluoroscopic)
Cinematography *(Cinematic,*
  Cinematographic)
Cinemicrography
Cineole
Cineraria
Cinerarin
Cinerin
Cineroentgenography
  (Cineroentgenographic)
Cingalobolus
Cinnamaldehyde
Cinnamate
Cinnamic
  X-Ref: Transcinnamic
Cinnamolide
Cinnamomum
Cinnamon
Cinnamosmolide
Cinnamoyl
Cinnamoylhistamine
Cinnamyl
Cinnamylphenols
Cinquefoil
Cintractia
Ciodrin
Cionus
Circadian
Circaeaster
Circobotys
Circocylliba
Circulation *(Circulated,*
  Circulating, Circulations,
  Circulatory,
  Microcirculatory)
  X-Ref: Microcirculation,
    Recirculation
Circulifer
Circumanal *see* Anus
Cirina
Cirphis
Cirrhaea
Cirrhosis
Cirrus
Cirsium
Cisidae
  X-Ref: Ciidae
Cissampelos
Cissus
Cistaceae
Cisternae
  X-Ref: Intercisternal
Cistron *(Cistrons)*
Cistus

Citellus
Citheroniidae
Citragil
Citral
Citrange
Citrates *(Citrate, Citrated)*
Citreoviridin
Citrex
Citric
Citrinin
Citrobacter
Citromycetin
Citronella
Citronellol
Citronellylamine
Citrons *(Citron)*
Citropsis
Citrulline
Citrullus
Citrus
Civilization *(Civilized)*
Cixiidae *(Cixiid)*
Cladietum
Cladina
Cladioneura
Cladobotryum
Cladocera *(Cladoceran)*
Cladonia
Cladoniaceae
Cladophora
Cladophoraceae
Cladopus
Cladosiphon
Cladosporium
Cladrastis
Clams
Clania
Clarifier
Clarkia
Classroom *see* Education
Clasterosporium
Clastes
Clastoptera
Clathridium
Claudberries *(Claudberry)*
Claudopus
Clausena
Clausenia
Clausia
Claustrum
Clavaria
Clavariaceae
  (Clavariaceous)
Clavatol
Claviceps
Clavicle
  X-Ref: Subclavian
Clavicornia
Clavigerinae
Clavigralla
Clavine
Clavulinopsis
Claws *(Claw)*
Clay *(Argillic, Clayey,*
  Claying, Clayish, Clays)
  X-Ref: Argillaceous

Claypan
Claytonia
Cleaning *(Clean,*
  Cleanability, Cleaned,
  Cleaner, Cleaners,
  Cleanliness, Cleanness,
  Cleans, Cleansers,
  Cleansing, Cleanup)
Clearcutting *(Clearcut,*
  Clearcuts, Clearcuttings,
  Clearfelling)
Clearings *(Clearing)*
Cleavage *(Cleavages, Cleave,*
  Cleaved, Cleaves,
  Cleaving)
  X-Ref: Precleavage
Cleft
Cleistocactus
Cleistocalyx
Cleistogamy *(Cleistogamous,*
  Cleistogenes,
  Cleistogenous)
Cleistoiodophanus
Cleistothecia
  (Cleistocarpous,
  Cleistocarps,
  Cleistothecial,
  Cleistothecium)
Clematis
Clematoside
Cleome
Cleomeprenols
Cleomis
Cleonini
Cleonus
Cleopus
Cleora
Clepsis
Cleptodromia
Cleridae *(Clerid)*
Clerodane
Clerodendrum
  (Clerodendron)
Clethrionomys
Cletus
Climacia
Climaciella
Climacium
Climate *(Agroclimatic,*
  Agroclimatological,
  Bioclimates, Bioclimatic,
  Bioclimatology, Climates,
  Climatic, Climatical,
  Climatically,
  Climatography,
  Climatological,
  Climatology,
  Climatonomy,
  Climosequence)
  X-Ref: Agroclimatology,
    Bioclimate,
    Homoclimates
Climbers *(Climber,*
  Climbing)
Clinicopathology
  (Clinicopathologic,

  Clinicopathological)
Clinohelea
Clinopodium
Clinoptilolite
Clinostats *(Clinostat)*
Clinteria
Clintonia
Clioxanide
Clipping *(Clipper, Clippings)*
Clitocybe
Clitopilus
Clitoria
Clitoris *(Clitoral)*
Clivia
Clivina
Cloaca *(Cloacal)*
Cloches *(Cloche, Cloched)*
Clods *(Clod, Cloddiness)*
Cloeon
Clogging
Clogmia
Clomiphene
Clone *(Clonal, Clonally,*
  Cloned, Clones, Cloning)
  X-Ref: Intraclonal,
    Subclones
Clonidine
Clonorchis
Clophen
Clopidol *see* Meticlorpindol
Cloprostenol
Closterium
Closterocerus
Closterothrix
Clostridium *(Clostridia,*
  Clostridial)
Clothing *(Clothes)*
  X-Ref: Apparel,
    Garments
Cloths *see* Textiles
Clotted *see* Coagulation
Cloudberries *(Cloudberry)*
Clouds *(Cloud, Cloudy)*
  X-Ref: Cumulus
Clover *(Clovers)*
  X-Ref: Senji, Sweetclover
Cloves *(Clove)*
Cloxacillin
Clubiona
Clubionidae
Clubroot
Clubs
Clunio
Clusia
Clusiaceae
Cluster *(Clustered,*
  Clustering, Clusters)
Clutch
Clytra
Clytrinae
Clytus
Cnaphalocrocis
Cnephasia
Cnephia
Cnicus
Cnidoscolus

Cnidospora
Coacervation
Coagulants *(Anticoagulant,*
   Coagulant, Coagulator)
   X-Ref: Anticoagulants
Coagulase
Coagulation
   (Anticoagulative, Clot,
   Clotpromoting, Clotting,
   Coagula, Coagulability,
   Coagulated, Coagulating,
   Coagulative, Coagulum)
   X-Ref: Anticoagulation,
   Clotted
Coagulothrombocytograms
Coal *(Coalfield, Coalfields,*
   Coals, Colliery)
Coalescence
Coarctation
Coating *(Coat, Coated,*
   Coatings, Coats,
   Uncoated)
   X-Ref: Uncoating
Coattractants *see*
   Attractants
Coaxial *see* Axis
Cobalamins
Cobalt *(Cobaltous)*
Cobamides *(Cobamide)*
Coban
Cobex
Cobinamides
Cobnuts
Cobra
Cobs *see* Corncobs
Coca
Cocaine
Coccidae *(Cocci, Coccid,*
   Coccids)
Coccidia *(Coccidial,*
   Coccidian)
Coccidiascus
Coccidioides
Coccidioidomycosis
Coccidiosis
Coccidiostat *(Anticoccidial,*
   Coccidiostatic,
   Coccidiostatics,
   Coccidiostats)
   X-Ref: Anticoccidials
Coccidiphila
Coccidium *see* Eimeria
Coccinella
Coccinellidae *(Coccinellid,*
   Coccinellids)
Coccinellina
Coccinia
Coccochloris
Coccoidea *(Coccoidean)*
Coccolithineae
Coccolithophoridae
   (Coccolithophorid,
   Coccolithophorids)
Coccolithus
Coccomyces
Cocconeis

Coccophagus
Coccopilatus
Coccotrypes
Cocculus
Coccus
Coccyx *(Coccygeal)*
Cochineal
Cochlea *(Cochlear)*
Cochleanthes
Cochlearia
Cochliobolus
Cochliomyia
Cochlospermum
Cockchafers *(Cockchafer)*
Cockerels *(Cockerel)*
Cockleburs *(Cocklebur)*
Cockroaches *(Cockroach)*
Cocks *(Cock)*
Cocksfoot *see* Orchardgrass
Cocktails *(Cocktail)*
Cocles
Cocoa
Coconuts *(Coco, Coconut)*
Cocoons *(Cocoon,*
   Cocooning)
Cocos
Cod *(Cods)*
Codeine
Codiaeum *(Codiaeums)*
Codinaeopsis
Coding *(Code, Coded,*
   Coder, Coders, Codes,
   Codex, Codification,
   Codifying, Encoded,
   Encoder, Encoders)
   X-Ref: Encoding
Codium
Codon
   X-Ref: Anticodon
Codonocarpus
Codonopsis
Coecum *see* Cecum
Coelaenomenodera
Coelalysia
Coelanthe
Coelastrum
Coeliac *see* Abdomen
Coelidia
Coelidiinae
Coeliopsis
Coelogyne *(Coelogynes)*
Coeloides
Coelom *(Coelomic)*
   X-Ref: Intracoelomal
Coelomomyces
Coelomomycetaceae
Coelomycetes *(Coelomycete)*
Coelomycidium
Coelopa
Coelophora
Coemansia
Coenagrion
Coenagrionidae
   (Coenagriidae,
   Coenagrionid)
Coenocyte *(Coenocytic)*

Coenomyia
Coenomyiidae
Coenonycha
Coenonympha
Coenosia
Coenotic *see* Cenosis
Coenurosis
Coenurus
Coenzymes *(Coenzyme)*
Coevolution *see* Evolution
Coffee *(Coffea, Coffees)*
Caffeine *see* Caffeine
Cognac *see* Brandies
Cohabitation
Cohesion *(Cohesive)*
   X-Ref: Noncohesive
Coir
Coitus *see* Copulation
Coix
Coke
Cokeromyces
Cola *(Kola)*
   X-Ref: Kolanut
Colacium
Colaspini
Colaspis
Colastes
Colcemid
Colchamine
Colchicine *(Colchicin)*
Colchicum
Colchiploids *(Colchiploid)*
Colchitetraploids
   (Colchitetraploidy)
Cold *(Colder)*
   X-Ref: Subzero
Cole
Coleochaeta
Coleochaete
Coleogyne
Coleomegilla
Coleophoma
Coleophora
Coleophoridae
Coleoptera *(Coleopteran,*
   Coleopterological,
   Coleopterous)
Coleopterists
Coleoptiles *(Coleoptile)*
Coleosoma
Coleosporium
Coleus
Colias
Colibacillosis *(Colibacillary,*
   Colibacteriosis)
Colibacillus *(Colibacillic)*
Colibacteria *see* Coliforms
Colic
Colicins *(Colicin, Colicine,*
   Colicines, Colicinogenic)
Colienteritis
Colienterotoxemia
Coliforms *(Coliform)*
   X-Ref: Colibacteria
Coligranuloma *see*
   Granulomas

Colinus
Colipase
Coliphages
Colisepticaemia *see*
   Septicemia
Colitis
Colladonus
Collagen *(Collagenolytic,*
   Collagenous, Collagens)
   X-Ref: Procollagen
Collagenase
Collagenosis
Collard *see* Kale
Collateral *(Collaterals)*
Collecephalus
Collections *(Collection)*
Collectives *(Collective,*
   Collectivism,
   Collectivization)
   X-Ref: Intercollective
Collectors *(Collector)*
Colleges *see* Education
Collema
Collemataceae
Collembola *(Collembolan)*
Collenchyma
Colleterium *(Colleterial)*
Colletidae
Colletodiol
Colletotrichin
Colletotrichum
Colliculus *(Collicle,*
   Collicular, Colliculi)
Collie *(Collies)*
Colliguaja
Collinsia
Colloids *(Colloid, Colloidal)*
Collops
Colloquims *see* Conferences
Collybia
Collyriclum
Collyrini
Colmatage
Colneleic
Colnelenic
Colobopterus
Colocasia
Colocleora
Cololejeunea
Colon *(Colonic)*
Colonies *(Colonial, Colony)*
   X-Ref: Microcolony
Colonization *(Colonisation,*
   Colonising, Colonists,
   Colonize, Colonized,
   Colonizing)
   X-Ref: Recolonization
Colopha
Colophony *see* Rosin
Color *(Coloration,*
   Colorations, Colored,
   Coloring, Colorings,
   Colorless, Colors, Colour,
   Colouration, Coloured,
   Colouring, Colours)
   X-Ref: Multicolor

Coloradoa
Colorants *(Colorant)*
Colorimetry *(Colorimeter,*
  Colorimeters,
  Colorimetric)
Colostomy *(Colostomised)*
Colostrum *(Colostral)*
  X-Ref: Precolostral
Colotis
Colpidium
Colpocephalum
Colpoda
Colpomenia
Coltricia
Colts *(Colt)*
Colubridae
Colubrina
Columba
Columbicola
Columbiformes
  (Columbiform)
Columellia
Columelliaceae
Columnea
Columnocystis
Colutea
Colydiidae
Colymbetes
Colza
Coma
Comandra
Combines *(Combine)*
Combo
Combretaceae
Combretum
Combs
Combustion *(Combustibility,*
  Combustible)
  X-Ref: Incombustibility,
  Noncombustible
Cometabolism *see*
  Metabolism
Comfrey
Comiso *(Comisana)*
Commelina
Commelinaceae
Commensalism
  (Commensal,
  Commensalic)
Commerce *(Commercial,*
  Commercialization,
  Commercially)
Comminution
  (Comminuted)
Commiphora
Commodities *(Commodity)*
Communication
  (Communicate,
  Communicating,
  Communications,
  Communicators)
  X-Ref:
  Telecommunications
Communism *(Communist,*
  Communists)
Communities *(Community)*

Compaction *(Compact,*
  Compacted, Compacting,
  Compactness)
Comparettia
Compendium
Compensation *(Compensate,*
  Compensatory)
Complexometry
  (Complexometric)
Complexonometry
  (Complexonometric)
Compositae
Compost *(Composted,*
  Composting, Composts)
Compression *(Compressed,*
  Compressibility,
  Compressing,
  Compressional,
  Compressions,
  Compressive)
Compsobata
Compsocerocoris
Compsomelissa
Compsopogon
Compsothrips
Comptonia
  X-Ref: Sweetfern
Computers *(Computer,*
  Computerisation,
  Computerised,
  Computerization,
  Computerized,
  Computerizing)
  X-Ref: Minicomputers
Comstockaspis
Comstockiella
Conalbumin
Conanthalictus
Concanavalin
  (Concanavaline)
Concentrates *(Concentrate,*
  Concentrating)
Concentric
Conceptus *see* Fetus
Conchylis
Concrete *(Concretes)*
Condalia
Condiments
Conditioning *(Conditioned,*
  Conditioner, Conditioners)
  X-Ref: Preconditioning
Conductivity *(Conductance,*
  Conductances,
  Conductibility,
  Conducting, Conduction,
  Conductive, Conductor,
  Conductors)
Conductometry
  (Conductimetric,
  Conductometric)
Conduits *(Conduit)*
Condyle
  X-Ref: Intercondylar
Cones *(Cone, Conelets,*
  Conical)
  X-Ref: Strobili

Confectionery
  (Confectionary,
  Confectioneries,
  Confectioners,
  Confections)
Conferdione
Conferences *(Colloquim,*
  Congress, Convention,
  Seminar, Session)
  X-Ref: Colloquims,
  Congresses,
  Conventions, Seminars,
  Sessions
Conferin
Conferol
Conferone
Confinement *(Confine,*
  Confined, Confines,
  Confining)
  X-Ref: Unconfined
Conformaton
Congelation
Congenital *(Congenitally)*
Congestion *(Congestive)*
Conglomerates
  (Conglomerate,
  Conglomeration)
Conglutinin
Congresses *see* Conferences
Coniceine
Conidiobolus
Conidiophores
  (Conidiophore)
Conidium *(Conidia,*
  Conidial, Conidian,
  Conidiation,
  Conidiogenesis,
  Conidiospore,
  Conidiospores,
  Microconidiogenesis)
  X-Ref: Aconidial,
  Microconidiating
Coniella
Coniferae *(Conifer,*
  Coniferous, Conifers)
Coniferaldehyde
Coniferin
Coniferyl
Coniochaeta
Coniophora
Coniopterygidae
Coniopteryx
Conioscinella
Coniosporium
Coniosternum
Coniothecium
Coniothyrium
Conium
Conjugase
Conjugate *(Conjugates,*
  Conjugation,
  Conjugations)
  X-Ref: Unconjugated
Conjunctiva *(Conjunctival)*
Conjunctivitis
Connaraceae

Conocephalidae
Conocephalinae
Conocephalum
Conocephalus
Conocybe
Conoderus
Conomorium
Conomyrma
Conopharyngia
Conophor
Conophorus
Conophthorus
Conophytum
Conopidae *(Conopinae)*
Conospermum
Conostigmus
Conotrachelus
Conotrema
Conotylidae
Conradina
Consanguinity
Consciousness *(Conscious)*
Conservation *(Conservancy,*
  Conservational, Conserve,
  Conserved, Conserves,
  Conserving)
Conservationists
  (Conservationist)
Consistometry
  (Consistometric)
Consolida
Constipation
Constriction
Consultation *(Consultant,*
  Consultants, Consultative)
Consumers *(Consumer,*
  Consumerism)
Consumption *(Consumable,*
  Consumed, Consumes,
  Consuming,
  Consumptions,
  Consumptive)
Containerization
  (Containerisation,
  Containerized)
Containers *(Carton,*
  Cartoned, Container)
  X-Ref: Cartons
Contamination
  (Contaminant,
  Contaminants,
  Contaminated,
  Contaminating,
  Contaminations,
  Monocontaminated)
  X-Ref:
  Monocontamination,
  Unsanitary
Contarinia
Contortae
Contours *(Contour,*
  Contouring)
Contrabithorax
Contraceptives
  (Contraceptive)
  X-Ref: Antifertility

Contractions *(Contractile,*
Contractility, Contraction,
Contracture)
Contracts *(Contract,*
Contractor, Contractors,
Contractual)
Convalescence
(Convalescent,
Convalescing)
Convallaria
Convection *(Convective,*
Convector, Convectors)
Conventions *see*
Conferences
Convergence *(Convergent,*
Converging)
Conveyors *(Conveyance,*
Conveyer, Conveying,
Conveyor, Conveyorized)
Convoluta
Convolution *(Convolutions)*
Convolvulaceae
Convolvulus
Convulsions *(Convulsion,*
Convulsive)
X-Ref: Anticonvulsant
Conwentzia
Conyza
Cookeina
Cookery *(Cook, Cooked,*
Cooker, Cooking, Cooks,
Precooking)
X-Ref: Culinary,
Precooked
Cookies *(Cookie)*
Cookware
Cool *(Chilled, Chiller,*
Chilling, Chills, Cooled,
Cooling, Prechilling,
Supercooled)
X-Ref: Chill, Prechill,
Precooling, Subcooling,
Superchill,
Supercooling
Coolers *(Cooler)*
Cooperatives *(Coop,*
Cooperate, Cooperating,
Cooperation, Cooperative,
Cooperatively,
Cooperativism,
Cooperativization,
Cooperators, Coops)
X-Ref: Intercooperative
Cooperia
Copablepharon
Copepoda *(Copepod,*
Copepods)
Copernicia
Copiapoa
Copidosoma
Copigments *see* Pigments
Copolymers *(Copolymer,*
Copolymeric,
Copolymerisation,
Copolymerization)
Copper

X-Ref: Cuprous
Coppices *(Coppice,*
Coppiced, Copse)
Copra
Coprecipitation *see*
Precipitation
Coprinaceae
Coprinae
Coprine
Coprinus
Copromyza
Coprophagy *(Coprophagous,*
Coprophilous)
Coproporus
Coproscopy *(Coproscopic)*
Coprostanol
Coptoformica
Coptosapelta
Coptosoma
Coptotermes
Copulation *(Coital,*
Copulating, Copulatory)
X-Ref: Coitus, Postcoital,
Postcopulation,
Precopulatory
Coquillettidia
Cora
Corallina
Corallinaceae
Coranus
Corbicula
Corchorus
Corcyra
Cordacin
Cordaitaceae
Cordaites
Cordia
Cordulia
Corduliidae *(Corduliid)*
Cordycepin
Cordyceps
Cordyla
Cordylanthus
Cordyline
Coreidae *(Coreid, Coreids)*
Coremia *(Coremium)*
Coreoidea *(Coreoid)*
Coreomycetopsis
Coreopsidinae
Coreopsis
Corethra
Coriandrum *(Coriander)*
Coriaria
Corilagin
Corimelaena
Corimelaenidae
Coriolus
Coris
Corisella
Corispermum
Corium *see* Skin
Corixa
Corixidae *(Corixids)*
Cork *(Corks, Corky)*
Cormocephalus
Cormophyta *(Cormophytes,*

Cormophytic)
Corms *(Corm, Cormous,*
Protocorm)
X-Ref: Protocorms
Corn *see* Maize
Cornaceae
Corncobs *(Cob, Corncob)*
X-Ref: Cobs
Cornea *(Corneal)*
Cornfield *see* Maize
Cornflakes
Cornicles
Cornicularia
Cornish
Cornmeal
Cornstalks *(Cornstalk)*
Cornstarch
Cornuaspis
Cornulaca
Cornus
Corolla *(Corollas)*
Coronaridine
Coronary *(Coronaries)*
Coronaviruses *(Coronaviral,*
Coronavirus)
Coronilla
Coronophorales
Coronopus
Coroxon
Corpora *(Corpus)*
Corporations *(Corporation,*
Corporative)
Corpuscles *(Corpuscle)*
Corrals *(Corral)*
Corriedale
Corrigia
Corrodentia *see* Psocoptera
Corrosion *(Anticorrosive,*
Corrosive, Corrosiveness,
Corrosivity)
X-Ref: Anticorrosives,
Biocorrosion
Corrugation *(Corrugated,*
Corrugating)
Cortaderia
Cortex *(Cortical, Cortically,*
Cortices, Corticofugal)
X-Ref: Adrenocortical,
Neocortex
Corthylus
Corticarina
Corticiaceae
Corticium
Corticoids
(Adrenocorticoids,
Adrenocorticosteroids,
Corticoid,
Corticoidogenesis,
Corticosteroid,
Corticosteroidogenesis,
Corticosteroids)
X-Ref: Adrenocorticoid,
Adrenocorticosteroid
Corticolous
Corticosterone
Corticotrophic

(Adrenocorticotrophic,
Corticotropic)
X-Ref:
Adrenocorticotropic
Corticotrophs
Corticotropin
(Adrenocorticotropin,
Corticotrophin)
Corticum
Cortinariaceae
Cortinarius
Cortinellus
Cortisol *see* Hydrocortisone
Cortisone
Cortodera
Cortusa
Coryanthes
Corydalidae
Corydalis
Corydalus
Corylaceae
Corylin
Corylus
Corymbites
Corymboporella
Corymbs
Corynanthe
Coryne
Corynebacteriaceae
Corynebacterium
(Corynebacteria,
Corynebacterial,
Coryneform)
Corynepteris
Corynespora
Coryneum
Corynocarpus
Corynoline
Coryphantha
Coryphista
Corytenchine
Corytenchirine
Corythucha *(Corythuca)*
Coryza
Coscinium
Coscinodiscaceae
Coscinodiscus
Cosmarium
Cosmetics *(Cosmetic,*
Cosmetology)
Cosmia
Cosmophila
Cosmopolites
Cosmopterygidae
(Cosmopterigidae)
Cosmopteryx *(Cosmopterix)*
Cosmos
Cossettes *(Cossette)*
Cossidae *(Cossid)*
Cossoninae
Cossus
Cost *(Costing, Costings,*
Costs)
Costal
X-Ref: Intercostal
Costelytra

Costunolide
Costus
Cosynthetase see
    Synthetases
Cotherapy see Therapy
Cotinis
Cotinus
Cotoneaster (Cotoneasters)
Cotoran see Fluometuron
Cottage
Cottendorfia
Cotterellia
Cotton (Cottons)
Cottonseed (Cottonseeds)
Cottontail see Rabbits
Cottonwood (Cottonwoods)
Coturnix
Cotylanthera
Cotyledons (Cotyledon,
    Cotyledonal,
    Cotyledonary)
    X-Ref: Intercotyledonary
Cotylenins
Cotylenol
Couchgrass
Couepia
Cough (Coughing)
    X-Ref: Antitussive
Coulometry (Coulometric)
    X-Ref: Microcoulometric
Coulters (Coulter)
Coumaphos
Coumarate
Coumaric
Coumarins (Bicoumarin,
    Coumarin)
    X-Ref: Bicoumarins,
    Cumarin
Coumaryl
Coumarylglucose
Coumarylquinic
Coumestan (Coumestanes,
    Coumestans)
Coumestrol (Coumestrols)
Counselors (Counseling,
    Counsellor, Counsellors,
    Counselor)
Couplings
Courts (Court)
Courtship see Mating
Courvoisiella
Cousinia
Covariance see Variance
Covers (Cover, Coverage,
    Coverages, Covering,
    Coverings)
Cowania
Cowberries (Cowberry)
Cowdria
Cowdung see Dung
Cowhouses see Barns
Cowmen (Cowhand,
    Cowhands, Cowman)
Cowpeas (Cowpea)
Cowpox
    X-Ref: Antivaccina,

Vaccina
Cows (Cow)
Cowsheds see Sheds
Coxa
    X-Ref: Subcoxa
Coxiella
Coxofemoral
Coxsackieviruses
    (Coxsackie,
    Coxsackievirus)
Coyden
Coyotes (Coyote)
Coypus (Coypu)
    X-Ref: Myocastor, Nutria
Crabapples (Crabapple)
Crabgrass
Crabronidae (Crabroninae)
Crabs (Crab)
Crackers
Cracking (Crack, Cracked,
    Cracks)
Crafts (Craft,
    Craftsmanship,
    Craftsmen)
Crambe
Crambidae (Crambids,
    Crambinae)
Crambini
Crambus
Cramptonomyia
Cranberries (Cranberry)
Cranes (Crane)
Cranesbill
Craniology (Craniological)
Cranium (Cranial,
    Intracranially)
    X-Ref: Intracranial,
    Neurocranium
Crapemyrtle (Crapemyrtles)
Craspedolepta
Craspedophyceae
Craspedorrhynchus
Craspeduchus
Craspidospermine
Craspidospermum
Crassinodine
Crassostrea
Crassula (Crassulas)
Crassulaceae (Crassulacean)
Crataegolic
Crataegus
Crataeva
Craterocercus
Crates (Crate)
Cratna
Cratospila
Cratoxylon
Crawlers (Crawler)
Crayfish (Crawfish)
Cream (Creamed,
    Creaming, Creams)
Creameries (Creamery)
Creatine (Creatinine)
Credit (Crediting, Credits)
Creeping (Creep, Creeper,
    Creepers)

Crematogaster
Cremona
Crenulaspidiotus
Creophilus
Creosote (Creosoted)
Crepes
Crepidotus
Crepis
Crepuscular
Crescentieae
Cresol (Cresols)
Cress
Cretaceous
Cricetus
Crickets (Cricket)
Criconema
Criconematidae
Criconematinae
Criconematoidea
Criconemoides
Cricotopus
Crimidine
Criminal
Crinitol
Crinkle
Crinocerus
Crinum (Crinums)
Criocephalus
Crioceras
Criocerinae
Crioceris
Criollo (Criollos)
Criorrhina
Cristacortis
Cristulariella
Crithidia
Crociflorinone
Crocin
Crocinae
Crocothemis
Crocus
Croesia
Croesus
Crofters (Crofter)
Cromadurine
Crombrugghia
Cronartium
Croneton
Cronicus
Cronolone
Crops (Crop, Cropland,
    Croplands, Cropped,
    Cropping, Forecrop)
    X-Ref: Aftercrops,
    Forecrops, Noncrop,
    Uncropped
Crossandra
Crossbreeding (Backcrosses,
    Backcrossing, Backrossed,
    Cross, Crossability,
    Crossbred, Crossbreds,
    Crossbreed, Crossbreeds,
    Crossed, Crossing,
    Intercrosses, Intercrossing)
    X-Ref: Backcross,
    Crossings, Doublecross,

Intercross, Outcrossing,
    Polycross
Crosscutting (Crosscut)
Crossidium
Crossings see Crossbreeding
Crosslinking (Crosslinked,
    Crosslinks)
Crossocerus
Crossosoma
Crossosomataceae
Crossovers (Crossover)
Crossties
Crotalaria (Crotalarias)
Crotepoxide
Crotilin
Croton
Crotonate
Crotonin
Crotonylidene
Crowberries (Crowberry)
Crowding
Crowns (Crown)
Crownvetch
Crows (Crow)
Cruciata
Cruciferae (Crucifer,
    Cruciferous, Crucifers)
Cruelty
Crufomate
Crumenula
Crumpets
Crushing (Crush, Crushed,
    Crusher, Crushers,
    Crushings)
Crustacea (Crustacean,
    Crustaceans, Crustaceous)
Crustecdysone
Crusts (Crust, Crusted,
    Crusting)
    X-Ref: Anticrustants
Crymodes
Cryobiology (Cryobiological,
    Cryobiotic)
Cryobius
Cryocleavage
Cryofracture
    (Cryofractured)
Cryogenics (Cryogenesis,
    Cryogenic, Cryogenically,
    Cryogeny)
Cryology (Cryologic)
Cryophilic
Cryopreservation
Cryoprotection
    (Cryoprotectants,
    Cryoprotected,
    Cryoprotective)
Cryoscopy (Cryoscope,
    Cryoscopic)
Cryosurgery (Cryosurgical)
Cryotherapy
Cryphalini
Cryphalus
Cryptantha
Cryptarcha
Crypthecodinium

Cryptic
Cryptinae
Cryptobiosis
Cryptoblabes
Cryptocanthon
Cryptocarya
Cryptocellus
Cryptocentrum
Cryptocephalinae
(Cryptocephaline)
Cryptocephalus
Cryptocercus
Cryptochironomus
Cryptococcaceae
Cryptococcosis
Cryptococcus (Cryptococcic)
Cryptocoryne
Cryptodesmidae
Cryptodiscus
Cryptoechinuline
Cryptogamia (Cryptogamic,
Cryptogams)
Cryptognathidae
Cryptognathus
Cryptolaemus
Cryptolestes
Cryptomeria
Cryptomonas
Cryptomyces
Cryptonemia
Cryptonemiales
(Cryptonemiacean)
Cryptophagidae
Cryptophagus
Cryptophilus
Cryptophlebia
Cryptophoranthus
Cryptophyceae
Cryptopleura
Cryptopogon
Cryptorchidism
(Cryptorchid,
Cryptorchids)
X-Ref:
Pseudocryptorchidism
Cryptorhynchinae
Cryptorhynchus
(Cryptorrhynchus)
Cryptoserphus
Cryptoses
Cryptospora
Cryptosporella
Cryptosporidiosis
Cryptosporidium
Cryptosporiopsis
Cryptostegia
Cryptostemma
Cryptostigmata
(Cryptostigmatid)
Cryptostroma
Cryptostylis
Cryptotaenia
Cryptotermes
Cryptothallus (Cryptothalli)
Crystallite
Crystallization

(Crystalization,
Crystalliferous,
Crystallisation,
Crystalliser, Crystallized,
Crystallizer, Crystallizing)
X-Ref: Noncrystallizing,
Recrystallization
Crystallochemistry
Crystallography
(Crystallographic)
Crystals (Crystal, Crystallin,
Crystalline, Crystallinity,
Crystallins, Crystalloid,
Crystalloids, Monocrystal)
X-Ref: Intracrystalline,
Microcrystalline,
Microcrystals,
Monocrystals
Crytopeltis
Crytotermes
Ctenicera
Ctenidae
Ctenitis
Ctenizidae
Ctenocephalides
Ctenocladus
Ctenolepisma
Ctenomyces
Ctenomys
Ctenopharyngodon
Ctenophthalminae
Ctenophthalmus
Ctenoplectrella
Ctenoxylocopa
Ctenuchidae
Ctenus
Cuauhtemone
Cubes (Cube, Cubed,
Cubing)
Cubital see Elbows
Cubitermes
Cuckoo
Cucujidae
Cucujoidea (Cucujoid)
Cucujomyces
Cucullaria
Cucullia
Cuculliinae
Cucullus
Cucumbers (Cucumber)
Cucumis
Cucurbita
Cucurbitaceae (Cucurbit,
Cucurbitaceous,
Cucurbits)
Cucurbitacins
(Cucurbitacin)
Cudonia
Cudrania
Cuerna
Cuisine
Culex
Culicidae (Culicid, Culicids,
Culicinae, Culicine)
Culicini
Culicoides

Culinary see Cookery
Culiseta
Culling (Culled)
Cullulase
Culm see Stem
Culmites
Cultivars (Cultivar)
X-Ref: Intercultivar
Cultivators (Cultivator)
Culverts (Culvert)
Cumambrin
Cumarin see Coumarins
Cumbria
Cumene
Cumin
Cuminal
Cuminum
Cumulus see Clouds
Cunabdella
Cunaxa
Cunaxidae
Cunninghamella
Cunninghamia
Cunoniaceae
Cupferron
Cuphea
Cuphocera
Cupiennius
Cuprates
Cupressaceae
Cupressus
Cupric
Cuprimyxin
Cuprizone
Cuprous see Copper
Cupule
Cura
Curare
Curasol
Curatives (Curative)
Curculigo
Curculio
Curculionidae (Curculid,
Curculionid,
Curculionids)
Curculionoidea
Curcuma
X-Ref: Zedoaria
Curcumin
Curdling (Curdle, Curdled)
Curds (Curd)
Curettage
Curing (Cure, Cured,
Cures)
X-Ref: Uncured
Curium
Curl (Curled, Curly)
Currants (Blackcurrant,
Currant)
X-Ref: Blackcurrants
Curriculum (Curricula,
Curricular)
Curry (Curried, Currying)
Curtonotum
Curvularia
Cuscohygrine

Cuscuta
Cuscutaceae
Cusparia
Cussonia
Custards (Custard)
Customers (Customer)
Customs (Custom)
Cutaneous see Skin
Cuterebra
Cuterebridae
Cuticles (Cuticle, Cuticular)
X-Ref: Intracuticular,
Transcuticular
Cuticulin
Cutins (Cutin)
Cutleria
Cutleriales
Cutlery
Cutlets
Cutters (Cutter)
Cuttings (Cut, Cutting)
X-Ref: Noncutting,
Overcutting, Precutting
Cutworms (Cutworm)
Cyamopsis
Cyanacethylhydrazone
Cyanamide
Cyanandrium
Cyanates (Cyanate)
Cyanatryn
Cyanazine
Cyanelles
Cyanides (Cyanide)
Cyanidins (Cyanidin)
X-Ref: Procyanidins
Cyanidium
Cyanins (Cyanin)
Cyanoacetic
Cyanoalanine
Cyanobacteria
(Cyanobacterium)
Cyanocobalamin
Cyanoethyl
Cyanoethylation
(Cyanoethylated)
Cyanogen (Cyanogenesis,
Cyanogenetic, Cyanogenic,
Cyanogens)
Cyanoglucosides
Cyanoguanidine
Cyanoimidodithiocarbonate
Cyanophages (Cyanophage)
Cyanophenyl
Cyanophos
X-Ref: Cyanox
Cyanophyceae
Cyanophyta (Cyanophytan,
Cyanophyte)
Cyanopsis
Cyanox see Cyanophos
Cyanuric
Cyasterone
Cyathea
Cyatheaceae
Cyathin
Cyathium

Cyathocline
Cyathodinium
Cyathodium
Cyathostoma
Cyathula
Cyathus
Cyazone
Cybernetics *(Cybernetic)*
X-Ref: Biocybernetic
Cybister
Cybocephalidae
Cybocephalus
Cycadaceae *(Cycad.*
Cycads)
Cycadales
Cycadophytae
(Cycadophytes)
Cycas
Cychrini
Cychropsis
Cyclamates *(Cyclamate)*
Cyclamen
Cyclanthera
Cyclas
Cyclase *(Cyclases)*
Cyclitols
Cycloartenol
Cycloate *(Ronite)*
X-Ref: Ronit
Cyclobutane
Cyclocephala
Cycloconium
Cyclocreatine
Cyclodextrins
Cyclodictyon
Cyclodiene
Cycloeucalenol
Cycloheptenone
Cyclohexadiene
Cyclohexane
Cyclohexaneacetate
Cyclohexanecarboxylate
Cyclohexanone
Cyclohexene
Cycloheximide
X-Ref: Actidione
Cyclohexyl
Cyclohexylamine
Cyclohexylhydroxylamine
Cyclohexylindan
Cyclohydrolase
Cycloleucine
Cyclolobium
Cyclommatus
Cycloneda
Cyclonerodiol
Cyclonerotriol
Cyclones *(Cyclone)*
Cyclopenin
Cyclopent
Cyclopenta
Cyclopentadiene
(Cyclopentadien)
Cyclopeptides
(Cyclopeptide)
Cyclophanes

Cyclophosphamide
X-Ref: Endoxan
Cyclophyllidea
Cyclopia
Cyclopropane
Cyclopropanecarboxamide
Cyclopropanic
Cyclopropanone
Cyclopropenoid
Cyclopropyl
Cyclops *(Cyclopoid)*
Cyclorrhapha
(Cyclorrhaphous)
Cyclosis
Cyclotella
Cycocel
X-Ref: TUR
Cydia
Cydnidae
Cydonia
Cygon *see* Dimethoate
Cylapinae
Cylas
Cylinder *(Cylinders.*
Cylindrical)
Cylindrocapsa
Cylindrocarpon
Cylindrocladium
Cylindroiulus
Cylindrospermum
Cylindrosporium
Cylindrotettix
Cylindrotheca
Cymatopsocus
Cymbidium *(Cymbidiums)*
Cymbopogon
(Cymbopogons)
Cymene
Cyminae
Cymindis
Cymodocea
Cymophorus
Cymopterus
Cynanchum
Cynara
Cynareae
Cynipidae *(Cynipid.*
Cynipids)
Cynipoidea
Cynips
Cynkotox *see* Zineb
Cynodon
Cynoglossum
Cynomoriaceae
Cynthia
Cyolane *see* Phosfolan
Cyperaceae
Cyperacites
Cyperorchis
Cyperus
Cyphelophorus
Cyphicerini
Cyphioidae
Cyphomandra
Cyphon
Cypress

Cyprinus
Cypripedilinae
Cypripedium
Cyproterone *(Cyproteron)*
Cyrilla
Cyrillaceae
Cyrtandra
Cyrtanthus
Cyrtidae
Cyrtidium
Cyrtochilum
Cyrtodiopsis
Cyrtolaus
Cyrtomium
Cyrtopeltis
Cyrtophora
Cyrtosperma
Cystadenoma
Cystathionine
Cystathioninuria
Cysteamine *(Cystamine)*
Cystectomy
Cysteic
Cysteine *(Cystein)*
Cysteinylvaline
Cysticercoids *(Cysticercoid)*
Cysticercosis *(Cysticerciasis)*
Cysticercus *(Cysticerci)*
Cystidia
Cystine
Cystitis
Cystocaulus
Cystoclonium
Cystopteris
Cystoseira
Cystoseiraceae
Cysts *(Cyst. Cystic)*
Cytauxzoonosis
Cythioate
Cytidine
Cytinus
Cytisus
Cytoarchitectonics
(Cytoarchitectonic.
Cytoarchitectural)
Cytochalasin
Cytochemistry
(Cytochemical)
Cytochromes *(Cytochrome)*
Cytocontraction
Cytodifferentiation
Cytoecetes
Cytoembryology
Cytogenetics *(Cytogenetic.*
Cytogenetical)
Cytogeography
(Cytogeographic.
Cytogeographical)
Cytohistochemistry
(Cytohistochemical)
Cytohistology
(Cytohistological)
Cytokinesis
Cytokinins *(Anticytokinin.*
Cytokinin)
X-Ref: Anticytokinins

Cytology *(Cytologic.*
Cytological. Cytologically)
Cytolysis *(Cytolytic)*
Cytomegaloviruses
(Cytomegalovirus)
Cytomixis
Cytomorphogenesis
Cytomorphology
(Cytomorphological)
Cytopathology *(Cytopathic.*
Cytopathogenic.
Cytopathogenicity.
Cytopathological.
Cytopathy)
Cytophaga
Cytophotometry
(Cytophotometer.
Cytophotometric.
Cytophotometrical)
Cytophysiology
(Cytophysiological)
Cytoplasm *(Cytoplasma.*
Cytoplasmatic.
Cytoplasmic.
Cytoplasmically.
Cytoplasms)
X-Ref: Acytoplasmic,
Intracytoplasmic
Cytoquinines
Cytosine
Cytosol *(Cytosolic. Cytosols)*
Cytospora
Cytosporina
Cytostatic
Cytosterile *see* Sterile
Cytotaxonomy
(Cytotaxonomic.
Cytotaxonomical)
Cytotherapy
Cytotoxicity *(Cytotoxic)*
X-Ref: Noncytotoxic
Cyttaria
Daboecia
Dachshund *(Dachshunds)*
Dacinae
Dacini
Dacninae
Daconil
Dacrydium
Dacrymyces
Dacrymycetales
Dactylaria
Dactylella
Dactyliandra
Dactylis
Dactylium
Dactyloctenium
Dactylopiidae
Dactylopius
Dactylorchis
Dactylorhiza
Dactylosphaera
Dactylus
Dactynotus
Dacus
Daedalea

Daemonorops
Daffodils *(Daffodil)*
Dahlias *(Dahlia)*
Daidzin
Dairy *(Dairies, Dairying)*
  X-Ref: Nondairy
Dairymen *(Dairyfarmer,*
  Dairyfarmers, Dairyman)
Daisies *(Daisy)*
Daktulosphaira
Dalapon
  X-Ref: Dowpon
Dalbergia
Dalbulus
Dallisgrass
Dalmatian *(Dalmatians)*
Damalinia
Damascenone
Damascone
Damascus
Damietta
Daminozide *see* Alar
Dammarane
Damping *(Damp)*
Dams *(Dam)*
Damselflies *(Damselfly)*
Damsons
Danaidae *(Danaid,*
  Danainae)
Danaus
Danazol
Dancing *(Dance, Dances)*
Dandelions *(Dandelion)*
Dandruff
Dane
Danothrips
Dansyl
Danthonia
Dantrolene
Daphne
Daphnetoxin
Daphnia
Daphnilactone
Daphnin
Daphniphyllaceae
Daphniphyllum
Daphnis
Darlingtonia
Darluca
Darvasoline
Dasheen
Dasybasis
Dasychira
Dasycladaceae
  (Dasycladacean)
Dasycladales
Dasyhelea
Dasylophia
Dasyneura *(Dasineura)*
Dasypogoninae
Dasypus
Dasyscyphus
Dasysyrphus
Dasyuridae *(Dasyurid)*
Datana
Datisca

Datura
Daucic
Daucus
Daughters *(Daughter)*
Daunorubicin
Davaineidae
Davallia
Davidius
Dawsonia
Daylight
Daylilies *(Daylily)*
Dazomet
  X-Ref: Basamid, Mylone,
  Thiazone
DDD
DDE
DDT
DDVP *see* Dichlorvos
Deacetoxycephalosporin
Deacetylation
  (Deacetylated)
Deacetylcephalosporin
Deacetylmatricarin
Deacidification
  (Deacidifying)
Deactivation *(Deactivating)*
Deacylation *(Deacylating)*
Deaeration
Deafferentation
Dealers *(Dealer)*
Dealkylation
Deamidation *(Deamidated)*
Deaminase *(Deaminases)*
Deamination *(Deaminated)*
Death *(Dead, Deaths)*
  X-Ref: Antemortem,
  Deceased, Demise,
  Died, Dying,
  Moribund, Mortem
Debarking
Debaryomyces
Debeaking *(Debeak)*
Debilitation *(Debilitated)*
Deblossomed *see* Flowers
Deboning *(Deboned,*
  Deboners)
Debraining
Debris
Debromoaplysin
Debt
Debudding
Decadienoic
Decaffeination
  (Decaffeinated)
Decahydrate
Decaisnea
Decalcification *(Decalcified,*
  Decalcifying)
Decalin
Decalone
Decanoate
Decanoic
Decantation *(Decanting)*
Decapeptide
Decapitation *(Decapitated)*
Decarbonylation

Decarboxylases
  (Decarboxylase)
Decarboxylation
  (Decarboxylating)
Decarboxymethylation
Decaschistia
Decay *(Decayed, Decaying,*
  Decays)
Deccanocarpon
Deceased *see* Death
Decene
Decerebration *(Decerebrate)*
Dechlorination
Deciduomata
Deciduous *(Deciduary,*
  Deciduously,
  Deciduousness)
Decolorization
  (Decoloration, Decolorize,
  Decolorizing,
  Decolourization)
Decomplementation
Decomposition
  (Decomposable,
  Decomposed, Decomposer,
  Decomposing)
Decompression
  (Decompressive)
Decontamination
  (Decontaminants,
  Decontaminate,
  Decontaminated,
  Decontaminating)
Decoquinate
Decoration *(Decorative)*
Decorosiphon
Decortication
  (Decorticating,
  Decorticator)
Decoys *(Decoy)*
Decrystallization
  (Decrystallized)
Decticinae
Dedeckera
Deepfreezing
Deer
Deerflies *(Deerfly)*
Defatting *(Defatted)*
Defaunation
Defecation
Defectoscopy *(Defectoscope)*
Defeminization
Defense *(Defence, Defenses,*
  Defensive)
Defibration *(Defibrator)*
Deficiency *(Deficient)*
Deficits *(Deficit)*
Deflation
Deflection
Deflectometer
Deflectors *(Deflector)*
Defleecing *see* Dehairing
Deflocculation
Defluorination
  (Defluorinated)
Defoliation *(Defoliant,*

Defoliants, Defoliate,
  Defoliated, Defoliating,
  Defoliations, Defoliator,
  Defoliators)
Deforestation *(Deforested)*
Deformation *(Deformable,*
  Deformations, Deformed)
Deformities *(Deformity)*
Defrosting *(Defreezing,*
  Defrost)
Degeneration *(Degenerate,*
  Degenerated,
  Degenerating,
  Degenerative,
  Degenerescence)
  X-Ref: Nondegenerate
Degerminator
Deglaciation
Deglanded *see* Glands
Deglutition
Degradation *(Degradability,*
  Degradable, Degradative,
  Degrade, Degraded,
  Degrading)
  X-Ref: Autodegradation
Degranulation
Degreening
Degumming
Degustation *(Degustator,*
  Degustators)
Dehairing
  X-Ref: Defleecing
Dehardening *see* Hardening
Dehelminthization
Dehiscence *(Dehiscent,*
  Indehiscent)
  X-Ref: Indehiscence
Dehorning
Dehulling *(Dehull,*
  Dehulled)
Dehydrase
Dehydratases *(Dehydratase)*
Dehydration *(Dehydrated,*
  Dehydrating)
Dehydrators *(Dehydrator)*
Dehydroabietic
Dehydroacetate
Dehydroacetic
Dehydroagapanthagenin
Dehydroalanine
Dehydroascorbic
Dehydrobromination
Dehydrochlorinase
Dehydrocholesterol
  (Dehydrocholesterols)
Dehydrocorydaline
Dehydroepiandrosterone
Dehydrofreezing
Dehydrogenases
  (Dehydrogenase)
Dehydrogenation
Dehydronerol
Dehydropolymers
Dehydroproline
Dehydropyrrolizidine
Dehydroquinase

Deicing
Deightoniella
Deilephila
Deinocerites
Delactosed
Deladenus
Delafondia
Delaine
Delamination
Deleatidium
Delesseriaceae
Delia
Delias
Delicatessen *(Delicacies)*
Delignification *(Delignified)*
Deliming
Delinting *(Delinted)*
Delipidation
Delisea
Delitschia
Delonix
Delphacidae
Delphinine
Delphinium *(Delphiniums)*
Delphoside
Deltapine
Deltas *(Deltaic, Deltal)*
Deltocephalidae
Deltocephalinae
    (Deltocephaline)
Deltocephalus
Delvotest
Demalon
Demand *(Demands)*
Dematiaceae
    (Dematiaceous)
Demercuration
Demetallization
Demethylase
Demethylation
    (Demethylate,
    Demethylated)
Demethylgardneramine
Demeton
    X-Ref: Systox
Demetrius
Demineralization
    (Demineralized)
Demise *see* Death
Democracy
Demodex *(Demodectic)*
Demodicidae
Demodicosis
Demography *(Demographic)*
Demyelination
    (Demyelinating)
Denaturation *(Denaturants,*
    Denaturating, Denatured,
    Denaturing,
    Denaturization)
    X-Ref: Undenatured
Dendriscocaulon
Dendrites
Dendrobacillin
Dendrobaena
Dendrobangia

Dendrobine
Dendrobium *(Dendrobiums)*
Dendrocalamus
Dendrocereus
Dendrochronology
    (Dendrochronological)
Dendroctonus
Dendrocygna
Dendrogram
Dendroides
Dendrolasin
Dendrolasma
Dendrolimus
Dendrology *(Dendrologic,*
    Dendrological)
Dendrometry *(Dendrometer,*
    Dendrometers,
    Dendrometric,
    Dendrometrical,
    Dendrometrics)
Dendropanax
Dendrophilic
    (Dendrophilous)
Dendrophoma
Dendrophthoe
Dendrosicus
Dendrosoter
Dendryphiella
Denervation *(Denervated)*
    X-Ref: Chemodenervated
Dengue
Denitrification
    (Denitrifying)
Dennstaedtia
Dennstaedtiaceae
Dennyus
Densitometry *(Densimetric,*
    Densitometer,
    Densitometric)
Density *(Dense, Densities)*
Densonucleosis
Dental
Dentin *(Dentinal, Dentine)*
Dentistry
Dentition
Dentures
Denudation *(Denuded)*
Deodar
Deodorization *(Deodoriser,*
    Deodorized, Deodorizers,
    Deodorizing)
Deoiling *(Deoiled)*
Deontology
Deoxidation *(Deoxidant)*
Deoxyadenosine
Deoxycholate
Deoxycorticosterone
    X-Ref:
    Desoxycorticosterone
Deoxycortisol
Deoxycytidine
Deoxygluconate
Deoxyglucose
Deoxyherqueinone
Deoxyloganic
Deoxynucleosides

    (Deoxynucleoside)
Deoxynucleotide
Deoxypeganine
Deoxypentosone
Deoxypyridoxine
Deoxypyridoxol
Deoxyribonucleases
    (Deoxyribonuclease)
    X-Ref:
    Desoxyribonuclease,
    DNase
Deoxyribonucleic
    (Desoxyribonucleic)
Deoxyribonucleohistone
Deoxyribonucleoproteins
    (Deoxyribonucleoprotein)
Deoxyribonucleosides
    (Deoxyribonucleoside)
Deoxyribonucleotides
    (Deoxyribonucleotide)
Deoxyribosyl
Deoxyriboviruses
    (Deoxyribovirus)
Deoxythymidine
Deoxyuridine
Depancreatized *see*
    Pancreatectomy
Depectenization
    (Depectinized)
Dephosphorylation
Depigmentation
Depilation
Deplasmolysis *see*
    Plasmolysis
Depletion *(Depleted)*
Depolarization *(Depolarized,*
    Depolarizer, Depolarizing)
Depolymerase
Depolymerization
Depolyploidization
Depopulation
Depreciation
Depression *(Depressant,*
    Depressed, Depressing,
    Depressional, Depressions)
Depressors *(Depressor)*
Deprivation *(Deprived)*
Deproteinization
    (Deproteinised,
    Deproteinized)
Depsides *(Depside)*
    X-Ref: Tetradepside
Depsidone
Depsipeptides *(Depsipeptide)*
Depuration *see* Purification
Deraeocoris
Derbesia
Derbesiales
Derbidae
Derivatives *(Derivative)*
Derivatization
Dermacarus
Dermacentor
Dermal *see* Skin
Dermanyssidae
    (Dermanyssid)

Dermanyssus
Dermaptera
Dermatitis
Dermatocarpon
Dermatology
    (Dermatological,
    Dermatologically)
Dermatomycosis
    (Dermatomycoses)
Dermatophagoides
Dermatophilosis
Dermatophilus
Dermatophyta
    (Dermatophytes,
    Dermatophytic)
Dermatophytoses
    (Dermatophytosis)
Dermatoplasty
    (Dermatoplastic,
    Dermoplastic,
    Dermoplasty)
Dermatoses *(Dermatosis)*
Dermestes
Dermestidae *(Dermestid)*
Dermis *see* Skin
Dermocystidium
Dermoid *see* Skin
Dermojet
Dermolepida
Dermonecrosis
Deroceras
Derocrania
Derodontidae
Deronectes
Derris
Desacetyl
Desalination
    (Desalinization, Desalted,
    Desalting)
Desaturase
Desaturation *(Desaturate,*
    Desaturated)
Deschampsia
Descurainia
Desensitization
Deserts *(Desert,*
    Desertification)
    X-Ref: Semidesert
Desiccants *(Antidesiccant,*
    Desiccant)
    X-Ref: Antidesiccants
Desiccation *(Desiccated,*
    Desiccating)
Desilaging *see* Silage
Desilicified *see* Silica
Desmarestia
Desmedipham
Desmethyl
Desmia
Desmidiaceae *(Desmid,*
    Desmids)
Desmidiales
Desmodium
Desmodontidae
    (Desmodontid)
Desmodus

Desmosomes
Desmotomy
Desorption
Desoxycorticosterone *see*
  Deoxycorticosterone
Desoxyribonuclease *see*
  Deoxyribonucleases
Desserts *(Dessert)*
  X-Ref: Sweets
Destun
Desugarization
Desulfotomaculum
Desulfovibrio
Desulfurization
Desynapsis *(Desynaptic)*
Detasseling
Detergents *(Detergency,*
  Detergent)
Deterioration *(Deteriorated)*
Dethiobiotin
Dethiomethylation
Detoxication *(Detoxicant,*
  Detoxicate, Detoxification,
  Detoxified, Detoxifier,
  Detoxify, Detoxifying)
Deudorix
Deuteration *(Deuterated)*
  X-Ref: Hexadeuterated
Deuterium
Deuteroheme
Deuteromycetes
Deuteromycotina
Deuterophoma
Deuterosminthurus
Deutocerebrum
Deutzia
Devaluation
Deviation *(Deviant)*
Devocalizing
Devrinol
Dew *(Dewfall)*
Dewatering *(Dewater)*
Dewberries *(Dewberry)*
Dewclaws
Deworming *(Dewormings)*
Dexamethasone
Dexon *see* Fenaminosulf
Dexter
Dextran *(Dextrans)*
Dextranase
Dextranglucosidase
Dextransucrase
Dextrinases *(Dextrinase)*
Dextrins *(Dextrin,*
  Dextrine)
Dextrose *see* Glucoses
Dextrostix
Deyeuxia
Dhomoandrost *see* Androst
Dhurrin
Diabetes *(Diabetic,*
  Diabetics, Diabetogenic)
Diabrotica
Diacetal
Diacetate
Diacetoxyscirpenol

Diacetyl
Diacetylmonoxime
Diacetyltartaric
Diachlorini
Diacrisia
Diacyl
Diacylglycerols
Diadegma
Diadema
Diadocidia
Diaecoderus
Diaeretiella
Diagenesis
Diagnosis *(Diagnose,*
  Diagnosed, Diagnoses,
  Diagnosing, Diagnostic,
  Diagnostical,
  Diagnosticians,
  Diagnostics)
Diakinesis
Dialdehydes *(Dialdehyde)*
Dialene
Dialeurodes
Dialictus
Dialineura
Dialium
Dialkyl
Dialkylnaphtamidine
Dialkylphthalate
Diallate
  X-Ref: Avadex
Diallels *(Diallel, Diallele,*
  Diallelic)
Dialuric
Dialysis *(Dialysate,*
  Dialyzable)
Diamesa
Diamesinae
Diameton
Diamide
Diamines *(Diamine)*
Diaminetetraacetate
Diaminoandrostane
Diaminobenzidine
Diaminobiphenyl
Diaminobutyric
Diaminonaphthalene
Diaminooxydase
Diaminopimelic
Diaminopropionic
Diaminopurine
Diaminosilanes
Diaminosuccinic
Diaminoxydase
Diammonium
Diamphenethide
Diamphidia
Dianthera
Dianthidium
Dianthrimide
Dianthus
Diapause *(Diapausing)*
  X-Ref: Nondiapausing,
  Postdiapause,
  Prediapause
Diapensiaceae

Diaphania
Diapheromera
Diaphonia
Diaphorase
Diaphorina
Diaphragm *(Diaphragmatic)*
Diaphysis
Diaporthaceae
Diaporthe
Diaprepes
Diapriidae
Diarrhea *(Diarrheal,*
  Diarrheic, Diarrhoea,
  Diarrhoeas)
Diarthrosis *(Diarthrodial)*
Diaryl
Diarylpropanes
  (Diarylpropane)
Diaspididae
Diaspidini
Diaspidoidea
Diaspores *see* Spores
Diastase *(Diastatic)*
Diastatidae
Diastereoisomers
  (Diastereoisomer)
Diastole *(Diastolic)*
Diathaea
Diathesis
Diatoma
Diatomaceae *(Diatom,*
  Diatomic, Diatoms)
Diatomaceous *(Diatomite)*
  X-Ref: Kieselguhr
Diatomeae
Diatraea
Diatraeophaga
Diatrypaceae
Diatrype
Diazepam
Diazinon *(Diazinone)*
Diazoacetic
Diazoben *see* Fenaminosulf
Diazocarbonyl
Diazoketones *(Diazoketone)*
Diazomethane
Diazonium
Diazoxide
Dibasic
Dibenzodioxin
Dibenzofuran
  (Dibenzofurans)
Dibenzoylmethane
Dibenzyloxycarbonyl
Dibiomycin
Dibolia
Dibotryon
Dibrachoides
Dibrom *see* Naled
Dibromide *(Dibromides)*
Dibromochloropropane
  X-Ref: Fumazone
Dibromoethane
Dibromothymoquinone
Dibromovinyl
Dibutylhydroxytoluene

Dibutyryl
Dicaelus
Dicaffeoyl
Dicalcium
Dicamba
  X-Ref: Banvel
Dicanthium
Dicarbonate
Dicarbonic
Dicarbonyl
Dicarboximide
Dicarboxylates
  (Dicarboxylate)
Dicarboxylic
Dicellomyces
Dicellostyles
Dicentra
Dichaetomyia
Dichanthium
Dichapetalaceae
Dichapetalum
Dichelacera
Dichillus
Dichlobenil
Dichlofluanid
  X-Ref: Euparen
Dichlone
Dichloride
Dichloroaniline
Dichlorobenzaldoxime
Dichlorobenzene
Dichlorobenzoic
Dichlorobenzonitrile
Dichlorobiphenyl
Dichlorocyclopropanes
Dichlorodifluoromethane
Dichlorodiphenyl
Dichloroethane
Dichloroethylene
Dichlorofluorescein
Dichloroisocyanurate
Dichloromethane
Dichloromethyl
Dichlorophen
  (Dichlorophene)
  X-Ref: Preventol
Dichlorophenol
Dichlorophenoxyacetate
Dichlorophenoxyacetic
Dichlorophenyl
Dichlorophenylamino
Dichloropicolinic
Dichloropropane
Dichloropropenes
  (Dichloropropene)
  X-Ref: Telone
Dichloropropionanilide
Dichlorovinyl
Dichlorvos *(Dichlorovos)*
  X-Ref: DDVP, Nuvan,
  Vapona
Dichocrocis
Dichondra
Dichoteleas
Dichotomosiphonales
Dichroism *(Dichroic)*

Dichromate
Dichromatism
Dichromothrips
Dichroplus
Dicksonia
Dicobalt
Dicofol
    X-Ref: Kelthane
Dicotex
    X-Ref: Dikotex
Dicotyledons *(Dicot, Dicots,*
    Dicotyledon,
    Dicotyledonous)
Dicoumarin *(Dicoumarol,*
    Dicumarin, Dicumarol)
Dicranum
Dicranura
Dicresyl
Dicroceliasis
Dicrocoeliidae
Dicrocoelium
Dicroidium
Dicrotendipes
Dicrotophos
Dictamnus
Dictionaries *(Dictionary)*
Dictynidae
Dictyocaulosis
Dictyocaulus
Dictyococcus
Dictyopanus
Dictyopharidae
Dictyoploca
Dictyoptera
Dictyopteris
Dictyosiphonales
Dictyosomes *(Dictyosome)*
Dictyostelium
Dictyota
Dictyotales
Dicumyl
Dicyandiamide
    (Dicyanodiamide)
Dicyclohexylcarbodiimide
Dicyphus
Didiereaceae
Didymella
Didymium
Didymocarpus
Didymuria
Dieback
Died *see* Death
Dieffenbachia
Dieldrin
    X-Ref: Photodieldrin
Dielectric *(Dielectrical)*
Dielocerus
Diels *(Diel)*
    X-Ref: Nondiel
Dielsiochloa
Dienamide
Diencephalon *(Diencephalic)*
Dienes *(Dien, Diene)*
Dienol
Dienolide
Diepoxides *(Diepoxide)*

Diepoxybutane
Diesel
Diesters *(Diester)*
Diestogyna
Diestrus
    X-Ref: Dioestrus
Dietan
Diethanolamides
Diethylacetal
Diethylamide
Diethylamine
Diethylaminoethyl
Diethylcarbamazine
Diethyldithiocarbamate
Diethylene
Diethylenetriamine
Diethylnitrosamine
Diethylstilbestrol
    (Stilboestrol)
    X-Ref: Stilbestrol
Diethylsulphate
Dietitians *(Dietitian)*
Diets *(Diet, Dietaries,*
    Dietary, Dietetic,
    Dietetics, Dieting,
    Dietotherapeutic,
    Dietotherapy,
    Monodietetic, Monodiets)
    X-Ref: Monodiet
Difenzoquat
Diffraction
Diffractograms
    (Diffractogram)
Diffractometry
    (Difractometric)
Diffusates *(Diffusate)*
Diffusion *(Diffusable,*
    Diffuse, Diffused,
    Diffuser, Diffusers,
    Diffusible, Diffusing,
    Diffusional, Diffusions,
    Diffusive, Diffusivity)
    X-Ref: Microdiffusion
Difil
Diflubenzuron
    X-Ref: Dimilin
Difluorobenzoyl
Digamasellus
Digenea
Digera
Digestion *(Digest, Digesta,*
    Digestability, Digested,
    Digester, Digestibilities,
    Digestibility, Digestible,
    Digesting, Digestions,
    Digestive, Digests)
    X-Ref: Microdigestion
Digging *(Dig, Digger,*
    Diggers)
Digital *(Digitized)*
Digitalis
Digitaria
Digitivalva
Digitonin
Digitorebutia
Digitoxin

Diglucosides *(Diglucoside)*
Diglycerides *(Diglyceride)*
Diglycosides *(Diglycoside)*
Diglyphus
Digoxin
Dihaloalkanes
    (Dihaloalkane)
Dihalovinyl
Dihaploids *(Dihaploid)*
Dihybrid
Dihydrate
Dihydroacridine
Dihydroasparagusic
Dihydrobenzofuranyl
Dihydrocadambine
Dihydrochalcone
    (Dihydrochalcones)
Dihydrochloride
Dihydrochrysanthemolacton
Dihydrocinnamic
Dihydrocleavamine
Dihydroconiferyl
Dihydrocubebin
Dihydroergotoxin
Dihydroestafiatone
Dihydroflavonols
Dihydrofolate
Dihydrofolic
Dihydrofuran
Dihydrofurocoumarin
    (Dihydrofurocoumarins)
Dihydrogen
Dihydrohomalicine
Dihydroisodrin
Dihydrokaempferol
Dihydrolanosterol
Dihydromaltol
Dihydroneopterin
Dihydronicotinamide
Dihydrophaseic
Dihydroprogesterone
Dihydroprostaglandin
Dihydroprotoberberine
Dihydropyridines
Dihydropyrimidine
Dihydropyrrolizine
Dihydrorhizobitoxine
Dihydrosamidin
Dihydrostreptomycin
Dihydrotestosterone
Dihydroxyacetone
Dihydroxyacetophenone
Dihydroxybenzenes
    (Dihydroxybenzene)
Dihydroxybenzoate
Dihydroxybenzoic
Dihydroxycholecalciferol
Dihydroxycoumarins
    (Dihydroxycoumarin)
Dihydroxyfumaric
Dihydroxyhexadecanoic
Dihydroxyphenethylamine
Dihydroxyphenylalanine
Dihydroxyphenylserine
Dihydroxypiperidine
Dihydroxypregn

Dihydroxyvitamin
Dihydrozeatin
Diimidoesters
Diisobutyrate
Diisocyanate
Diisoeugenol
Diisopropyl
Diisopropylthiocarbamate
Diisothiocyanates
Dikaryons *(Dikaryotic,*
    Dikaryotization)
Dikaryophase
Dikes *(Dike, Diked)*
Diketene
Diketones *(Diketone)*
Dikinase
Dikotex *see* Dicotex
Dikraneura
Dikraneurini
Dilactol
Dilactone
Dilation *(Dilatation,*
    Dilatatory, Dilated,
    Dilations, Dilator)
Dilatometry *(Dilatometric)*
Dilatotarsa
Dilauryl
Dilignol
Dill
Dillapiole
Dilleniaceae
Dilleniales
Dilophospora
Dilor
Dilta
Diltiazem
Diludin
Dilution *(Diluent, Diluents,*
    Dilute, Diluted, Diluter,
    Diluting)
    X-Ref: Undiluted
Dima
Dimecron
Dimedrol
Dimelaena
Dimer *(Dimeric,*
    Dimerisation,
    Dimerization, Dimerizing,
    Dimers)
Dimercaprol
Dimeriella
Dimethoate
    X-Ref: Cygon,
    Phosphamide, Rogor
Dimethoxybenzene
Dimethoxyphosphinothioyl
Dimethrin
Dimethyl
Dimethylalkanes
Dimethylallyltryptophan
Dimethylamine
    (Dimethylamino)
Dimethylaminobenzaldehyde
Dimethylaminobenzenediazo
Dimethylaminoethoxy

Dimethylaminophenol
Dimethylarginine
Dimethylarsinate
Dimethylarsinic
Dimethylbenzimidazolyl
Dimethylcurine
Dimethylcyclopentyl
Dimethyldithiocarbamate
Dimethylepifukiic
Dimethylether
  (Dimethylethers)
Dimethylethyl
Dimethylformamide
Dimethylformamidine
Dimethylfukiic
Dimethylhydantoin
Dimethylhydrazides
  (Dimethylhydrazide)
Dimethylhydrazine
Dimethylmorpholinium
Dimethylnitrosamine
Dimethylnona
Dimethylnonyl
Dimethylpyrazol
Dimethylpyrimidinole
Dimethylsulfate
Dimethylsulfoxide
Dimethyltriazene
Dimethylurea
Dimetridazole
Dimidin
Dimilin see Diflubenzuron
Dimorphosciadium
Dimorphotheca
Dimorphs see
  Polymorphism
Dimorphus
Dimyristoyl
Dinarmus
Dineura
Dineutes
Dining (Dinner, Dinners)
Dinitramine
Dinitroaniline
  (Dinitroanilines)
Dinitrocresol
Dinitrogen
Dinitroorthocresol
Dinitrophenol
  (Dinitrophenolic,
  Dinitrophenols)
Dinitrophenyl
Dinitrophenylhydrazones
  (Dinitrophenylhydrazone)
Dinitrothiophene
Dinobryon
Dinobuton
Dinocap
Dinochernes
Dinochloa
Dinocras
Dinoderus
Dinoflagellata
  (Dinoflagellate,
  Dinoflagellates)
Dinophyceae (Dinophycean)

Dinoseb
  X-Ref: Premerge
Dinothrombium
Dinucleotides (Dinucleotide)
Dioctophyma
Dioctophymatoidea
Dioctophymidae
Dioctria
Dioctyl
Diodyrrhynchus
Dioecism (Dioecious,
  Dioecy, Unisexual)
  X-Ref: Unisexuality
Dioestrus see Diestrus
Diols (Diol)
Dionaea
Diones (Dione)
Diopsis
Diopsiulus
Dioptidae
Dioryctria
Dioscorea (Dioscoreas)
Dioscoreaceae
Dioscoreophyllum
Diosgenin (Diosgenine)
Diosindigo
Diosmeae
Diosmetin
Diosmin
Diospyros
Dioxane
Dioxathion
Dioxide
Dioxins (Dioxin)
Dioxolane
Dioxopiperazine
Dioxygenase
Diparopsis
Dipcadi
Dipentene
Dipeptidase (Dipeptidases)
Dipeptides (Dipeptide)
Dipetalonema
Dipetalonematidae
Diphasia
Diphasium
Diphenadione
Diphenamid
Diphenhydramine
Diphenol
Diphenolase
Diphenoloxidase
Diphenyl see Biphenyls
Diphenylacetamide
Diphenylamide
Diphenylamine
Diphenylcarbazone
Diphenylethers
Diphenylguanidine
Diphenylhydantoin
Diphenylmethane
Diphenylolpropane
Diphenylpyrazolium
Diphenylsulfone
Diphenylthiocarbazone see
  Dithizone

Diphenylthiourea
Diphenyltin
Diphenylurea
Diphosphatase
  (Diphosphatases,
  Pyrophosphatases)
  X-Ref: Pyrophosphatase
Diphosphate (Diphosphates,
  Pyrophosphates)
  X-Ref: Pyrophosphate
Diphosphoglucose
Diphosphoglycerate
Diphosphoinositide
Diphosphonate
Diphosphopyridine
Diphtheria
  X-Ref: Antidiphtheria
Diphyllin
Diphyllobothriidae
Diphyllobothrium
Diphyllum
Diphyscium
Dipivaloylmethanate
Diplandroid
Diplatys
Diplazon
Diplazontinae
Diplectrona
Diplocardia
Diplocarpon
Diplocentridae
Diplocentrus
Diplococcus (Diplococci,
  Pneumococcal,
  Pneumococci)
  X-Ref:
  Antipneumococcal,
  Pneumococcus
Diplodia
Diplodialide
Diplodinium
Diplodoma
Diplogasteridae
Diplogyniidae
Diploids (Diploid, Diploidal,
  Diploidization,
  Diploidized, Diploidizing,
  Diploidy)
Diplolophium
Diplopappus
Diplophlyctis
Diploplectron
Diplopoda (Diplopod,
  Diplopods)
Diploptera
Diploschistes
Diplostomidae
Diplotaxis
Diplotene
Diplotriaenoidea
Diplura
Dipotassium see Potassium
Dipping
Diprenorphine
Diprion
Diprionidae (Diprionid)

Dipropionate
Dipropylthiocarbamate
  (Dipropylthiolcarbamates)
Dipsacaceae (Dipsacus)
Diptera (Dipteran,
  Dipterological, Dipterous)
  X-Ref: Microdiptera
Dipterex see Trichlorfon
Dipterocarpaceae
  (Dipterocarp,
  Dipterocarps)
Dipterocarpus
Dipterosiphonia
Dipteryx
Dipylidium
Dipyridamole
Dipyridinium
Dipyridyl
Diquat
Directories
Dirian
Dirofilaria
Dirofilariasis
Disaccharidases
  (Disaccharidase)
Disaccharides
  (Disaccharide)
Disacidification see
  Acidulation
Disbudding
Discaria
Discestra
Discocactus
Discoidin
Discolomidae
Discoloration (Discolored,
  Discolouration)
Discomycetes (Discomycete,
  Discomycetous)
Discophora
Discophrya
Discosia
Dishes (Dish)
Dishwashing (Dishroom,
  Dishwasher)
Disincentives see Incentives
Disinfection (Disinfect,
  Disinfectant,
  Disinfectants, Disinfected,
  Disinfecting)
Disinfestation (Disinfested)
Disinsectization
Disintegration
  (Disintegrated,
  Disintegrating,
  Disintegrator)
Disks (Disc, Discing, Discs,
  Discus, Disk, Disking)
Dislocation (Dislocations)
Dismutase (Dismutases)
Dismutation
Disocactus
Disodium
Disomics (Disomic)
Disophenol
  X-Ref: Ancylol

Disparlure
Dispensary
Dispensers *(Dispenser)*
Dispersion *(Dispersant,*
 Dispersants, Dispersibility,
 Dispersing, Dispersions,
 Dispersity, Dispersive)
Displacement *(Displaced)*
Disposal *(Disposable,*
 Disposing)
Dissection *(Dissectible,*
 Dissecting)
 X-Ref: Microdissection
Dissemination
 (Disseminated,
 Disseminators)
Dissipation
Dissodinium
Dissolution *(Dissolved)*
Distantiella
Distemonanthus
Distemper
Distichlis
Distichophyllum
Distillation *(Distillate,*
 Distillates, Distillative,
 Distilled, Distilling)
 X-Ref: Microdistillation
Distilleries *(Distillery)*
Distillers *(Distiller)*
Distomatosis
Distomum
Distributors *(Distributor)*
Disturbance *(Disturbances,*
 Disturbed)
 X-Ref: Undisturbed
Disulfates *(Disulfate,*
 Disulphates, Pyrosulphate)
 X-Ref: Pyrosulfate
Disulfides *(Disulfide,*
 Disulphide)
Disulfiram
Disulfonate
Disulfonic
Disulfoton
Ditches *(Ditch)*
Diterpenes *(Diterpene,*
 Diterpenic)
 X-Ref: Bisditerpene
Diterpenoids *(Diterpenoid)*
Dithane
Dithioacetals
Dithiocarbamates
 (Dithiocarbamate)
Dithiocarbamic
Dithiodinicotinic
Dithiolane
Dithionite
Dithiophosphate
Dithiothreitol
Dithizone *(Dithizon)*
 X-Ref:
 Diphenylthiocarbazone
Ditrazine
Ditriaena
Ditrichum

Ditylenchus
Ditylum
Diurea
Diuretics *(Diuresis,*
 Diuretic)
 X-Ref: Antidiuretic
Diurnal
Diuron
 X-Ref: Karmex
Diversification *(Diversified,*
 Diversifies, Diversify,
 Diversifying)
Diversoside
Diverticulum *(Diverticula,*
 Diverticular)
Divinyl
DMSO
DNA *(DNAs)*
 X-Ref: rDNA
DNase *see*
 Deoxyribonucleases
Doassansia
Dobsonflies *(Dobsonfly)*
Dociostaurus
Dockage
Docking
Docks *(Dock)*
Docosahexaenoic
Docosenoic
Documentation
 (Documentalists)
Dodder
Dodecadien
Dodecadienoates
Dodecane
Dodecatheon
Dodecenoic
Dodecenyl
Dodecyl
Dodecylsulfate
 (Dodecylsulphate)
Dodine
Dodonea
Dodonidia
Doe
Dogs *(Bitch, Canine, Dog,*
 Hound, Pup, Puppy,
 Pups)
 X-Ref: Bitches, Canines,
 Hounds, Puppies,
 Watchdogs
Dogwoods *(Dogwood)*
Dolerite
Dolicheremaeus
Dolichoderidae
Dolichoderinae
Dolichoderus
Dolichodorus
Dolichol
Dolicholon
Dolichomiris
Dolichomins
Dolichopetalum
Dolichopeza
Dolichopoda
Dolichopodidae

Dolichos
Dolichothele
Dolichovespula
Dolichyl
Doliopria
Dollarspot
Dolomite *(Dolomites,*
 Dolomitic)
Dolphins *(Dolphin)*
Domatia
Domestication
 (Domesticated)
 X-Ref: Nondomesticated
Domiati
Dominance *(Dominant)*
 X-Ref: Overdominance
Donkeys *(Donkey)*
Donors *(Donor)*
Dontostemon
Donus
Doona
Dopa
Dopamine
Doppler
Dorcadion
Dorcatoma
Dorema
Dormancy *(Dormancies,*
 Dormant)
 X-Ref: Nondormant
Dormin *see* Abscisin
Doronicum
Dorper
Dorset
Dorsiventrality
Dorstenia
Doryclus
Dorycnium
Doryctinae
Doryctinus
Dorylaimida
Dorylaimina
Dorylaimoidea
Doryline
Doryllium
Dorylus
Dorymerus
Doryphora
Dorypteryx
Dorytomus
Dosimetry *(Dosimeter,*
 Dosimeters, Dosimetric)
Dothichiza
Dothidella
Dothiorella
Dothistroma
Dothistromin
Dotyophycus
Doublecross *see*
 Crossbreeding
Dough *(Doughs)*
Doughnuts
Dovenix *see* Nitroxynil
Doves *(Dove)*
Dowco
Dowex

Downpours *see* Rain
Dowpon *see* Dalapon
Doxantha
Doxapram
DPX
Draba
Dracaena
Dracunculus
Draeculacephala
Drafting *(Draft, Drafted,*
 Drafts)
Dragonflies *(Dragonfly)*
Drainage *(Drain, Drainable,*
 Drainages, Drained,
 Draining, Drains,
 Underdrained)
 X-Ref: Underdrainage,
 Undrained
Draintile
Drakes *(Drake)*
Dranosan
Drassodes
Drat *see* Chlorophacinone
Drawings *(Drawing)*
Drazoxolon
Drechslera
Dredging *(Dredge, Dredger,*
 Dredgers, Dredgings)
Dregs
Drenching *(Drench,*
 Drenches)
Drepanacra
Drepanidae
Drepanocerus
Drepanolejeunea
Drepanopeziza
Drepanosiphum
Drepanotermes
Dressers *(Dresser)*
Dresses *(Dress)*
Dressing *(Dressings)*
Dreyfusia
Drifting *(Drift, Driftless,*
 Drifts)
Drilling *(Drill, Drilled,*
 Drillings, Drills)
 X-Ref: Overdrilled
Drinking *(Drink, Drinks)*
Drivers *(Driver)*
Driving *(Drive, Driven,*
 Drives)
Dromaius
Dromedaries *(Dromedary)*
Dromotropism
 (Dromotropic)
Drones *(Drone)*
Droperidol
Dropouts *see* Education
Droppings
Dropsy
Drosera
Droseraceae
Drosophila
Drosophilella
Drosophilidae *(Drosophilids)*
Drosophilin

Drosophyllum
Drosopterins
Droughts *(Drought,
Droughting, Drouth)*
Drugs *(Drug)*
Drupacin
Drupanin
Drupe
Drupelet
Drusus
Dry *(Dryness)*
Dryadanthe
Dryadula
Dryas
Dryers *(Drier, Driers,
Dryer)*
Drying *(Dried, Dries,
Dryed)*
X-Ref: Redrying,
Undried
Dryinidae
Drylands *(Dryland)*
Drylot
Drymaplaneta
Drymaria
Drymonia
Drymusa
Dryocoetes
Dryocosmus
Dryocrassol
Dryocrassyl
Dryopidae
Dryopoidea
Dryopteris
Drysdale
Duboisia
Duboscqia
Duchesnea
Ducks *(Duck, Duckling,
Ducklings)*
Duckweed *(Duckweeds)*
Ducts *(Duct, Ductal,
Ductus)*
X-Ref: Intraductal
Duff
Dufoureinae
Dugaldia
Dugesia
Dugesiella
Duikers *(Duiker)*
Dulcitol *see* Galactitol
Dulichium
Dulinius
Dumortiera
Dumping *(Dump, Dumped,
Dumps)*
Dumplings *(Dumpling)*
Dunaliella
Dunes *(Dune)*
Dung
X-Ref: Cowdung
Duodenum *(Duodenal)*
X-Ref: Intraduodenal
Duplaspidiotus
Dupontia
Durability *(Durable)*

Durcupan
Durio
Duroc
Duroquinone
Durotomy
Durra *(Jowars)*
X-Ref: Cholam, Jowar
Durum
Duseniella
Dust *(Dustfree, Dusting,
Dusts, Dusty)*
Dustbathing
Dusters *(Duster)*
Duter
Duties *(Duty)*
Duvatrienediols
Dwarfism *(Dwarf, Dwarfed,
Dwarfing, Dwarfness,
Dwarfs)*
Dwayalomella
Dyads
Dyckia
Dyers *(Dyer)*
Dyes *(Dye, Dyed, Dyeing,
Dyestuff, Dyestuffs)*
Dyfonate *see* Fonofos
Dying *see* Death
Dymondia
Dynamometry
*(Dynamometer,
Dynamometers,
Dynamometric)*
Dynastes
Dynastidae *(Dynastinae)*
Dyrene
Dysaletria
Dysaphis
Dyschondroplasia
Dyscritomyia
Dysdercus
Dysentery *(Dysenterial,
Dysenteric, Dysenteries)*
Dysfunctions *(Dysfunction)*
Dysgalactia
Dysgerminoma
Dysglobulinemia
Dysmachus
Dysmicoccus
Dysmictocarabus *see*
Carabus
Dyspepsia *see* Indigestion
Dysphagia
Dysplasia *(Dysplastic)*
Dyspnea *(Dyspnoea)*
Dyssodia
Dystaenia
Dystocia *(Dystocias)*
Dystrophy *(Dystrophic,
Dystrophogenic)*
Dytiscidae *(Dytiscid)*
Dytiscus
Eacles
Eagles *(Eagle)*
Earheads *(Earhead)*
Earias
Earlywood *see* Springwood

Earnings *see* Income
Ears *(Ear)*
X-Ref: Aural
Earth *(Earths)*
Earthenware
Earthquake *(Earthquakes)*
Earthworks *(Earthwork)*
Earthworms *(Earthworm)*
Earwigs *(Earwig)*
Earworms *(Earworm)*
Easements *(Easement)*
Eating *(Eat, Eaten, Eaters,
Eats)*
Eatonia
Ebenaceae
Ebony
Eburnamine
Eburnamonine
Ecballium
Ecclitura
Eccremocactus
Eccrine
Ecdyonurus
Ecdysis *(Ecdysial)*
X-Ref: Postecdysis
Ecdysone *(Ecdyson,
Ecdysones)*
Ecdysterone
Eceriferum
ECG *see*
Electrocardiography
Echeveria *(Echeverias)*
X-Ref: Thompsonella
Echidnophaga
Echinacea
Echinatine
Echinocereus
Echinochloa
Echinochondrium
Echinococcosis
Echinococcus
Echinocystis
Echinocyte *(Echinocytic)*
Echinodermata
(Echinoderms)
Echinodontium
Echinofossulocactus
Echinolaena
Echinomastus
Echinopharyphium
Echinops
Echinopsis
Echinostelium
Echinostoma
Echinostomatidae
Echis
Echitoideae
Echium
Echmepteryx
Echocardiography
(Echocardiographic)
Echthistus
Echthromorpha
Eciton
Ecklonia
Eclipophleps

Eclipses *(Eclipse)*
Eclipta
Eclosion
X-Ref: Posteclosion
Ecofallow *see* Fallowing
Ecogeography
*(Ecogeographic,
Ecogeographical)*
Ecologists *(Ecologist)*
Ecology *(Agroecological,
Agroecosystem,
Bioecological, Ecologic,
Ecological, Ecologically,
Ecologies, Ecosystem)*
X-Ref: Agroecology,
Agroecosystems,
Bioecology, Ecosystems
Econazol
Econometrics *(Econometric)*
Economics *(Agroeconomic,
Economic, Economical,
Economically, Economico,
Economies, Economy)*
X-Ref: Agroeconomy
Economists *(Economist)*
Economize *(Economise,
Economised, Economising,
Economization,
Economizing)*
Ecophysiology
(Ecophysiological)
Ecosphere
Ecosystems *see* Ecology
Ecotone *(Ecotones)*
Ecotoxicology
Ecotypes *(Ecotype,
Ecotypic)*
X-Ref: Agroecotypes
Ecribellatae *(Ecribellate)*
Ectemnius
Ectendomycorrhizae
*(Ectendomycorrhiza,
Ectendomycorrhizal,
Ectendomycorrhizas)*
ECTHAM
Ecthyma
Ectocarpales
Ectocarpus
Ectoderm *(Ectodermal)*
Ectodesmata
Ectoedemia
Ectomycorrhizae
*(Ectomycorrhiza,
Ectomycorrhizal,
Ectomycorrhizas)*
Ectomyelois
Ectoparasites *(Ectoparasite,
Ectoparasitic,
Ectoparasiticide)*
Ectophytes *(Ectophytic)*
Ectopia *(Ectopic)*
Ectoplasm
Ectopsocidae
Ectopsocus
Ectosymbiosis
(Ectosymbiont,

Ectosymbiotic)
Ectotrophic *(Ectotropic)*
Ectromelia
Ectropion
Ectropis
Eczema
Edam *(Edamer)*
Edaphic *see* Soils
Edaphus
Eddies *(Eddy)*
Edeine
Edelweiss
Edema *(Hydropsy)*
  X-Ref: Hydrops, Oedema
Edgaria
Edible *(Edibility)*
Edicts
Edifenphos
  X-Ref: Hinosan
Edilbaev
Edithocolea
EDTA
Education *(Educate,*
  Educated, Educating,
  Educational,
  Educationally, Educative,
  Graduate, Graduated,
  Pedagogic, Teach,
  Teaches, Teachings)
  X-Ref: Classroom,
  Colleges, Dropouts,
  Graduates, Pedagogy,
  Taught, Teaching,
  Tutorial
Educators *(Educator,*
  Teacher)
  X-Ref: Teachers
Edwardsiana
Edwardsiella
EEG *see*
  Electroencephalography
Eelgrass
Eels *(Eel)*
Eelworms *(Eelworm)*
Efferia
Efflorescence *see* Flowering
Effluent *(Effluence,*
  Effluents)
Efulensia
Egeria
Egesta
Eggplant *(Brinjals,*
  Eggplants)
  X-Ref: Brinjal
Eggs *(Egg, Egging)*
Eggshells *(Eggshell)*
Eggyolk
Egregia
Ehrlichia
Ehrlichiosis
Eichhornia *(Eichornia)*
Eicosapeptide
Eigenvector
Eilica
Eimeria
  X-Ref: Coccidium

Eimeriidae
Einkorn
Eisenia
Ejaculation *(Ejaculate,*
  Ejaculated, Ejaculates,
  Ejaculating, Ejaculations,
  Ejaculatory)
Ejection *(Ejected, Ejecting,*
  Ejectors, Ejects)
Ekatin *see* Thiometon
Ekebergia
EKG *see*
  Electrocardiography
Elachertus
Elachiptera
Elachista
Elaeagnaceae
Elaeagnus
Elaeis
Elaeocarpaceae
Elaeocarpus
Elaeophora
Elaidic
Elands *(Eland)*
Elasmolomus
Elasmopalpus
Elasmucha
Elasmus
Elastase
Elasticity *(Elastic,*
  Elasticities)
  X-Ref: Inelastic
Elastin *(Elastins)*
Elastomers *(Elastomer,*
  Elastomeric)
Elateridae *(Elaterid)*
Elaters
Elatine
Elatobium
Elatophilus
Elbaue
Elbows *(Elbow)*
  X-Ref: Anconeal, Cubital
Eldana
Elderberries *(Elderberry)*
Elderly
  X-Ref: Centenarians,
  Oldsters
Elecampane
Electrapis
Electricians *(Electrician)*
Electricity *(Electric,*
  Electrical, Electrically,
  Electrics)
Electrification *(Electrified,*
  Electrify)
Electroanalysis
  (Electroanalytical)
Electroanesthesia
Electroantennograms
  (Electroantennogram)
Electrocardiography
  (Electrocardiogram,
  Electrocardiograms,
  Electrocardiograph,
  Electrocardiographic,

Electrocardiographical,
  Electrocardiographs)
  X-Ref: ECG, EKG
Electrocardiology
Electrocautery *see*
  Cauterization
Electrochemistry
  (Electrochemical)
Electrochromatography
  (Electrochromatograph)
Electrocoagulation
Electrocution *(Electrocutor,*
  Electrocutors)
Electrodes *(Electrode)*
Electrodiagnostics
Electrodialysis
  (Electrodialytic)
Electroejaculation
Electroencephalography
  (Electroencephalogram,
  Electroencephalograms,
  Electroencephalographic)
  X-Ref: EEG
Electrofiltration
Electroflotation
Electrofocusing
  X-Ref: Isoelectrofocusing
Electrogenesis
  (Electrogenic)
Electrogrammetry
Electrography
  (Electrograms)
Electrohydraulics
  (Electrohydraulic)
Electroimmunoassay *see*
  Immunoassay
Electrokinetics
  (Electrokinetic)
Electrolysis *(Electrolytic)*
Electrolytes *(Electrolyte)*
Electromagnetism
  (Electromagnet,
  Electromagnetic)
Electromechanics
  (Electromechanical)
Electrometry *(Electrometer,*
  Electrometric)
Electromyography
  (Electromyogram,
  Electromyograms,
  Electromyographic)
  X-Ref: EMG
Electronics *(Electronic,*
  Electronically)
Electronmicroscopy
  (Electronmicroscopic,
  Electronmicroscopical)
Electroosmosis
  (Electroosmotic)
Electrophoregrams
  (Electrophoregram)
Electrophoresis
  (Electrophoretic,
  Electrophoretical,
  Electrophoretically)
Electrophotoluminescence

Electrophysics
  (Electrophysical)
Electrophysiology
  (Electrophysiologic,
  Electrophysiological)
Electropotentials
  (Electropotential)
Electroretinography
  (Electroretinogram,
  Electroretinograms,
  Electroretinographic)
Electroshock
  (Electroconvulsive)
Electroslag
Electrostatics *(Electrostatic)*
Electrostimulation
Electrotherapy
Electrotransport
  (Electrotransporting)
Eledoisin
Elemanolides
Elenchidae
Elenchus
Eleocharis
Eleodes
Eleostearic
Elephantiasis
Elephantopus
Elephants *(Elephant)*
Elettaria
Eleusine *(Eleucine)*
Eleutherococcus
Elevators *(Elevator)*
Eliaea
Eliminase
Elks *(Elk)*
Ellagic
Ellagitannins
Ellimenistes
Elliptochthoniidae
Elliptocytes *see*
  Erythrocytes
Ellobiidae
Elmidae
Elminthidae
Elms *see* Ulmus
Elodea
Elongatin
Elongation *(Elongated,*
  Elongating)
Elphinstonia
Elsinoe
Eltinor
Eltonella
Elution *(Eluent, Eluted)*
Eluviation *(Eluvial)*
Elymana
Elymoclavine
Elymus
Elysia
Elysius
Elytra *(Elytral, Elytron,*
  Elytrum)
Elytrigia
Elytrurus
Emasculation *see* Castration

Embargo *(Embargoes)*
Embedding *(Embedded)*
Embelia
Embelin
Embergeria
Embidobiini
Embiidae *(Embiid)*
Embioptera
Emblica
Embolanthera
Embolectomy
Embolemidae
Embolemus
Embolism *(Emboli,*
  Embolic, Embolization)
Embryo
  X-Ref: Postembyonic
Embryogeny
  (Embryogenesis,
  Embryogenic)
  X-Ref: Proembryogenesis
Embryology *(Embryologic,*
  Embryological)
Embryos *(Embryo,*
  Embryoid, Embryoids,
  Embryoless, Embryon,
  Embryonal, Embryonate,
  Embryonated,
  Embryonating,
  Embryonic, Embryony,
  Postembryonal,
  Postembryonic)
  X-Ref: Proembryo
Embryotoxicity
  (Embryotoxic)
Emendation *(Emended)*
Emericellin
Emerin
Emesinae
Emetics *(Emetic)*
Emetine
Emex
EMG *see* Electromyography
Emigrants *(Emigrant)*
Emigration *(Emigrations)*
Emilia
Emmental *(Emmentaler,*
  Emmenthal,
  Emmenthaler)
Emmer
Emmonsia
Emmotins
Emmotum
Emodin
Empetraceae
Empetrum
Emphoropsis
Emphysema
  (Emphysematous)
Empididae
Employment *(Employ,*
  Employed, Employee,
  Employees, Employer,
  Employers, Employing,
  Employments, Employs,
  Job, Jobless)

X-Ref: Jobs,
  Unemployment
Empoasca
Empyema
Emulsifiers *(Emulsifier)*
Emulsion *(Emulsifiable,*
  Emulsification,
  Emulsified, Emulsify,
  Emulsifying, Emulsions)
Enallagma
Enamel *(Enamelled)*
Enamide
Enamine
Enantia
Enantiomer *(Enantiomeric,*
  Enantiomers)
Enantiomorphism
  (Enantiomorph,
  Enantiomorphic,
  Enantiomorphous)
Enaphalodes
Enarmonia
Encalypta
Encapsulation
  (Encapsulated,
  Encapsulating,
  Microencapsulated)
  X-Ref:
  Microencapsulation
Encarsia
Encelia
Encephalartos
Encephalitis
  (Encephalitogenic)
  X-Ref: Panencephalitis
Encephalitozoon
Encephalomalacia
Encephalomyelitis
Encephalomyocarditis
Encephalon
Encephalopathy
  X-Ref:
  Hepatoencephalopathy
Enchenopa
Enchlaena
Enchytraeidae
Encoding *see* Coding
Encyclia
Encyclopedias
  (Encyclopedia)
Encyrtidae *(Encyrtid,*
  Encyrtids, Encyrtinae)
Encyrtus
Encystment *(Encysted,*
  Encysting)
Endemic *(Endemism)*
Endeostigmata
Endive
  X-Ref: Witloof
Endocarditis
Endocardium *(Endocardial)*
Endocarp
Endocellular *see* Cells
Endocrine *(Endocrinal,*
  Endocrines,
  Endocrinologic,

Endocrinological,
Endocrinology,
Endocrinous)
Endocronartium
Endocuticle
Endoderm *(Endodermal,*
  Endodermic)
Endogeny *(Endogenous,*
  Endogenously)
Endoglucanase
Endogonaceae
Endogone
Endolithic
Endomannanase
Endometritis
Endometrium *(Endometrial)*
Endomitosis *see* Mitosis
Endomyces
Endomycetaceae
Endomychidae
Endomycopsis
Endomycorrhizae
  (Endomycorrhiza,
  Endomycorrhizal,
  Endomycorrhizas)
Endonuclease
  (Endonucleases)
Endoparasites
  (Endoparasite,
  Endoparasitic,
  Endoparasitism)
Endopeptidases
  (Endopeptidase)
Endophragmia
Endophytes *(Endophyte,*
  Endophytic)
Endoplasm *(Endoplasmatic,*
  Endoplasmic)
Endopolyploids
  (Endopolyploid,
  Endopolyploidy)
Endoreduplication
Endoscopy *(Endoscope)*
Endoskeleton *(Endoskeletal)*
Endosperm *(Endospermal,*
  Endospermic,
  Endospermous,
  Endosperms)
  X-Ref: Nonendosperm
Endospores *(Endospore)*
Endostigme
Endosulfan
  X-Ref: Thiodan
Endosymbiosis
  (Endosymbionts,
  Endosymbiotic)
Endothall *(Endothal)*
Endothecium
Endothelioma
Endothelium *(Endothelial)*
  X-Ref: Interendothelial
Endothia
Endothiella
Endothion
Endotoxemia
Endotoxins *(Endotoxic,*

Endotoxin)
Endotracheal *see* Tracheal
Endoxan *see*
  Cyclophosphamide
Endozoic
Endrin
Endromis
Endymion
Energy *(Energies,*
  Energized)
Enflurane
Engelhardia
Engineering *(Engineered)*
Engineers *(Engineer)*
Engines *(Engine)*
Englerastrum
Enhydra
Enhydrin
Enicocephalidae
Enicospilus
Enkianthus
Enlinia
Enneothrips
Ennominae
Ennomos
Enoate
Enochrus
Enoclerus
Enol *(Enolic)*
Enolase
Enolate
Enolide
Enology
Enolophosphates
Enones
Enophthalmos
  (Enophthalmia)
Enoplognatha
Ensete
Ensifera
Ensilage *see* Silage
Entada
Entamoeba
Entandrophragma
Enteral *(Enterally)*
Enteric
Enteritis
Enteroanastomosis *see*
  Anastomoses
Enterobacter
Enterobacteriaceae
  (Enterobacteria)
Enterobius
Enterochromaffin
Enterococcus *(Enterococci)*
Enterocolitis
Enterocytes
Enterohepatitis
  (Enterohepatic)
Enterolithiasis
Enteromorpha
Enteronephritis
Enteropathy
  (Enteropathogenetic,
  Enteropathogenic,
  Enteropathogenicity,

Enteropathogens)
Enterotomy
Enterotoxemia
(Enterotoxaemia)
Enterotoxins *(Enterotoxic,*
Enterotoxigenic,
Enterotoxigenicity,
Enterotoxin)
Enterotropic
Enteroviruses *(Enterovirus)*
Enthalpimetry
(Enthalpimetric)
Enthalpy
Entisols *(Entisol)*
Entobacterin *(Entobacterins,*
Entobakterin)
Entodinium
Entodon
Entomobryidae
Entomofauna *see* **Insects**
Entomogenous
Entomologists
(Entomologist)
Entomology *(Entomological)*
Entomopathogenic
Entomophaga
(Entomophage,
Entomophages,
Entomophagous)
Entomophagy
Entomophily *(Entomophilic,*
Entomophilous)
Entomophthora
Entomophthoraceae
(Entomophthoraceous)
Entomophthorales
Entomopoxviruses
(Entomopoxvirus)
Entomoscelis
Entomoxan *see*
Hexachlorocyclohexane
Entophlyctis
Entrepreneurs
(Entrepreneur,
Entrepreneurial,
Entrepreneurship)
Entropion
Entropy *(Entropies)*
Entyloma
Entypus
Enucleation *(Enucleated,*
Enucleations)
Environment
(Bioenvironmental,
Environmental,
Environmentally,
Environments)
X-Ref: **Bioenvironment**
Enzootic
Enzymes *(Apoprotein,*
Enzymatic,
Enzymatically, Enzyme,
Enzymic, Enzymically,
Exoenzyme, Nonenzymic)
X-Ref: **Apoenzyme,**
Apoproteins,

Exoenzymes,
Multienzyme,
Nonenzymatic,
Photoenzymatic
Enzymology
(Enzymological)
Eocene
Eoeurysa
Eolian
Eomenacanthus
Eosentomidae
Eosentomoidea
Eosentomon
Eosin
Eosinophilia
Eosinophils *(Eosinophil,*
Eosinophile, Eosinophiles,
Eosinophilic)
Eotetranychus
Epacridaceae
Epallage
Epanerchodus
Ependyma *(Ependymal)*
Ependymitis
Epeoloides
Epermeniidae
Eperythrozoon
Eperythrozoonosis
Ephedra
Ephedrine
Ephedrus
Ephelis
Ephemerals *(Ephemera,*
Ephemeral, Ephemeron,
Ephemers)
Ephemerella
Ephemerellidae
Ephemeridae
Ephemeroidea
Ephemeropsis
Ephemeroptera
(Ephemerida)
Ephemoptera
Ephestia
Ephippigerida
Ephydra
Ephydridae
Epibenthos *see* **Benthos**
Epiblast
Epiblema
Epicardium *(Epicardial)*
Epicarp
X-Ref: **Exocarp**
Epicauta
Epichloe
Epichlorohydrin
Epichoristodes
Epicobalamin
Epicoccum
Epicormic
Epicotyls *(Epicotyl)*
Epicuticle *(Epicuticular)*
Epidemiology *(Epidemic,*
Epidemical, Epidemics,
Epidemiologic,
Epidemiological)

X-Ref:
Microepidemiology,
Seroepidemiologic
Epidendrum
Epidermatitis
Epidermis *(Epidermal,*
Epidermic, Epidermoid,
Epidermology)
X-Ref: Intraepidermal
Epidermophyton
Epidermoptidae
Epididymides
Epididymis *(Epididymal)*
Epididymitis
Epidinium
Epidural
Epifagus
Epifauna *see* **Fauna**
Epigaeic
Epigenesis *(Epigenetic)*
Epiglottis *(Epiglottic)*
Epilachna
Epilachninae
Epilampridae
Epilamprinae
Epilepsy *(Epileptic,*
Epileptiform)
Epilithic
Epilobium
Epimedium
Epimer *(Epimeric,*
Epimerization)
Epimerases *(Epimerase)*
Epimetasia
Epimyodex
Epinasty
Epinephrine
Epinotia
Epiornithic *see* **Epornitic**
Epipactis
Epiphyas
Epiphyll *(Epiphyllous)*
Epiphyllum
Epiphysiolysis
Epiphysis *(Epiphyseal,*
Epiphyses)
Epiphysitis
Epiphytes *(Epiphyte,*
Epiphytic)
Epiphytology *(Epiphytotic,*
Epiphytotics)
Epipogium
Epipsocidae
Epipsocopsis
Epipyrops
Epirhyssa
Episcia
Episcleritis
Episiotomy
Episome *(Episomal)*
Episomic
Epistasy *(Epistasis,*
Epistatic)
Epistaxis
Epistemology
(Epistemological)

Epitettix
Epithelioma
Epithelium *(Epithelia,*
Epithelial,
Epitheliogenesis,
Epitheliums)
Epithem
Epitoxasia
Epitrimerus
Epitritus
Epitrix
Epizoa *(Epizoic, Epizoon)*
Epizootiology *(Epizootic,*
Epizootics, Epizootiologic,
Epizootiological)
Epornitic
X-Ref: Epiornithic
Epoxides *(Epoxide)*
X-Ref: Monoepoxides
Epoxy *(Epoxidation,*
Epoxidized)
Epoxycarotenoids
Epoxygeranyl
Epoxygermacranolides
Epoxylathyrol
Epoxypropyl
Epoxysuccinic
Eptam *see* **EPTC**
EPTC
X-Ref: Eptam
Eptesicus
Epulis
Epuration *see* **Purification**
Equator *(Equatorial)*
Equidae *(Equid, Equids)*
Equilibration *(Equilibrated,*
Equilibrating)
Equines *see* **Horses**
Equipment *(Equipments)*
Equisetaceae
Equisetales
Equisetolic
Equisetum
Equithesin
Equus
Eradication *(Eradicant,*
Eradicants, Eradicate,
Eradicated, Eradicating)
Eragrosteae
Eragrostis
Eragrostoideae
Erannis
Eranthemum
Eranthis
Erebia
Eremanthine
Eremantholide
Eremanthus
Eremocarpus
Eremochloa
Eremocitrus
Eremophila
Eremophilane
Eremopoa
Eremopsocus
Eremosphaera

Eremotes
Eremothecium
Eremurus
Eresidae
Eretmapodites
Eretmocerus
Ereynetidae
Ergastoplasm
   (Ergastoplasmic)
Ergocalciferol
   X-Ref: Calciferol
Ergocornine
Ergocryptine
Ergoline
Ergometrine
Ergonomics *(Ergonomic,*
   Ergonomical, Ergonomy)
Ergost
Ergosterol
Ergot *(Ergotism)*
Ergotamine
Ergothioneine
Erianthus
Erica
Ericaceae *(Ericaceous)*
Ericales
Ericameria
Erigeron
Erigonidae
Erigorgus
Erinacea
Erinaceus *see* Hedgehogs
Erineum
Erinnyis
Erinose *(Erinoses)*
Eriobotrya
Eriocampidea
Eriocaulaceae
Eriocaulon
Eriocereus
Eriococcidae
Eriococcus
Erioderma
Eriogonum
Erioischia
Erionota
Eriopeltis
Eriophorum
Eriophyes
Eriophyidae *(Eriophyid,*
   Eriophyinae)
Eriophyllum
Eriophyoidea *(Eriophyoid)*
Eriosoma
Eriostemon
Eriostomum
Eriozona
Eriphus
Eriphyes
Eriplanus
Erisoma
Eristalis
Eritadenine
Ermines *(Ermine)*
Erodium
Erosion *(Erode, Eroded,*

Erodibility, Erodible,
   Eroding, Erosive,
   Erosiveness)
X-Ref: Antierosion,
   Nonerodible,
   Nonerosive
Erotylidae *(Erotylid)*
Eruca
Erucic
Eructation *(Eructated)*
   X-Ref: Belching
Eruption *(Erupt, Erupted,*
   Erupting, Eruptive,
   Erupts)
Ervum
Erwinia *(Erwinias)*
Eryngium
Erynniopsis
Eryphanis
Erysimoside
Erysimum
Erysipelas
Erysipelothrix
Erysiphaceae
Erysiphales
Erysiphe
Erythorbic
Erythraeidae
Erythrea
Erythrina
Erythrinan
Erythritol
   X-Ref: Anhydroerythritol
Erythroblastosis
Erythroblasts
Erythrocytes *(Erythrocyte,*
   Erythrocytic)
   X-Ref: Elliptocytes,
   Intraerythrocytic
Erythrodiol
Erythrogenesis *see*
   Erythropoiesis
Erythrogram
Erythroid
Erythroleukemia
Erythroleukosis
Erythromycin
Erythroneura
Erythronium
Erythrophagocytosis
Erythrophleum
Erythropoiesis
   (Erythropoietic)
   X-Ref: Erythrogenesis
Erythropoietin
Erythroxylaceae
Erythroxylon
Erythroxylum
Erythrulose
Esastigmatobius
Escalerina
Escalloniaceae
Escherichia
Escherichieae
Eschscholzia
Esculin

Eskimos
Esophagitis
   X-Ref: Oesophagitis
Esophagogastric
Esophagostomy
Esophagotracheal
Esophagus *(Esophageal,*
   Oesophageal)
   X-Ref: Megaesophagus,
   Oesophagus,
   Paraoesophageal,
   Subesophageal,
   Suboesophageal
Espalier *(Espaliered)*
Essences *(Essence)*
Esterases *(Esterase)*
   X-Ref: Antiesterase
Esterification *(Esterified,*
   Esterifying,
   Interesterified)
   X-Ref: Interesterification,
   Nonesterified,
   Unesterified
Esters *(Ester)*
Estigmene
Estivation *(Aestivating,*
   Estival, Estivating)
   X-Ref: Aestivation
Estola
Estradiol
   X-Ref: Oestradiol
Estrogen *(Antiestrogenic,*
   Estrogenic, Estrogenically,
   Estrogenicity,
   Estrogenization,
   Estrogenized, Estrogens,
   Oestrogeneous,
   Oestrogenic,
   Oestrogenicity,
   Oestrogens)
   X-Ref: Antiestrogens,
   Oestrogen
Estrone
Estrus *(Anestrous,*
   Anoestrous, Estral,
   Estrous, Estrual, Estruses,
   Oestral, Oestrous,
   Oestrual, Oestrum)
   X-Ref: Anestrus,
   Anoestrus, Oestrus,
   Polyestrus
Estuaries *(Estuarine,*
   Estuary)
Eteobalea
Ethane
Ethanediol *see* Glycol
Ethanol
Ethanolamine
Ethanolic
Ethaverine
Ethazol
Etheostoma
Ethephon
   X-Ref: Ethrel
Ethers *(Ether, Etherization)*
Ethics *(Ethic, Ethical)*

Ethidium
   X-Ref: Homidium
Ethion
Ethionine
Ethirimol
Ethmia
Ethmiidae
Ethmoid *(Ethmoidal)*
Ethnic
Ethnobotany
   (Ethnobotanical)
Ethnography
Ethnosociology
Ethofumesate
Ethology *(Ethological)*
Ethoxycaffeine
Ethoxycarbonyl
Ethoxylupanin
Ethoxyquin
   X-Ref: Santoquin
Ethrel *see* Ethephon
Ethylacetate
Ethylamide
Ethylamine
Ethylation *(Ethylated)*
Ethylbenzimidazole
Ethylcarbamate
Ethylcyclohexyl
Ethyldithiophosphate
Ethylene *(Ethylenebis)*
   X-Ref: Antiethylene
Ethylenediamine
Ethyleneglycols
   (Ethyleneglycol)
Ethylenethiourea
Ethylenimine
   (Ethyleneimine)
Ethylethylenimine
Ethylmalemide
Ethylmercury
   (Ethylmercuric)
Ethylmethane
Ethylmethanesulfonate
   (Ethylmethanesulphonate)
Ethylphenoxy
Ethylphenyl
Ethylthiocarbamate
Ethylurea
Etidocaine
Etiella
Etiochloroplasts
   (Etiochloroplast)
Etiocholanolone
Etiolation *(Etiolated)*
Etiology *(Aetiological,*
   Etiologic, Etiological)
   X-Ref: Aetiology
Etiopathogenesis *see*
   Pathogenesis
Etioplasts *(Etioplast)*
Etorphine
Etymology
Euandrena
Euarestoides
Eubelidae
Eublaberus

Eublemma
Euborellia
Eucallipterus
Eucalyptus *(Eucalypt,*
Eucalypts)
Eucaryote *(Eucaryotes,*
Eucaryotic, Eukaryota,
Eukaryote, Eukaryotes,
Eukaryotic)
Eucera
Euceraphis
Eucerine
Eucharis
Eucheuma
Euchlaena
Euchloe
Euchromatin
Euchromius
Euclea
Euclidium
Eucnemidae
Eucoilidae
Eucoilinae
Eucommia
Eucondylops
Eucosma
Eucosmidae *see*
Olethreutidae
Eucosoma
Eucryphia
Eucryphiaceae
Euderma
Eudesmane
Eudia
Eudiplogaster
Eudorina
Eudusbabekia
Euetheola
Eugaurax
Eugenia
Eugenol
Euglena
Euglenales
Euglenophyceae
Euglenophyta
Euglobulin
Euglossa
Euglyphidae
Eugregarina *(Eugregarine,*
Eugregarines)
Euhamitermes
Eukoenenia
Eulachnus
Eulaelaps
Eulecanium
Eulophidae *(Eulophid,*
Eulophids)
Eumargarodes
Eumastacidae
(Eumasticinae)
Eumastacoidea
Eumedonia
Eumenes
Eumenidae
Eumerus
Eumichtis

Eumicromus
Eumolpidae *(Eumolpinae)*
Eumycetes
Eunidia
Eunotus
Euodia
Euolalin
Euonymus
Eupackardia
Eupafolin
Eupaformonin
Euparagia
Euparen *see* Dichlofluanid
Eupatilin
Eupatolide
Eupatorieae
Eupatorium
Eupelmidae
Euphaedra
Euphalangium
Euphasiopteryx
Euphausia
Euphausiidae
Euphorbetin
Euphorbia *(Euphorbias)*
Euphorbiaceae
(Euphorbiaceous)
Euphorbieae
Euphoriana
Euphorinae
Euphotic
Euphrasia
Euphydryas
Eupithecia
Eupitheciinae
Eupithecini
Euplectrus
Euplilis
Euploids *(Euploid,*
Euploidy)
Euplusia
Eupoecila
Eupomatenoid
Eupomatia
Euproctis
Euprosthenops
Eupsilia
Eupteromalus
Eupterote
Eupterotidae
Eupteryx
Euptoieta
Eureiandra
Eurhynchium
Euriphene
Eurois
Europium
Eurosta
Eurotia
Eurotium
Eurya
Eurycantha
Eurycanthinae
Eurycotis
Eurydema
Eurygaster

Euryglossa
Euryglossina
Euryglossinae
Euryhaline
Eurytoma
Eurytomidae
Eurytrema
Euscelini
Euscelis
Euscepes
Euschistus
Euscorpius
Euselasia
Eusiderin
Eusideroxylon
Eusimulium
Eusminthurus
Eustachian
Eustigmatophyceae
Eustrongylides
Eutectic
Euterpe
Eutetranychus
Euthanasia
Euthyrhynchus
Eutogenes
Eutolmus
Eutricharea
Eutrichomelina
Eutrombicula
Eutrombidium
Eutrophication *(Eutrophic,*
Eutrophicated)
Eutypa
Eutypella
Euura
Euxanthe
Euxanthellus
Euxoa
Euxylophora
Euzophera
Evacuation *(Evacuate,*
Evacuated, Evacuates,
Evacuating)
Evagination *(Evaginated,*
Evaginations)
Evaporation *(Evaporated,*
Evaporating, Evaporative,
Evaporativity, Evaporator,
Evaporators)
Evaporimetry
(Evaporimeter,
Evaporimeters)
Evapotranspiration
Eveque
Evergestis
Evergreens *(Evergreen)*
Evernia
Everninomicin
Evetria
Evisceration *(Eviscerated,*
Eviscerating)
Evodia
Evodinus
Evolution *(Evolutional,*
Evolutionary, Evolutive,

Evolve, Evolved, Evolving)
X-Ref: Coevolution,
Overevolution
Evonoside
Ewes *(Ewe)*
Exanthema
Excavation *(Excavate,*
Excavated, Excavates,
Excavating, Excavations,
Excavators)
Excelsine
Exchange *(Exchangeability,*
Exchangeable,
Exchanged, Exchanger,
Exchangers, Exchanges)
Exciccatae
Excision *(Excised)*
Excitation *(Excitability,*
Excitable, Excitations,
Excitatory, Excited,
Excitement, Exciting)
Excrement *(Excrements,*
Excreta)
Excretion *(Excreted,*
Excreting, Excretions,
Excretory)
Excystment *(Excystation,*
Excysted)
Executives *(Executive)*
Exencephaly
Exenterus
Exercise *(Exercised,*
Exercises, Exercising)
Exetasis
Exfoliation *(Exfoliated,*
Exfoliative)
Exhalation *(Exhalates,*
Exhaled, Exhales,
Exhaling, Expiratory)
Exhaustion *(Exhausted,*
Exhausting)
Exhausts *(Exhaust)*
Exhibits *(Exhibit,*
Exhibiting, Exhibition,
Exhibitions)
Exidia
Exidiopsis
Exilispora
Exine *(Exines)*
Exitianus
Exobasidiales
Exobasidiellum
Exobasidium
Exocarp *see* Epicarp
Exocarpos
Exocellular *see* Cells
Exocortis
Exocrine
Exocuticle
Exocytosis
Exodeoxyribonuclease
Exodermis
Exoenzymes *see* Enzymes
Exogonium
Exometabolites
Exoneurella

Exonuclease
Exophiala
Exophthalmos
  (Exophthalmus)
Exopolysaccharide
Exoprotease
Exorista
Exoskeleton (Exoskeletal)
Exosmosis see Osmosis
Exostemma
Exostosis (Exostoses)
Exoteleia
Exotherm (Exothermic)
Exotoxin
Expeditions (Expedition)
Expellers (Expeller)
Expenditures (Expenditure)
Explants (Explant)
Exploitation (Exploit,
  Exploited, Exploiting,
  Exploits)
Explosion (Exploding,
  Explosions, Explosiveness)
Explosives (Explosive)
Exporters (Exporter)
Exports (Export,
  Exportation, Exportations,
  Exported, Exporting)
Exposure (Exposures)
Expoxide
Expropriation
  (Expropriated,
  Expropriations)
Exsanguination
Exserohilum
Exsicatae
Extenders (Extender)
Extension
Extensometer
Extermination
  (Exterminate)
Exterminators
  (Exterminator)
Extinction (Extinct)
Extinguishers
  (Extinguishing)
Extraaxonal see Axons
Extracellular see Cells
Extracerebral see Brain
Extrachromosomal see
  Chromosomes
Extracorporeal
Extraction (Extractability,
  Extractable, Extractant,
  Extractants, Extracted,
  Extractible, Extracting,
  Extractions, Extractive,
  Extractives, Extractor)
  X-Ref: Unextracted
Extracts (Extract)
Extrafloral
Extrahepatic see Hepatic
Extramedullary see Medulla
Extramitochondrial see
  Mitochondria
Extraneuronal see Neurons

Extranuclear see Nucleus
Extraocular see Eyes
Extraperoxisomal see
  Peroxisomes
Extrapulmonary see
  Pulmonary
Extrarenal see Kidneys
Extraretinal see Retina
Extraskeletal see Skeleton
Extrauterine see Uterus
Extravascular see Vascular
Extremities (Extremity)
Extrusion (Extruded,
  Extruder, Extruders,
  Extruding)
Exudates (Exudate,
  Exudation, Exudative)
  X-Ref: Antiexudative
Exuviella
Exyra
Eyepiece
Eyes (Eye, Eyeball,
  Eyeballs, Eyelashes,
  Eyeless, Eyelid, Eyelids)
  X-Ref: Binocular,
  Extraocular,
  Intraocular
Eyespots (Eyespot)
Eyeworms (Eyeworm)
Eylais
Fabaceae (Fabaceous)
Faboideae
Fabraea
Fabrics (Fabric)
Faceflies (Facefly)
Factories (Factory)
Fads (Fad, Faddism,
  Faddists)
Fagaceae
Fagara
Fagarine
Fagonia
Fagopyrum
Fagus (Beech)
  X-Ref: Beeches
Fairs (Fair, Fairgrounds)
Fairways (Fairway)
Falconiformes
Falcons (Falcon)
Falibetan
Fall (Falling)
Fallopian
Fallout
Fallowing (Fallow,
  Fallowed, Fallows)
  X-Ref: Ecofallow
Fallugia
Famines (Famine)
Famophos see Famphur
Famphur
  X-Ref: Famophos
Fanleaf
Fannia
Fans (Fan, Fanning)
Farcy see Glanders
Farina (Farinaceous)

Farinocystis
Farinograph
Farmeria
Farmers (Farmer)
  X-Ref: Husbandman
Farmhouses (Farmhouse)
Farming (Farmed)
Farmlands (Farmland)
Farms (Farm, Farmsteads)
  X-Ref: Interfarm,
  Intrafarm, Nonfarm
Farmsteads (Farmstead)
Farmworkers (Farmhands,
  Farmworker)
Farmyards (Farmyard)
Farnesane
Farnesate
Farnesene
  X-Ref: Alphafarnesene
Farnesenic
Farnesiferol
Farnesoate
Farnesols (Farnesol)
Farnesyl
Farriers (Farrier)
Farrowing (Farrow,
  Farrowed, Farrowings)
Farsetia
Fasciations (Fasciated,
  Fasciation)
Fasciola
Fascioliasis (Fasciolasis)
Fasciolicides (Fasciolicidal,
  Fasciolicide)
Fascioloides
Fastening (Fasteners,
  Fastenings)
Fastigium (Fastigial)
Fasting (Fasted)
Fatal
Father (Fathers)
Fatigue (Fatiguing)
Fatliquoring
Fatoua
Fats (Fat, Fatty, Nonfatty)
  X-Ref: Nonfat
Fatteners (Fattener)
Fattening (Fatness, Fatted,
  Fatten, Fattened,
  Fattiness)
Fauna (Faunal, Faunas,
  Faunistic, Faunistical,
  Faunistics)
  X-Ref: Epifauna,
  Macrofauna
Faurelina
Favism
Favus
Fawns (Fawn)
Fayoumi
Feathergrass
Feathers (Feather,
  Feathered, Feathering,
  Feathery)
  X-Ref: Plumage
Febrile see Fever

Feces (Faecal, Faeces,
  Fecal, Stool)
  X-Ref: Stools
Fecund see Fertility
Fedex
Feed (Feedingstuff,
  Feedingstuffs, Feeds,
  Feedstuff, Feedstuffs)
Feeders (Feeder)
Feedgrains (Feedgrain)
Feeding (Fed, Feedings)
  X-Ref: Overfeeding
Feedlots (Feedlot, Feedyard)
  X-Ref: Feedyards
Feedyards see Feedlots
Fees (Fee)
Fegatella
Feijoa
Feldspars
Felicia
Felidae (Felids)
Felis (Feline)
Felling see Logging
Felt (Felted, Felting, Felts)
Feltia
Feltriidae
Female (Femaleness,
  Females, Feminine)
Feminization
Femur (Femoral)
Fenac
Fenaminosulf
  X-Ref: Dexon, Diazoben
Fenbendazole (Fenbendazol)
Fences (Fence, Fenced,
  Fencing)
Fenchlorphos see Ronnel
Fenestration
Fenfluramine
Fenitrooxon (Fenitrooxone)
Fenitrothion
  X-Ref: Accothion,
  Folithion, Metathion,
  Metation, Sumithion
Fennel
Fenoprop
Fensulfothion
Fentanyl
Fenthion (Phenthione)
  X-Ref: Baytex, Lebaycid,
  Phenthion
Fentin
Fenugreek
Fenuron
Fenusa
Feral
Ferbam
Fermentation (Ferment,
  Fermentability,
  Fermentations,
  Fermentative, Fermented,
  Fermenting, Ferments,
  Unfermented)
  X-Ref:
  Heterofermentative,
  Overfermented,

Prefermentation,
Refermentation,
Unfermentable
Fermentors *(Fermentator,*
Fermenter, Fermenters,
Fermentor)
Ferns *(Fern)*
Ferocactus *(Ferocacti)*
Ferralization *(Ferralitic)*
Ferredoxin *(Ferredoxins)*
Ferrets *(Ferret)*
Ferric
X-Ref: Monoferric
Ferricrocin
Ferricyanide
Ferricytochrome
Ferrihemoglobin
Ferrimyoglobin
Ferrite
Ferritin
X-Ref: Apoferritin
Ferrochelatase
Ferroconcrete
Ferrocyanide
Ferrodextran
Ferrokinetics
Ferrous
Ferruginol
Ferruginous *see* **Iron**
Fertility *(Fecundation,*
Fecundity, Fertile,
Fertilities)
X-Ref: Autofertility,
Fecund, Nonfertile,
Subfertility
Fertilization *(Fertilisation,*
Fertilised, Fertilising,
Fertilizable, Fertilizations,
Fertilize, Fertilized,
Fertilizes, Fertilizing)
X-Ref: Autofertilization,
Unfertilized
Fertilizers *(Fertiliser,*
Fertilisers, Fertilizer)
Ferula
Ferulate *(Ferulates)*
Ferulic
Ferutin
Ferutinin
Fescue *see* **Festuca**
Festuca *(Fescues)*
X-Ref: Fescue
Festuclavine
Feta
Feterita
Fetlock
Fetoprotein *(Fetoproteins)*
X-Ref: Foetoprotein
Fetuin
Fetus *(Conceptuses, Fetal,*
Feto, Fetuses, Foetal,
Foeto, Foetuses)
X-Ref: Conceptus,
Foetus, Superfetation
Feudalism *(Feudal,*
Feudalistic)

Feulgen
Fever *(Antipyretics,*
Feverish, Fevers,
Hyperthermic)
X-Ref: Antipyretic,
Febrile, Hyperthermia
Fiberboard *(Fiberboards,*
Fibreboard, Fibreboards,
Paperboards)
X-Ref: Cardboard,
Paperboard
Fiberglass *(Fibreglass)*
Fiberiser
Fiberoptics *(Fiberoptic)*
Fibers *(Fiber, Fibered,*
Fibre, Fibres, Fibrous)
X-Ref: Interfiber
Fibrillation
Fibrils *(Fibril, Fibrillar,*
Fibrillary)
Fibrin
Fibrinogen
Fibrinoid
Fibrinolysis *(Fibrinolytic)*
Fibrinopeptide
Fibroblasts *(Fibroblast,*
Fibroblastic)
Fibrocartilage
(Fibrocartilaginous)
Fibrocyst *(Fibrocystic)*
Fibroins *(Fibroin)*
Fibroma
Fibrosarcoma
(Fibrosarcomas)
Fibrosis
Fibula *(Fibulae, Fibular)*
Ficalbia
Ficins
Ficoll
Ficus
Figs *(Fig)*
Filaments *(Filament,*
Filamentary, Filamentous,
Microfilaments)
X-Ref: Microfilament
Filaria *(Filarial)*
Filariasis *(Filariosis)*
Filariata
Filaricides
Filariidae *(Filariid,*
Filariids)
Filarioidea
Filariomyces
Filaroides
Filberts *(Filbert, Hazelnut,*
Hazels)
X-Ref: Hazelnuts
Filia
Filicales
Filicineae *(Filicinae)*
Filipendula
Fillers *(Filler)*
Fillies *(Filly)*
Films *(Film, Filmed,*
Filming, Filmstrip,
Filmstrips)

Filobasidiella
Filopodia
Filters *(Filter)*
X-Ref: Biofilters
Filth
Filtrates *(Filtrate)*
Filtration *(Filterability,*
Filterable, Filtered,
Filtering, Filtrability,
Filtrable, Filtrated)
X-Ref: Nonfilterable,
Prefiltration,
Thermofiltration
Fimbriae
Fimbristylis
Fimbristylus
Finance *(Financed,*
Finances, Financial,
Financiers, Financing,
Funding)
X-Ref: Fiscal, Funds
Fineness *(Fine)*
X-Ref: Ultrafine
Finewooled
Finfish
Fingerjoints
Fingerlings *(Fingerling)*
Finishes *(Finish, Finishing)*
Finlaya
Fiorinia
Fire *(Fires, Wildfire)*
X-Ref: Backfires,
Wildfires
Fireblight
Firebrats *(Firebrat)*
Firebreaks
Firefighting *(Firefighter,*
Firefighters, Fireline)
Fireflies *(Firefly)*
Fireplaces *(Fireplace)*
Fireproofing *(Fireproof)*
X-Ref: Flameproofing
Firewood *see* **Fuelwood**
Firs *see* **Abies**
Fiscal *see* **Finance**
Fischerula
Fish *(Fishes, Fishy)*
Fisheries *(Fishery)*
Fishing
Fishmeal *(Fishmeals)*
Fishponds *(Fishpond)*
Fissidens
Fissidentaceae
Fistula *(Fistulae, Fistulas,*
Fistulated, Fistulation,
Fistulization)
Fistulina
Fixation *(Fix)*
Fixatives *(Fixative)*
Flabellospora
Flacherie
Flacourtiaceae
Flagellariaceae
Flagellate *(Flagella,*
Flagellar, Flagellatae,
Flagellated, Flagellates,

Flagellation, Flagellum)
X-Ref: Aflagellate
Flagyl *see* **Metronidazole**
Flakeboard *see* **Chipboard**
Flaking *(Flaked, Flakers,*
Flakes)
Flameproofing *see*
Fireproofing
Flames *(Flame, Flameless,*
Flaming)
Flamingoes *(Flamingo)*
Flammability *(Flammable)*
X-Ref: Inflammability
Flammulina
Flammutoxin
Flank
Flatfish
Flatidae
Flatlands *(Flatland)*
Flatus *(Flatulence)*
Flatwoods *(Flatwood)*
Flavanols
Flavanones *(Flavanone)*
Flavans *(Flavan)*
Flaveria
Flavescence
Flavines *(Flavin, Flavine,*
Flavinogenesis, Flavins)
Flaviolin
Flaviviruses
Flavobacterium
(Flavobacteria)
Flavocytochrome
Flavodoxin
Flavomycin
Flavones *(Flavone, Flavonic)*
X-Ref: Bisflavones
Flavonoids *(Flavonoid,*
Flavonoidal)
Flavonols *(Flavanol,*
Flavonol)
Flavoproteins *(Flavoprotein)*
Flavor *(Flavored, Flavorful,*
Flavoring, Flavorings,
Flavors, Flavour,
Flavoured, Flavouring,
Flavourings, Flavours)
Flavorist *(Flavourist)*
Flavosorbin
Flax *(Flaxes)*
Flaxseed
Fleabeetles *(Fleabeetle)*
Fleahoppers *(Fleahopper)*
Fleas *(Flea)*
Fleckvieh *see* **Simmental**
Fleece *(Fleeces, Fleecy)*
Flemingia
Fleshflies *(Fleshfly)*
Flexure *(Flex, Flexion,*
Flexural)
Flickers *(Flicker)*
Flies *(Fly)*
Flights *(Flight)*
Flindersia
Flindersine
Floating

Flocculants *(Flocculant)*
Flocculation *(Flocculating)*
X-Ref: Autoflocculation,
Bioflocculation
Flocks *(Flock)*
Floodplains *(Floodland,
Floodplain)*
Floods *(Flood, Floodable,
Flooded, Flooding,
Floodings, Floodwater,
Floodwaters, Inundated)*
X-Ref: Inundation,
Underflooding,
Unflooded
Floors *(Floor, Flooring,
Floorings)*
X-Ref: Underfloor
Flora *(Floras)*
X-Ref: Megafloras
Floracarus
Florescence *see* Flowers
Floresianus
Floriculture *(Floricultural)*
Floride
Florideophyceae
Floridoside *see* Furcellaran
Florisil
Floristics *(Floristic,
Floristical)*
Florists *(Florist)*
Florule
Flotation
Flounder
Flour *(Flours, Floury)*
Flourensia
Flouridation *(Flouridating)*
Flowering *(Flowered)*
X-Ref: Anthesis,
Efflorescence,
Nightflowering
Flowerpots *see* Pots
Flowers *(Bloom, Bloomed,
Bloomers, Blooming,
Blooms, Blossoming,
Blossoms, Floral,
Florescences, Floret,
Florets, Flower, Flowery)*
X-Ref: Blossom,
Deblossomed,
Florescence,
Nonflowering,
Unifloral
Flowmeters *(Flowmeter)*
Fluctuadin
Fluctuanin
Flues *(Flue)*
Fluidics *(Fluidic)*
Fluidity
Fluidization *(Fluidised,
Fluidized)*
Fluids *(Fluid)*
Flukes *(Fluke)*
Flumes *(Flume)*
Flumethasone
Fluoborate
Fluometuron

X-Ref: Cotoran
Fluorapatite
Fluorene
Fluorescamine
Fluorescein
Fluorescence *(Fluorescent,
Fluorescing)*
X-Ref: Autofluorescence,
Nonfluorescent
Fluoridation *(Fluoridated)*
Fluorides *(Fluoride)*
Fluorination *(Fluorinated)*
Fluorine
Fluoro *(Fluor)*
Fluoroacetamide
Fluoroacetate
Fluoroborate
Fluorocarbon
(Fluorocarbons)
Fluorochemical
Fluorochromes
(Fluorochrome)
Fluorocitrate
Fluorodensitometry
(Fluorodensitometric)
Fluorodeoxyuridine
Fluorodifen
Fluorogestone
Fluorography
Fluorometry *(Fluorimetric,
Fluorimetry, Fluorometer,
Fluorometric)*
Fluorophenyl
Fluorophenylalanine
Fluorophores
Fluorophosphate
Fluoroprednisolone
Fluoropyrimidines
Fluoroscopy
Fluorosis
Fluorosugars
Fluorosulfonylbenzoyl
Fluorouracil
Fluorspar
Fluosilicate
Fluosilicic
Fluothane *see* Halothane
Flurogestone
Flurothyl
Flying
Flytrap
Foaling
Foals *(Foal)*
Foam *(Foamed, Foaming,
Foams, Foamy)*
X-Ref: Antifoaming
Fodder *(Fodders)*
Foeniculum
Foetidine
Foetoprotein *see*
Fetoprotein
Foetus *see* Fetus
Fog *(Fogs)*
Foils *(Foil)*
Folacin
Folate *(Folates)*

Foliage *(Foliar,
Unifoliolate)*
X-Ref: Foliose,
Interfoliar, Unifoliate
Foliation
Folic
Folicote
Foliicolous
Foliminine
Folin
Folioles
Foliose *see* Foliage
Folithion *see* Fenitrothion
Folklore *(Folkloric)*
Follicles *(Follicle,
Follicular, Follicules,
Folliculi)*
X-Ref: Intrafollicular
Folsomia
Fomannosin
Fomes
Fomitopsis
Fonofos
X-Ref: Dyfonate
Fonsecia
Fontinalis
Food *(Foods)*
Foodborne
Foodgrains *(Foodgrain)*
Foodplants *(Foodplant)*
Foodstuffs *(Foodstuff)*
Foot *(Feet)*
Football
Footbath
Foothills
Footwear
Forage *(Forages)*
Foraging *(Forager,
Foragers)*
Foramen
Foraminifera
Forbs *(Forb)*
Forcemeat
Forcipomyia
Forda
Forearm *see* Limbs
Forecast *(Forecasted,
Forecaster, Forecasters,
Forecasting, Forecasts,
Prospect)*
X-Ref: Outlook,
Prospects
Forecrops *see* Crops
Foregut
Foreleg *see* Legs
Forelia
Forelimbs *see* Limbs
Foremen
Foremilk *see* Milk
Forest *(Forested, Forests,
Nonforested)*
X-Ref: Nonforest,
Unforested
Forestation *see*
Afforestation
Foresters *(Forester)*

Forestomachs *see* Stomach
Forestry
Forficula
Forficulidae
Forging *(Forge, Forges)*
Formaldehyde
X-Ref: Formol
Formalin *(Formalinized)*
Formamidase
Formamide
Formamidines
(Formamidine)
Formate
Formetanate
Formic
Formica
Formicidae
Formicoidea
Formicomiris
Formiminoglycine
Formiminotransferase
Formol *see* Formaldehyde
Formononetin
Formosanine
Formothion
X-Ref: Antio
Formyl
Formylmethionyl
Formylstearate
Fornix
Forsythia *(Forsythias)*
Fortification *(Fortified,
Fortifier, Fortifying)*
Fortran
Fortunella
Fortuynia
Foschlor *see* Trichlorfon
Fossils *(Fossil, Fossilized,
Microfossil, Subfossils)*
X-Ref: Macrofossils,
Megafossil,
Microfossils, Subfossil
Fossombronia
Foulbrood
Fouquieria *(Fouquierias)*
Fouquieriaceae
Fowl *(Fowls)*
Fowlpox
Foxes *(Fox)*
Foxgloves *(Foxglove)*
Foxhounds
Foxtails *(Foxtail)*
Fozalon *see* Benzophosphate
Fractures *(Fracture,
Fractured)*
Fragaria
Fragilaria
Fragilariaceae
Fragipan *(Fragipans)*
Fragmentography
Fragrance *(Fragrances,
Fragrant)*
Frailea
Frames *(Frame, Framing)*
Franchises *(Franchise,
Franchisers, Franchising)*

Francisella
Francochthonius
Francolinus *(Francolin)*
Frangulanine
Frankenia
Frankfurters *(Frankfurter)*
Frankincense
    X-Ref: Olibanum
Franklinia
Frankliniella
Franseria
Frantzia
Frass
Fraud *(Frauds,*
    Fraudulence, Fraudulent)
Fraxidin
Fraxinus
Freelingyne
Freemartins *(Freemartin,*
    Freemartinism)
Freesia *(Freesias)*
Freezers *(Freezer)*
Freezing *(Freezability,*
    Freezable, Freeze,
    Freezed, Freezes)
    X-Ref: Refreezing,
    Subfreezing
Freight
Fremontodendron
Frenkelia
Frenulum
Freon
Frerea
Freshness *(Freshening)*
Freshwaters *(Freshwater)*
Fressia
Freycinetia
Freziera
Friction *(Frictional)*
    X-Ref: Antifriction
Fried *see* Frying
Friedelin
Friesea
Friesian *(Friesians)*
Fringilla
Frithia
Fritillaria *(Fritillaries,*
    Fritillary)
Fritschiella
Froesia
Frogeye
Frogs *(Frog)*
Fronds *(Frond)*
Frontalin
Frontalure
Frontinalis
Frost *(Frosted, Frosts)*
Frozen
    X-Ref: Refrozen,
    Unfrozen
Fructification *see* Fruiting
Fructofuranosidase
    (Fructofuranosidases)
Fructolysis
Fructosan *(Fructosans)*
Fructose

X-Ref: Levulose
Fructosediphosphatase
Fructosidase
Fruit *(Fruitless, Fruits)*
Fruitbodies *(Fruitbody)*
Fruitflies *(Fruitfly)*
Fruiting *(Fructifications,*
    Fruited, Fruitfulness)
    X-Ref: Fructification
Fruitworms *(Fruitworm)*
Frullania
Frustule
Frustulum
Fryers *(Fryer)*
Frying
    X-Ref: Fried
FSH
Fucaceae
Fucales
Fuchsias *(Fuchsia)*
Fuchsin
Fucogalactan
Fucoidan
Fucose
Fucosidases
Fucosterol
Fucosyl
Fucoxanthin
Fucus
Fuels *(Fuel, Fueling)*
Fuelwood
    X-Ref: Firewood
Fuirena
Fulani
Fulgoridae
Fulgoroidea
Fulleritermes
Fulvic
Fumagillin
Fumaramine
Fumarase
Fumarate
Fumaria
    X-Ref: Fumitory
Fumariaceae
Fumaric
Fumazone *see*
    Dibromochloropropane
Fumes *(Fume)*
Fumidil
Fumigants *(Fumigant)*
    X-Ref: Nonfumigant
Fumigation *(Fumigate,*
    Fumigated, Fumigating,
    Fumigations)
    X-Ref: Postfumigation
Fumitory *see* Fumaria
Funaria
Fundazol
Funds *see* Finance
Fundulus
Fundus
Fungi *(Fungal, Fungous,*
    Fungus, Microfungal)
    X-Ref: Microfungi,
    Micromycetes,

Mycoflora,
    Mycoorganisms
Fungicides *(Fungicidal,*
    Fungicide)
    X-Ref: Antifungal
Fungigrams
Fungistasis *(Fungistat,*
    Fungistatic, Fungistats)
Fungitoxicities *(Fungitoxic,*
    Fungitoxicant,
    Fungitoxicants,
    Fungitoxicity)
Fungivoridae
Funicles
Funiculus
Furacin
Furadan
Furan *(Furano, Furanoid,*
    Furans)
Furanacrylamide
Furanocoumarins
    (Furanocoumarin)
Furanoeremophilanes
Furanone
Furanoterpenes
    (Furanoterpenoid)
Furansequiterpene
    (Furanosesquiterpenes)
Furazolidone *(Furazolidon)*
    X-Ref: Furoxone, Neftin
Furbearing *(Furbearers,*
    Furred)
Furcaspora
Furcellaran
    X-Ref: Floridoside
Furcellaria
Furcraea
Furfurol *(Furaldehyde,*
    Furfural)
Furfuryl
Furfurylidene
Furidin
Furnaces *(Furnace)*
Furniture
Furoates *(Furoate)*
Furocoumarins
    (Furocoumarin)
Furosemide
    X-Ref: Lasix
Furosine
Furostan
Furostanol
Furoxone *see* Furazolidone
Furrows *(Furrow, Furrower,*
    Furrowing)
Furs *(Fur, Furring)*
Furunculosis
Furylfuramide
Fusarex *see*
    Tetrachloronitrobenzene
Fusaric
Fusariella
Fusariosis *(Fusarioses)*
Fusarium *(Fusaria,*
    Fusarial)
Fuscidea

Fuscuropoda
Fusea
Fusicladium
Fusicoccin
Fusicoccum
Fusidium
Fusiform
Fusiformis
Fusimotor
Fusion *(Fuse, Fused,*
    Fusions)
Fusobacterium
Futurology *(Futurological)*
Fynbos
Gabriola
Gaddi *see* Bhadarwah
Gadflies *(Gadfly)*
Gaeumannomyces
Gaffkya
Gagea
Gahnia
Gaillardia
Galactan
Galactitol
    X-Ref: Dulcitol
Galactofuranose
Galactofuranosyl
Galactoglucomannan
Galactography
Galactokinase
Galactolipase
Galactolipids *(Galactolipid)*
Galactomannans
    (Galactomannan)
Galactonic
Galactopyranose
Galactopyranoside
Galactopyranosyl
Galactosamine
Galactosaminogalactan
Galactose
Galactosemia
Galactosidases
    (Galactosidase)
Galactosides *(Galactoside)*
Galactosyls *(Galactosyl)*
Galactosyltransferases
    (Galactosyltransferase)
Galactoxyloglucans
Galacturonase
Galacturonic
Galanthus
Galantines
Galba
Galbanate
Galbanum
Galea
Galecron
Galega
Galenicals *(Galenical)*
Galeobdolon
Galeopsis
Galeopsis
Galerucella
Galerucidae
Galerucinae

Galinsoga
Galipea
Galium
Gall *(Galled, Galling, Galls)*
Gallae
Gallate
Gallbladder
X-Ref: Cholecyst
Galleria
Galleriidae
Gallic
Galliformes *(Gallinaceous)*
Galligonum
Gallimycin
Galline
Gallium
Gallotannins
Galloway
Gallstones *(Gallstone)*
Gallus
Galumna
Galumnidae *(Galumnids)*
Galvanization *(Galvanised, Galvanized)*
Galway
Gamasida
Gamasidae *(Gamasid, Gamasids, Gamasoid, Gamasoids)*
Gamasiphoides
Gamasoidea
Gambusia
Game
X-Ref: Metagame
Gametangia *(Gametangial, Gametangium)*
Gametes *(Gamete, Gametic, Microgamete)*
X-Ref: Microgametes, Semigametic
Gametocides *(Gametocidal, Gametocide)*
Gametocytes *(Gametocyte, Gonocyte)*
X-Ref: Gonocytes
Gametogenesis
X-Ref:
Macrogametogenesis,
Megagametogenesis,
Microgametogenesis
Gametogony
Gametophores *(Gametophore)*
Gametophytes *(Gametophyte, Gametophytic)*
Gammaglobulins *(Gammaglobulin)*
Gammarus
Gammexane
Gamones *(Gamone)*
Gamophylly
Gampsocoris
Gander
Gangaleoidin
Gangara

Ganglia *(Ganglion, Ganglionic)*
Ganglionectomy
Ganglioneuroblastoma
Ganglionitis
Gangliosides *(Ganglioside)*
Gangrene *(Gangrenous)*
Ganoderma
Gapeworm
Garbage
Garcinia
Gardencress
Gardeners *(Gardener)*
Gardenia *(Gardenias)*
Gardening
Gardens *(Garden)*
Gardneramine
Gardneria
Garlic
Garments *see* Clothing
Garnishing *(Garnish, Garnishes)*
Garrigues *(Garrigue)*
Garrya
Garypus
Gascardia
Gases *(Gas, Gaseous, Gassing)*
Gasoline
X-Ref: Petrol
Gasteracantha
Gasteracanthinae
Gasteria
Gasteromycetales
Gasteromycetes
Gasterophilidae
Gasterophilus
Gastina
Gastric *see* Stomach
Gastrimargus
Gastrin
Gastritis
Gastrocnemius
Gastrodia
Gastrodiscoides
Gastrodiscus
Gastroenteritis
Gastroenteropathy
Gastrointestinal *(Gastroduodenal, Gastroenteric)*
Gastrolobium
Gastromucoprotein
Gastropathy
Gastropexy
Gastrophilus
Gastrophysa
Gastropoda
Gastrostomy
Gastrothylax
Gastrotricha
Gastrulation
Gatnon
Gauge *(Gages, Gaging, Gauges, Gauging)*
Gaultheria

Gaura
Gautieria
Gaya
Gayal
Gaylussacia
Gayophytum
Gazania
Gazelles *(Gazelle)*
Gears *(Gear, Geared, Gearing)*
Geastrum
Geese *(Gosling)*
X-Ref: Goose, Goslings
Geissoloma
Geissolomataceae
Geitleribactron
Gelasinospora
Gelastocoridae
Gelastocoris
Gelatin *(Gelatine, Gelatins)*
Gelatinization *(Gelatinisation, Gelatinized, Gelatinous)*
X-Ref: Pregelatinized
Gelation *(Gell, Gelled, Gelling, Gells)*
Geldings *(Gelding)*
Gelechia
Gelechiidae *(Gelechidae, Gelechiids)*
Gelechioidea
Gelfiltration
Gelidiales
Gelidiella
Gelidium
Gelis
Gelonium
Gels *(Gel)*
Gelsemine
Gelsemium
Gemma *(Gemmae)*
Gemmule
Genealogy *(Genealogical)*
Genecology *(Genecological)*
Generation *(Generations)*
X-Ref: Multigeneration
Generators *(Generator)*
Generic
X-Ref: Infrageneric,
Intergeneric,
Intrageneric,
Subgeneric
Genes *(Gene, Genic, Polygenes, Polygeny)*
X-Ref: Intragenic,
Polygenic, Supergenes
Geneticists *(Geneticist)*
Genetics *(Androgenetic, Genetic, Genetical, Genetically)*
X-Ref: Androgenetics,
Chemogenetical
Geniculate
Genins *(Genin)*
Genista
Genisteae

Genitals *(Genital, Genitalia)*
X-Ref: Postgenital
Genkwanin
Genomes *(Genome, Genomic)*
X-Ref: Intergenome
Genotrophs
Genotypes *(Genotype, Genotypic, Genotypical)*
X-Ref: Intergenotypic
Gentamicin *(Gentamycin)*
Gentiana
Gentianaceae
Gentianine
Gentians *(Gentian)*
Gentiobiose
Geobiology
Geobotanists *see* Botanists
Geobotany *see* Botany
Geocalycaceae
Geochemistry *(Geochemical)*
Geocoris
Geodromicus
Geoelectric
Geogamasus
Geoglossaceae
Geographers *(Geographer)*
Geography *(Geographic, Geographical, Geographically)*
Geohydrology *(Geohydrologic, Geohydrological)*
Geoica
Geology *(Geologic, Geological)*
Geolycosa
Geomagnetism *(Geomagnetic)*
Geomedicine
Geometra
Geometridae *(Geometrid, Geometrids)*
Geometry *(Geometric, Geometrical)*
Geomorphology *(Geomorphic, Geomorphologic, Geomorphological)*
Geomydoecus
Geomyidae
Geomylichus
Geomys
Geonemy
Geoperception
Geophagy *(Geophagia)*
Geophilomorpha
Geophilous *(Geophilic)*
Geophylesis *(Geophyletic)*
Geophysics *(Geophysical)*
Georgefischeria
Geosmin
Geotaxis *see* Geotropism
Geotechnology *(Geotechnical)*

Geothermy *(Geothermal)*
Geotrichium
Geotrichosis
Geotrichum
Geotropism *(Geotactic,*
  Geotaxes, Geotaxies,
  Geotaxy, Geotropic,
  Geotropical, Geotropically)
  X-Ref: Geotaxis
Geraniaceae
Geraniol
Geranium *(Geraniums)*
Geranyl *(Geranylgeranyl)*
Gerbera
Gerbils *(Gerbil)*
Geriatrics *(Geriatric)*
Germacradienolide
Germacranolide
  (Germacranolides)
Germacrone
Germanium
Germicides *(Germicidal,*
  Germicide)
Germinal
Germination *(Germinability,*
  Germinated, Germinates,
  Germinating,
  Germinations,
  Germinative, Germinator)
  X-Ref: Nongerminating,
  Pregermination,
  Ungerminated
Germplasm *(Germplasms)*
Germs *(Germ, Germfree)*
Gerontology
  (Gerontological)
Gerrardanthus
Gerridae
Gerris
Gerthus
Gertrudis
Gesagard *see* Prometryne
Gesaprim *see* Atrazine
Gesatop *see* Simazine
Gesneria
Gesneriaceae *(Gesneriads)*
Gestagen
Gestation *(Gestating,*
  Gestational, Gestations)
Gestosis
Geum
Ghee
Gherkins *(Gherkin)*
Ghetto
Giallume
Giardia
Giardiasis
Gibasis
Gibberella
Gibberellanes
Gibberellate
Gibberellic
Gibberellins *(Gibberellin)*
  X-Ref: Antigibberellins
Gibbobruchus
Gibbosity

Gibbsite
Giblets
Gibsmithia
Giemsa
Giffordia
Gigantism
Gigantolaelaps
Gigartina
Gigartinales
Gigaspora
Gilbertiodendron
Gills
Gilpinia
Gilts *(Gilt)*
Gimnomera
Gin *(Ginned, Ginner,*
  Ginning, Gins)
  X-Ref: Ginners
Ginger
Gingiva *(Gingival)*
Ginkgo
Ginners *see* Gin
Ginseng
Ginsenosides
Gir
Giraudiella
Girdling *(Girdled)*
Girls *(Girl)*
Gisekia
Gitogenin
Gitoxigenin
Givotia
Gizzard *(Gizzards)*
Glacial *(Glacially, Glacier)*
  X-Ref: Interglacial,
  Postglacial
Glaciation *(Glaciated,*
  Glaciations)
Glades *(Glade)*
Gladiolus *(Gladioli)*
Glaebules
Glandectomy
Glanders
  X-Ref: Farcy
Glands *(Gland, Glanded,*
  Glandless, Glandulae,
  Glandular)
  X-Ref: Deglanded,
  Intraglandular
Glandularia
Glandulation
Glaphyrinae
Glaphyroptera
Glass *(Glasses, Glassiness,*
  Glassless)
Glasshouses *see*
  Greenhouses
Glassware
Glasswort
Glaucine
Glaucium
Glaucoma *(Glaucomatous)*
Glauconites *(Glauconite)*
Glaucopsyche
Glaux
Glazing *(Glaze, Glazed)*

Glaziovine
Gleaning *(Glean, Gleaned)*
Glebe
Glechoma
Gleditsia *(Gleditschia)*
Gleichenia
Gleicheniaceae
Gleistanthus
Glenniea
Glenodinium
Glenostictia
Gley *(Glei, Gleization,*
  Gleyed, Gleying, Gleys,
  Gleysolic, Gleysols,
  Pseudogleyed,
  Pseudogleys)
  X-Ref: Pseudogley
Glia *(Glial, Neuroglial)*
  X-Ref: Neuroglia
Gliadins *(Gliadin)*
Gliocephalotrichum
Gliocladium
Glischrochilus
Globaria
Globidium
Globins *(Globin)*
Globularia
Globules *(Globular,*
  Globule)
Globulins *(Antiglobulin,*
  Globulin)
  X-Ref: Antiglobulins
Glochidion
Gloeocapsa
Gloeocercospora
Gloeochaete
Gloeodendron
Gloeodontia
Gloeomonas
Gloeophyllum
Gloeosporium
Gloeotinia
Gloisiphonia
Glomerella
Glomeris
Glomerules *(Glomerule)*
Glomerulitis
Glomerulonephritis
Glomerulus *(Glomerular,*
  Glomeruli)
Glomus
Gloriosa
Glossa
Glossaries *(Glossary)*
Glossina
Glossinidae
Glossitis
Glossna
Glossoboralfs
Glossopteris *(Glossopterid)*
Glossosoma
Glossosomatidae
Gloxinia
Glucagon
Glucanase *(Glucanases)*
Glucanohydrolase

(Glucanohydrolases)
Glucans *(Glucan)*
Glucavamorin
  (Glucavamorine)
Glucides *(Glucide, Glucidic)*
Glucitol *see* Sorbitol
Glucoamylase
  (Glucoamylases)
Glucocerebroside
Glucocorticoids
  (Glucocorticoid)
Glucogenic
Glucokinase
Glucolipids
Glucomannans
  (Glucomannan)
Gluconate
Gluconeogenesis
  (Gluconeogenic)
Gluconic
Gluconobacter
Gluconolactone
Glucoprotein
Glucopyranose
Glucopyranosides
  (Glucopyranoside)
Glucopyranosyl
Glucoreceptor
Glucosamine
Glucosaminidase
Glucoseoxidase
Glucoseptanosides
Glucoses *(Gluco, Glucose)*
  X-Ref: Acylglucoses,
  Dextrose, Polyglucose
Glucosidases *(Glucosidase)*
Glucosides *(Glucoside,*
  Glucosidic)
Glucosinolates
  (Glucosinolate)
Glucosuria
Glucosyl
Glucosylation
Glucosyltransferases
  (Glucosyltransferase)
Glucuronic
Glucuronidase
  (Glucuronidases)
Glucuronide
Glucuronosyl
Glucuronosyltransferase
Glucuronoxylan
Glucuronyltransferase
Glue *(Glued, Glues, Gluing)*
Glugea
Glumes *(Glume, Glumed)*
Gluroside
Gluta
Glutamate *(Glutamates)*
Glutamic
Glutamine *(Glutamin)*
Glutaminyl
Glutamyl
Glutamyltransferase
Glutaraldehydes
  (Glutaraldehyde)

Glutaric
Glutathione
Glutathioneperoxidase
Glutelins (Glutelin)
Glutenins (Glutenin)
Glutens (Gluten)
Glutethimide
Glutinosone
Glutinous
Glycans (Glycan)
Glycemia (Glycemic)
Glyceraldehyde
Glycerate
Glyceria
Glycerides (Glyceride,
   Glyceridic)
Glycerine see Glycerol
Glycerokinase
Glycerol (Glycerin,
   Glycerinated, Glycerols)
   X-Ref: Glycerine
Glycerolipids (Glycerolipid)
Glycerolphosphate
Glycerophosphatases
   (Glycerophosphatase)
Glycerophosphate
Glyceryl
Glycidate
Glycidol
Glycidyl
Glycine
Glycinebetaine
Glycinin
Glycoalkaloids
   (Glycoalkaloid)
Glycobius
Glycocalyx
Glycocholate
Glycoconjugates
Glycoflavones
Glycogen (Glycogenic)
Glycogenolysis
   (Glycogenolytic)
Glycogenosis
Glycohydrolase
Glycol (Glycols)
   X-Ref: Ethanediol
Glycolate (Glycollate)
Glycolic
Glycolipids (Glycolipid)
Glycolysis (Glycolytic)
Glycopeptides
   (Glycomacropeptide,
   Glycopeptide)
Glycoperine
Glycophytes (Glycophytic)
Glycoproteins (Glycoproteic,
   Glycoprotein,
   Glycoproteinic)
Glycopyranoside
Glycopyranosyl
Glycosaminoglycans
   (Glycosaminoglycan)
Glycose
Glycosidases (Glycosidase)
Glycosides (Glycosidal,

Glycoside, Glycosidic)
Glycosmis
Glycosuria
Glycosyl
Glycosylases
Glycosylation (Glycosylated)
Glycosylflavones
Glycosylflavonoids
   (Glycosylflavonoid)
Glycosyltransferase
   (Glycosyltransferases)
Glycosylureas
Glycyphagidae
Glycyphagus
Glycyrrhiza
Glyodin
Glyoxal
Glyoxalase
Glyoxylate
Glyoxylic
Glyoxysomes (Glyoxysomal,
   Glyoxysome)
Glyphipterygidae
Glyphochilus
Glyphosate
   X-Ref: Roundup
Glyphosphates
   (Glyphosphate)
Glyphotaelius
Glyptholaspis
Glyptorhaestus
Glyptoscelis
Glyptotendipes
Gmelina
Gmelinol
Gnaphalium
Gnaphosidae
Gnathamitermes
Gnathosoma
Gnathostoma
Gnathotrichus
Gnats (Gnat)
Gnetaceae
Gnetopsida
Gnetum
Gnidia
Gnomonia
Gnophothrips
Gnorimoschema
Gnorimoschemini
Gnotobiotics (Gnotobiology,
   Gnotobiotic)
GNP
Goats (Goat)
Goatsrue
Godronia
Goebelia
Goera
Goethite
Goiter (Goitre, Goitrogenic,
   Goitrogens, Goitrous)
Goiterogen
Goitrin
Goldenrod (Goldenrods)
Goldfinch
Golenkinia

Golf
Golgi
Goliathus
Golofa
Gomesa
Gomori
Gomortega
Gomphidae
Gomphidius
Gomphrena
Gomphus
Gonadectomy
   (Gonadectomized)
Gonadotrophs
Gonadotropins
   (Gonadotrophic,
   Gonadotrophin,
   Gonadotrophins,
   Gonadotropic,
   Gonadotropin)
   X-Ref: Antigonadotropin,
   Progonadotropic
Gonads (Gonad, Gonadal,
   Gonadic)
Gonarthrosis
Gonatobotrys
Gonioanthela
Gonioctena
Goniodes
Gonioma
Goniopteris
Goniotrichales
Goniozus
Gonium
Gonocephalum
Gonocytes see Gametocytes
Gonodonta
Gonolobus
Gonomyia
Gonoplectus
Gonopods (Gonopod)
Gonyaulax
Gonyleptidae
Gonyleptomorphi
Gonystylus
Goodeniaceae
Goodia
Goodyera
Goose see Geese
Gooseberries (Gooseberry)
Goosefoot
Goosegrass
Gophers (Gopher)
Goplana
Gorgoderina
Gorse
Goslings see Geese
Gossypium
Gossyplure
Gossypol
   X-Ref: Hemigossypol
Gouania
Gouda
Gouinia
Goulash
Gourds (Gourd)

Gourmet
Gout (Gouty)
Government (Govern,
   Governed, Governing,
   Governmental,
   Governments)
   X-Ref: Intergovernmental
Governors (Governor)
Graafian
Graciela
Gracilaria
Gracilariidae
   (Gracillariidae)
Grackles
Grading (Grade, Graded,
   Grader, Graders, Grades,
   Gradings)
   X-Ref: Ungraded
Graduates see Education
Graeffea
Graellsia
Grafting (Graft, Grafted,
   Grafter, Graftings, Grafts)
   X-Ref: Regrafting,
   Topworking, Ungrafted
Grains (Grain, Grained)
Grama
Gramicidin
Graminae
Gramineae (Graminaceae,
   Graminaceous,
   Gramineous)
Graminella
Graminoid see Grasses
Grammatophyllum
Grammonota
Gramoxone see Paraquat
Grana
Granadilla
Granaries (Granary)
Grandisol
Grandlure
Granilin
Granite (Granites, Granitic)
Granosan (Granozan)
   X-Ref: Ceresan
Granulation (Granular,
   Granularity, Granulate,
   Granulated, Granulates,
   Granulating, Granulator,
   Granule, Granules,
   Microgranules)
   X-Ref: Intergranular,
   Microgranulated
Granulin
Granulocytes (Granulocyte,
   Granulocytic)
Granulomas (Granuloma,
   Granulomatous)
   X-Ref: Coligranuloma
Granulometric
Granulosis (Granuloses)
Grapefruit
Grapes (Grape, Grapeseed)
   X-Ref: Winegrape
Grapevines (Grapevine)

Graphidostreptus
Graphite
Graphium
Graphognatus
   (Graphognathus)
Grapholita
Grapholitha
Graphopterus
Graphosoma
Grasses (Graminoids, Grass,
   Grassy)
   X-Ref: Graminoid,
   Regrassing, Tallgrass
Grasshoppers (Grasshopper)
Grasslands (Grassland)
Gratiana
Graticules
Grating (Gratings)
Gravel
   X-Ref: Pebbles
Graveyards (Graveyard)
Gravidity see Pregnancy
Gravignost
Gravimetry (Gravimetric,
   Gravimetrical)
Gravimorphism
Gravity (Gravitation,
   Gravitational, Gravities)
Gravohormones
Gravy
Grayanotoxin
Grayia
Grazing (Graze, Grazed,
   Grazings)
   X-Ref: Nongrazed,
   Overgrazing, Ungrazed,
   Wintergrazing
Grease (Greasing, Greasy)
Greenbelt (Greenbelts)
Greenbugs (Greenbug)
Greeneria
Greenery
Greenfeed
Greenhouses (Glasshouse,
   Greenhouse, Hothouse,
   Hothoused)
   X-Ref: Glasshouses,
   Hothouses
Greening
   X-Ref: Regreening
Gregarina
Gregarinida (Gregarine)
Gregopimpla
Gremmeniella
Gretchena
Grevillea (Grevilleas)
Grewia
Greyhounds (Greyhound)
Griddles
Grids (Grid)
Griffithsia
Griffonia
Grimmia
Grindelia
Grinders (Grinder)
Grinding (Grind)

Griphosphaeria
Grippe
Grisein
Griseofulvin
Grits (Grit)
Groats (Groat)
Grobya
Groceries (Grocery)
Groin
   X-Ref: Inguinal
Gromphadorhina
Grooming (Groom)
Grooves (Groove)
Grossheimia
Grossulariaceae
Grotea
Groundhogs (Groundhog)
Groundnuts see Peanuts
Groundsel
Groundwater
   (Groundwaters)
Groundwood
Grouse
Groves (Grove)
Growers (Grower)
Grubs (Grub)
Grumusols (Grumusol,
   Smolnitzas, Smonitza)
   X-Ref: Smolnitsa
Gruyere
Gryllacridae
Gryllacrididae
Gryllacris
Gryllidae
Grylloblattidae
Grylloblattodea
Gryllodes
Grylloidea
Grylloptera
Gryllotalpa
Gryllotalpidae
Gryllus
Gryon
Guacamole
Guaiacolate
Guaiacols (Guaiacol)
Guaiacum
Guaiacyl
Guaiacylcarbinol
Guaianolides (Guaianolide)
Guanamine (Guanamines)
Guanethidine
Guanide (Guanides)
Guanidines (Guanidine)
Guanidinobenzoate
Guanine
Guano
Guanocytes
Guanosine
Guanyl
Guanylate
Guanylhydrazone
Guanylnucleotides
Guar
Guaran
Guatteria

Guavas (Guava)
Guayacanin
Guayin
Guernsey
Guggulu
Guidebooks (Guidebook)
Guignardia
Guilielma
Guillemots
Guilliermondella
Guindilia
Guizotia
Gulamentus
Gullies (Gully)
Gulls (Gull)
Gulonic
Gulose
Gum (Gums)
Gumbo see Okra
Gumboro
Gummosin
Gummosis (Gumming)
Gunnera
Guns (Gun)
Gur see Sugar
Gurleya
Gusathion see
   Azinphosmethyl
Gustatory see Taste
Gustometry
Gut (Guts)
Guthion see
   Azinphosmethyl
Gutierrezia
Guttiferae
Gutturomyces
Gutturomycosis
Guzerat (Guzera)
Gyalideopsis
Gymnadenia
Gymnaetron
Gymnamine
Gymnema
Gymnoascaceae
Gymnocalycium
   (Gymnocalyciums)
Gymnocarpium
Gymnocarpos
Gymnocladus
Gymnodinium
Gymnogramme
Gymnomitriaceae
Gymnomitrion
Gymnomyces
Gymnopilus
Gymnosperms
   (Gymnosperm,
   Gymnospermae)
Gymnosporangium
Gymnosporia
Gymnostomum
Gynaecomastia
Gynaikothrips
Gynandromorphism
   (Gynandromorph,
   Gynandromorphic,

Gynandromorphs)
Gynandropsis
Gynecium
Gynecology (Gynecological)
Gynodioecism
   (Gynodioecious)
Gynoecium (Gynoecious)
Gynogenesis
Gynostemium
Gynura
Gyoerffyella
Gyponana
Gyponinae
Gypsonoma
Gypsophila
Gypsum (Gypsic,
   Gypsiferous)
Gyrinidae
Gyrinus
Gyrodinium
Gyrogonites
Gyromitra
Gyromitrin
Gyrophora
Gyropidae
Haarupia
Habenaria (Habenarias)
Haberlea
Habitats (Habitat,
   Habitation, Microhabitat)
   X-Ref: Microhabitats
Habits (Habit, Habitual,
   Habituation)
Habranthus
Habrobracon
Habrocerus
Habronema
Habronemiasis
Hackberries
Hackelia
Hadacidin
Haddock
Hadena
Hadeninae
Hadrodactylus
Haemagglutination see
   Hemagglutination
Haemagglutinins see
   Hemagglutinins
Haemanthus
   X-Ref: Hemanthus
Haemaphysalis
Haematin see Hematin
Haematobia see Hematobia
Haematocrit see Hematocrit
Haematology see
   Hematology
Haematoma see Hematoma
Haematophagia see
   Hematophagia
Haematopinidae
Haematopinus
Haematopoiesis see
   Hematopoiesis
Haematopota
Haematoxenus

Haematuria *see* Hematuria
Haemin *see* Hemin
Haemobartonella
  (Hemobartonella)
Haemobartonellosis
  (Hemobartonellosis)
Haemochrome *see*
  Hemochrome
Haemocyanin *see*
  Hemocyanin
Haemocytes *see* Hemocytes
Haemodoraceae
Haemodynamics *see*
  Hemodynamics
Haemogamasus
Haemoglobins *see*
  Hemoglobins
Haemoglobinuria *see*
  Hemoglobinuria
Haemogregarinidae
Haemolymphs *see*
  Hemolymphs
Haemolysates *see*
  Hemolysates
Haemolysins *see*
  Hemolysins
Haemolysis *see* Hemolysis
Haemomelasma
Haemonchosis
Haemonchus
Haemophilus *(Hemophilus)*
Haemoprotein *see*
  Hemoproteins
Haemoproteus
Haemorrhage *see*
  Hemorrhage
Haemosporidia *see*
  Hemosporidia
Hafnia
Hahniidae
Hail *(Hailstorms)*
Hair *(Haired, Hairiness,*
  Hairing, Hairless, Hairs,
  Hairy)
Hairballs *(Hairball)*
Hake
Hakea
Hakiulus
Halacaridae
Halenia
Halesia
Haliaeetus
Haliaspis
Halibut
Halictidae *(Halictinae,*
  Halictine)
Halictophagus
Halictus
Halicystis
Halides *(Halide)*
Halimione
Halimodendron
Haliotis
Haliplidae
Haliplus
Halkendin

X-Ref: Isohalkendin
Hallikar
Halloysites *(Halloysite)*
Hallucinogens
  (Hallucinogen,
  Hallucinogenic,
  Hallucinogenically)
Haloalkylamines
Halobacterium
Halobates
Halocarbons
Halocellulose
Halocnemum
Halodule
Halogenides *(Halogenide)*
Halogens *(Halogen,*
  Halogenated,
  Halogenation)
Halogeton
Halophila
Halophilism *(Halophil,*
  Halophile, Halophilic)
Halophytes *(Halophyte,*
  Halophytic, Halophytism)
Haloprogin
Halopyrethroids
Haloragidaceae
Haloragis
Halothane
  X-Ref: Fluothane
Haloxon
Haloxylon
Haltica
Haltichella
Halticinae
Halticoptera
Halticus
Halyini
Ham *(Hams)*
Hamadryadinae
Hamadryas
Hamamelidaceae
Hamamelis
Hamartoma
Hamatine
Hamaxas
Hamburgers *(Hamburger)*
  X-Ref: Cheeseburgers
Hamelia
Hammada
Hammocks
Hammondia
Hampea
Hamsters *(Hamster)*
Hand *(Hands)*
Handbooks *(Handbook)*
Handelia
Handicapped
Handicraft
Handlooms *see* Looms
Handsheets *(Handsheet)*
Hannemania
Hansenula
Hapalia
Hapalopilus
Hapalosiphon

Haplamine
Haplaxius
Haplocladium
Haplodiploidy
Haplodiplosis
Haploesthes
Haplogonidae
Haploids *(Haploid,*
  Haploidization, Haploidy,
  Monohaploid)
  X-Ref: Monohaploids,
  Polyhaploids
Haplomitrium
Haplonts
Haplopappus
Haplophyllum
Haplosporella
Haplothrips
Haplothysanus
Hapten *(Haptenic)*
Haptoglobin *(Haptoglobins)*
Haptophyceae
Hardboard *(Hardboards)*
Hardening *(Hardened)*
  X-Ref: Dehardening
Hardiness *(Hardy)*
  X-Ref: Nonhardy
Hardware
Hardwoods *(Hardwood)*
Hares *see* Rabbits
Harmonia
Harness *(Harnessed)*
Harnischia
Harpactea
Harpagophoridae
Harpagoxenus
Harpalinae
Harpalini
Harpalus
Harpullia
Harringtonine
Harrowing *(Harrow,*
  Harrowed, Harrows)
Harvest *(Harvested,*
  Harvesting, Harvests,
  Preharvest)
  X-Ref: Preharvesting
Harvesters *(Harvester)*
Harvestmen *(Harvestman)*
Harveyella
Hasubanan
Hatcheries *(Hatchery)*
Hatching *(Hatch,*
  Hatchability, Hatched,
  Hatchling)
  X-Ref: Posthatch,
  Prehatch
Hauling *(Haul, Haulage,*
  Hauler)
Haulm *(Halm)*
Haustorium *(Haustoria,*
  Haustorial)
Hawkmoths *(Hawkmoth)*
Hawks *(Hawk)*
Hawkweed
Haworthia *(Haworthias)*

Hawthorns *(Hawthorn)*
Hay *(Hays)*
Haying *(Hayfields,*
  Haylands, Haymaking,
  Haystack)
Haylage *(Haylages)*
Hazel
Hazelnuts *see* Filberts
HCH *see*
  Hexachlorocyclohexane
Head *(Headed, Headless,*
  Heads)
Headlands *(Headland)*
Headrig
Headwaters *(Headwater)*
Healing
Health *(Healthful,*
  Healthier, Healthy)
Hearing
  X-Ref: Auditory
Heart *(Hearts)*
Heartwater
Heartwoods *(Heartwood)*
Heartworms *(Heartworm)*
Heat *(Heated, Heater,*
  Heaters, Heating, Heats,
  Hot, Preheating,
  Reheated, Reheating)
  X-Ref: Overheated,
  Preheated, Reheat,
  Superheated, Unheated
Heather *(Heathers)*
Heaths *(Heath, Heathland)*
Heaving *(Heave, Heaves)*
Hebeclinium
Hebeloma
Hebrus
Hecalapona
Hecalini
Hechtia
Hectares *(Hectare)*
Hectorella
Hectorellaceae
Hedeoma
Hedera
Hederagenin
Hedgehogs *(Hedgehog)*
  X-Ref: Erinaceus
Hedges *(Hedge,*
  Hedgecutting, Hedged,
  Hedger, Hedgerow,
  Hedgerows, Hedging)
  X-Ref: Unhedged
Hedophyllum
Hedya
Hedychium *(Hedychiums)*
Hedylepta
Hedyotideae
Hedysareae
Hedysarum
Heifers *(Heifer)*
Height *(Heights)*
Heimia
Helenalin
Helenieae
Helenium

Heleochloa
Heleomyzidae
Heleonastes
Heleritrine
Heliamphora
Heliangine
Heliangolide
Heliantheae
Helianthemum
Helianthocereus
Helianthus
Helichrysum
Helicidae
Helicobasidin
Helicobasidium
Helicocephalum
Helicoconis
Helicoid *(Helicoidal)*
Heliconia
Heliconiidae
Heliconiinae
Heliconius
Helicopters *(Helicopter)*
Helicosporidium
Helicotylenchus
Helicoverpa
Helictotrichon
Helina
Heliocereus
Heliococcus
Heliocopris
Helioctamenus
Heliolonche
Heliophilus
Heliophily
Heliopsis
Heliothis
Heliotridane
Heliotropism
Heliotropium
Helipterum
Helium
Helius
Helix *(Helical, Helices)*
Hellebore
Helleborus
Hellebrin
Hellula
Helmatac *see* Parbendazole
Helminthiasis
   (Helminthiases)
Helminthology
   (Helminthological)
Helminthosporiosis
Helminthosporium
Helminths *(Helminth,*
   Helminthic)
Helodes
Helodidae
Heloniopsis
Helopeltis
Helophorus
Helotiales
Helotium
Helvella
Helvellaceae

Helvolic
Helwingia
Helwingiaceae
Hemadsorption
Hemagglutination
   (Haemagglutinating,
   Hemagglutinability,
   Hemagglutinating)
X-Ref:
   Haemagglutination
Hemagglutinins
   (Antihaemagglutinins,
   Haemagglutinin,
   Hemagglutinin)
X-Ref:
   Antihemagglutinins,
   Haemagglutinins
Hemangioma
Hemangiopericytoma
Hemangiosarcoma
Hemanthus *see* Haemanthus
Hematic *see* Blood
Hematin
   X-Ref: Haematin
Hematite
Hematobia
   X-Ref: Haematobia
Hematocrit
   X-Ref: Haematocrit
Hematology *(Haematologic,*
   Haematological,
   Haematologically,
   Hematologic,
   Hematological)
X-Ref: Haematology
Hematoma *(Hematomas)*
   X-Ref: Haematoma
Hematophagia
   (Haematophagous,
   Hematophagous)
X-Ref: Haematophagia
Hematopoiesis
   (Antihemopoietic,
   Haematopoietic,
   Haemopoiesis,
   Haemopoietic,
   Hematogenesis,
   Hematopoietic,
   Hemopoiesis,
   Hemopoietic)
X-Ref:
   Antihematopoietic,
   Haematopoiesis
Hematoxylin
Hematozoa
Hematuria *(Haematuric)*
   X-Ref: Haematuria
Heme
   X-Ref: Nonheme
Hemerobiidae
Hemerobius
Hemerocallis
Hemerocampa
Hemiacetal
Hemiaminal
Hemiascomycetidae

Hemiberlesia
Hemicastration
Hemicellulases
   (Hemicellulase)
Hemicelluloses
   (Hemicellulose)
Hemicordulia
Hemicriconemoides
Hemicycliophora
Hemicycliophoridae
Hemideina
Hemidiscaceae
Hemieutreptia
Hemigossypol *see* Gossypol
Hemigossypolone
Hemigraphis
Hemilaminectomy
   (Hemilaminectomies)
Hemileia
Hemilepistus
Hemileuca
Hemin
   X-Ref: Haemin,
   Nonhemin
Hemiparasites
   (Hemiparasite,
   Hemiparasitic)
Hemiparesis
Hemiplegia
Hemiptera *(Hemipteran,*
   Hemipterans,
   Hemipterous)
Hemisphaeriales
Hemitarsocheylus
Hemitarsonemus
Hemitaxonus
Hemizygotes *(Hemizygote)*
Hemlock *see* Tsuga
Hemoblastoses
Hemochrome
   X-Ref: Haemochrome
Hemocirculation
   (Hemocirculatory)
Hemoclips
Hemocoel *(Hemocoele)*
Hemocyanin *(Hemocyanins)*
   X-Ref: Haemocyanin
Hemocytes *(Haemocyte,*
   Haemocytic, Hemocyte,
   Hemocytic)
X-Ref: Haemocytes,
   Prohemocytes
Hemocytoblasts
   (Hemocytoblastic)
Hemocytology
   (Hemocytological)
Hemocytopoietic
Hemodialysis
Hemodynamics
   (Haemodynamic,
   Hemodynamic)
X-Ref: Haemodynamics
Hemoglobins *(Haemoglobin,*
   Hemoglobin)
X-Ref: Haemoglobins
Hemoglobinuria

X-Ref: Haemoglobinuria
Hemogram *(Hemograms)*
Hemolymphs *(Haemolymph,*
   Haemolymphal,
   Hemolymph)
X-Ref: Haemolymphs
Hemolysates
   X-Ref: Haemolysates
Hemolysins *(Hemolysin)*
   X-Ref: Haemolysins
Hemolysis *(Haemolytic,*
   Hemolytic)
X-Ref: Haemolysis
Hemopexin *(Hemopexins)*
Hemophilia
Hemoproteins
   (Hemoprotein)
X-Ref: Haemoprotein
Hemorrhage *(Haemorragic,*
   Haemorrhages,
   Haemorrhagic,
   Hemorrhages,
   Hemorrhagic)
X-Ref: Haemorrhage
Hemorrhoids
Hemosporidia
   X-Ref: Haemosporidia
Hemostasis *(Hemostatic)*
Hemp
Hempa
Hempseed
Henbane
Hendersonia
Hendersonula
Heneicosen
Henhouse
Henicocephalidae
Henosepilachna
Henoticus
Hens *(Hen)*
Heodes
Heparan
   X-Ref: Heparitin
Heparin
Heparitin *see* Heparan
Hepatectomy
Hepatic *(Hepatis, Hepato)*
   X-Ref: Extrahepatic,
   Intrahepatic,
   Transhepatic
Hepaticae *(Hepatics,*
   Liverworts)
X-Ref: Liverwort
Hepatitis
Hepatocarcinoma
Hepatocellular
Hepatocytes *(Hepatocyte)*
Hepatoencephalopathy *see*
   Encephalopathy
Hepatojarakus
Hepatomas *(Hepatoma)*
Hepatopancreas
Hepatopathy
   (Hepatopathies)
Hepatosis
Hepatotoxicity

(Hepatotoxic, Hepatotoxins)
Hepatozoon
Hepatozoonosis
Hepes
Hepialidae
Hepialus
Heppia
Heptachlor
Heptachlorobornane
Heptadecatetraene
Heptafluorobutyric
Heptafluorobutyryl
Heptageniidae
Heptane
Heptanone
Heptatoma
Heptitol
Heptodonta
Heptose
Heptyl
Heracleum
Heratomin
Heratomol
Herbaceous *(Herbaceae)*
Herbage *(Herbages)*
Herbaria *(Herbarium,
Herbariums)*
Herberta
Herbicides *(Herbicidal,
Herbicide, Weedicide,
Weedkiller)*
X-Ref: Weedicides,
Weedkillers
Herbigation
Herbivores *(Herbivore,
Herbivorous)*
X-Ref: Antiherbivore
Herbs *(Herb, Herbal)*
Hercothrips
Hercules
Hercynine
Herdbook *(Herdbooks)*
Herders *(Herder)*
Herding
Herdoniini
Herds *(Herd)*
Herdsmen *(Herdsman)*
Herdwick
Heredity *(Hereditability,
Hereditary, Heritabilities,
Heritability, Heritable)*
Hereford *(Herefords)*
Heritiera
Hermannia
Hermaphroditism
*(Hermaphrodite,
Hermaphroditic,
Pseudohermaphrodite)*
X-Ref:
Pseudohermaphroditism
Hermatobates
Hermetic *(Hermetical,
Hermetically)*
Hernandia
Hernandiaceae
Hernia *(Hernias, Herniated,*

Herniation)
Herniaria
Herons *(Heron)*
Herpes *(Herpetic)*
Herpesviruses *(Herpesvirus)*
Herpetogramma
Herpetomonas
Herpotrichia
Herpyllus
Herring *(Herrings)*
Hersiliidae
Hesperethusa
Hesperetin
Hesperia
Hesperiidae *(Hesperid,
Hesperiid)*
Hesperinidae
Hesperinus
Hesperis
Hesperumia
Hesychia
Heterakidae
Heterakis
Heterapoides
Heteroagglutinins *see*
Agglutinins
Heteroauxin *see* **IAA**
Heterobasidion
Heterobeltiosis
Heteroblastic *(Heteroblasty)*
Heterocampa
Heterocarpy
Heterocaryons *see*
Heterokaryons
Heterocephalum
Heterocera
Heteroceridae
Heterochromatin
*(Heterochromatic,
Heterochromatization)*
Heterococcus
Heterocycles *(Heterocycle,
Heterocyclic,
Heterocyclics)*
Heterocysts *(Heterocyst,
Heterocystous)*
Heterodendron
Heterodera
Heteroderidae
Heterodoxus
Heterofermentative *see*
Fermentation
Heterogeneity
*(Heterogeneous,
Heterogenic,
Heterogenous)*
Heterogenesis
(Heterogenicity)
Heterogony *(Heterogonic,
Heterostyly)*
X-Ref: Heterostylous
Heterograft *(Heterografts,
Xenografts)*
X-Ref: Xenografting
Heterokaryons
(Heterocaryon,

*Heterokaryon,
Heterokaryosis,
Heterokaryotic,
Heterokaryotypes)*
X-Ref: Heterocaryons
Heterology *(Heterologous)*
Heteromeles
Heteromera
Heterometrus
Heteromorphism
(Heteromorphic)
Heteromyidae *(Heteromyid)*
Heteronychia
Heteronychus
Heteropappus
Heteropelma
Heteropeza
Heterophagous
Heterophyes
Heterophylly
(Heterophyllous)
Heteroploids *(Heteroploid,
Heteroploidy)*
Heteropoda
Heteropodidae
Heteropogon
Heteropoly
Heteroptera *(Heteropteran)*
Heterorhabditidae
Heterorhabditis
Heterosaccharide
Heterosexual
X-Ref: Nonheterosexual
Heterosides *(Heteroside)*
Heterosigma
Heterosis *(Heterotic)*
Heterosperm *see*
Spermatozoa
Heterosporium
Heterospory
(Heterosporous)
Heterostemon
Heterostigmata
Heterostylous *see*
Heterogony
Heterotanytarsus
Heterotardigrada
Heterotarsonemus
Heterotermes
Heterothallism
(Heterothallic)
Heterotheca
Heterotopia *(Heterotopic,
Heterotopies)*
Heterotransplantation
Heterotricha
Heterotrissocladius
Heterotrophy
*(Heterotrophic,
Heterotrophically)*
Heterotylenchus
Heteroxanthin
Heterozygotes
*(Heterozygosis,
Heterozygosities,
Heterozygosity,*

*Heterozygote,
Heterozygotic,
Heterozygous)*
Hevea
Hevein
Hexabromobiphenyl
Hexachloran *see*
Hexachlorocyclohexane
Hexachlorbutadien
Hexachloride
Hexachlorobenzene *see*
Hexachlorocyclohexane
Hexachlorobiphenyl
Hexachlorobutadiene
Hexachlorocyclohexane
(Hexachlorane)
X-Ref: BHC, Entomoxan,
HCH, Hexachloran,
Hexachlorobenzene,
Lindane
Hexachloroparaxylene
Hexachlorophene
Hexacola
Hexadecane
Hexadecen
Hexadecenoic
Hexadeuterated *see*
Deuteration
Hexaflurate
Hexagenia
Hexahydroindole
Hexametaphosphate
Hexamethonium
Hexamethyl
Hexamethylenetetramine
Hexamine
Hexamita
Hexamitiasis
Hexanal
Hexandrone
Hexane
Hexapeptide
Hexaphosphate
Hexaploids *(Autohexaploids,
Hexaploid)*
X-Ref: Autohexaploid
Hexatydeus
Hexenal
Hexene
Hexisea
Hexobarbital
Hexokinases *(Hexokinase)*
Hexosamine *(Hexosamines)*
Hexosaminidase
(Hexosaminidases)
Hexosemonophosphate
Hexoses *(Hexose)*
Hexosyl
Hexyl
Hexyleneglycol
Heyderia
Heza
Hibbertia
Hibernation *(Hibernating)*
Hibiscus
X-Ref: Abelmoschus

Hicklingia
Hickory *(Hickories)*
Hicoria *see* Carya
Hides *(Hide)*
Hidradenitis
Hieracium *(Hieracia)*
Hierarchy *(Hierarchical,*
Hierarchies)
Hierochloe
Hierodula
Hieroglyphus
Highlands *(Highland)*
Highveld *see* Veld
Highways *(Highway)*
Hiking *(Hike, Hikes)*
Hilara
Hilaria
Hilarimorpha
Hilarimorphidae
Hills *(Hill, Hillocks,*
Hillside, Hillslope,
Hilltop, Hilly)
Hilus *(Hilar, Hilum)*
Himantandraceae
Himantarium
Hindgut
Hindlegs *see* Legs
Hindlimb *see* Limbs
Hinges *(Hinge, Hinged)*
Hinosan *see* Edifenphos
Hip *(Hips)*
Hippeastrum
Hippelates
Hippoboscidae
Hippocampus
(Hippocampal)
Hippocrateaceae
Hippocrepis
Hippodamia
Hippomane
Hippomanes
Hippophae
Hippotion
Hippuric
Hippuris
Hircinoic
Hirschmanniella
(Hirschmaniella)
Hirstionyssus
Hirsutella
Hirudo
Hirundo
Hishimonus
Hisingerite
Hispidae *(Hispid, Hispids)*
Hispidin
Hispinae
Histaminase
Histamine *(Antihistaminic,*
Antihistaminics,
Histamines, Histaminic)
X-Ref: Antihistamines
Histeridae *(Histerid)*
Histidine
Histidinemia
Histidinol

Histiocytes
Histiogaster
Histiopteris
Histoacryl
Histoautoradiography
(Histoautoradiographic)
Histoblasts
Histochemistry
(Histochemical,
Histochemically)
X-Ref:
Ultrahistochemical
Histocompatibility
Histoenzymology
(Histoenzymological)
Histogenesis *(Histogenetic,*
Histogenetical,
Histogenetics, Histogenic)
Histology *(Histologic,*
Histological)
X-Ref: Histomorphology
Histolysis
Histometry *(Histometric)*
Histomonas
Histomoniasis
Histomorphology *see*
Histology
Histones *(Histone)*
Histopathology
(Histopathogenesis,
Histopathologic,
Histopathological)
Histophotometry *see*
Photometry
Histophysiology
(Histophysiological)
Histoplasma
Histoplasmosis
Historadiography
(Historadiograph)
Historians
Historiography
History *(Historic,*
Historical, Histories)
Hives *(Hive)*
Hobbies *(Hobby, Hobbyist)*
Hodotermes
Hodotermitidae
Hoeing *(Hoe, Hoed,*
Hoeings, Hoes)
Hofmeisterella
Hoggets *(Hogget)*
Hogpens
Hogs *see* Swine
Hoheria
Hoists *(Hoist, Hoisting)*
Hoja
Holacantha
Holarrhena
Holcostethus
Holcus
Holdings *(Holder, Holders)*
Holes *(Hole)*
Holidays *(Holiday)*
Holly *see* Ilex
Hollyhock

Holoacardius
Holobasidia *see* Autobasidia
Holocalyx
Holocarpic
Holocellulose
Holocene
Holocerina
Holochrome
Holography *(Holographic)*
Holometabola
(Holometabolous)
Holoparasitus
Holoprotein *see* Proteins
Holotricha
Holotrichia
Holotypes *(Holotype)*
Holstein *(Holsteins)*
Homadaula
Homalicine
Homalium
Homarus
Home *(Homes)*
Homemade
Homemakers *(Homemaker,*
Homemaking)
Homeopathy
Homeosis *(Homeotic,*
Homoeotic)
X-Ref: Homoeosis
Homeostasis *(Homeostatic)*
Homeowners
Homeria
Homesteads *(Homestead,*
Homesteaders,
Homesteading)
Homidium *see* Ethidium
Homing
Homoarginine
Homobasidiomycetes
(Homobasidiomycete)
Homocarnosine
Homocitrate
Homoclimates *see* Climate
Homocysteine
Homocytotropic
Homoeocerus
Homoeology
(Homoeologous)
Homoeosis *see* Homeosis
Homoeosoma
Homoerythrina
Homogenization
(Homogenate,
Homogenates,
Homogenisation,
Homogenised,
Homogenized,
Homogenizers,
Homogenizing)
Homogentisic
Homogony
X-Ref: Homostyly
Homografts *(Homograft)*
Homoionic
Homokaryon
(Homokaryotic)

Homology *(Homolog,*
Homologies, Homologous,
Homologue)
X-Ref: Nonhomologous
Homomevalonic
Homona
Homoneura
Homonucleoside
Homoptera *(Homopterous)*
Homorocoryphus
Homoserine
Homosexual
Homospermidine
Homospory *(Homosporous)*
Homostyly *see* Homogony
Homothallism
(Homothallic)
Homotropic
Homotropus
Homozygotes
(Homozygosity,
Homozygote,
Homozygous)
Honey *(Honeys)*
X-Ref: Melliferous
Honeybees *(Honeybee)*
Honeycombs *(Honeycomb,*
Honeycombing)
Honeydew *(Honeydews)*
Honeylocust
Honeypot
Honeysuckle
(Honeysuckles)
Hoofs *(Hoof, Hoofed,*
Hooves)
Hookeria
Hookeriaceae
Hookeriales
Hookworms *(Hookworm)*
Hoopoes
Hopea
Hoplia
Hoplitis
Hoplitoxenus
Hoplocampa
Hoplocerambyx
Hoplolaimidae
Hoplolaiminae
Hoplolaimus
Hoplopleuridae
Hoploteleia
Hoplothrips
Hoppers *(Hopper)*
Hops *(Hop, Hopping)*
Horama
Hordenine
Hordeum
Hordnia
Horehound
Hormidium
Hormones *(Hormonal,*
Hormonally, Hormone)
X-Ref: Antihormones,
Bihormonal
Hornbeam
Hornets *(Hornet)*

Horniolus
Horns *(Horn, Horned)*
Hornworms *(Hornworm)*
Hornwort
Horsebeans *(Horsebean)*
Horsechestnuts
  (Horsechestnut)
Horseflies *(Horsefly)*
Horsemanship
Horsemeat
Horsemen *(Horseman)*
Horseradish
Horses *(Equine, Horse,*
  Racehorses)
  X-Ref: Equines,
  Racehorse
Horseshoeing *(Horseshoers,*
  Horseshoes, Shoer)
  X-Ref: Shoeing
Horsesickness
Horsfieldia
Hortensia *see* Hydrangea
Hortia
Horticulture *(Horticultural)*
Horticulturists
  (Horticulturist)
Hose *(Hoses)*
Hospitals *(Hospital,*
  Hospitalized)
Hostathion *see* Triazophos
Hosts *(Host)*
  X-Ref: Nonhost
Hotbeds *(Hotbed)*
Hotels
Hothouses *see* Greenhouses
Hounds *see* Dogs
House *(Housed, Houses,*
  Housing)
Houseflies *(Housefly)*
Households *(Household)*
Housekeeping
Houseplants
Housewives *(Housewife)*
Hovenia
Howardula
Hoya
Huckleberries
  (Huckleberry)
Huernia
Hulling *(Hull, Hulled,*
  Huller, Hullers, Hulls)
  X-Ref: Unhulled
Hulsea
Hulupone
Human *(Hominids,*
  Humanized)
Humane *(Humanitarian)*
Humaria
Humariaceae
Humates *(Humate)*
Humectants *(Humectant)*
Humerobates
Humerus *(Humeral,*
  Humeri)
Humicola
Humidimeters

(Humidimeter,
  Humidimetre,
  Humidimetres)
Humidity *(Humid,*
  Humidification,
  Humidified, Humidifier,
  Humidifying, Humidities)
  X-Ref: Semihumid,
  Subhumid
Humins *(Humin)*
Humiriaceae
Hummingbird
Humor *(Humoral)*
Humulone
Humulus
Humus *(Humic,*
  Humification, Humified)
Hunger *(Hungry)*
Hunteracarus
Hunterellus
Hunteria
Hunting *(Hunt, Hunted,*
  Hunter, Hunters, Hunts)
Huntleya
Huperzia
Hura
Hurricanes *(Hurricane)*
Husband
Husbandman *see* Farmers
Husbandry
Husks *(Husk, Husker,*
  Husking, Huskless)
  X-Ref: Multihusked,
  Unhusked
Hutchinsia
Hyacinthus *(Hyacinth,*
  Hyacinths)
Hyadaphis
Hyadesia
Hyadesidae
Hyalesthes
Hyaline
Hyalinosis
Hyalochlora
Hyalodendron
Hyalomma
Hyalophora
Hyaloplasm
Hyalopterus
Hyalotia
Hyalotiella
Hyalotiopsis
Hyaluronate
Hyaluronic
Hyaluronidase
  (Hyaluronidases)
Hyblaea
Hybomitra
Hybosciara
Hybosorinae
Hybridization *(Hybrid,*
  Hybridisation,
  Hybridising, Hybridity,
  Hybridizability, Hybridize,
  Hybridized, Hybridizer,
  Hybridizers, Hybridizing,

Hybridogenic, Hybrids,
  Polyhybrid)
  X-Ref: Monohybrid,
  Polyhybrids
Hycanthone
Hydantoin *(Hydantoins)*
Hydatella
Hydatellaceae
Hydaticus
Hydatidosis
Hydatids *(Hydatid)*
Hydnaceae
Hydnocarpus
Hydnoraceae
Hydnum *(Hydnums)*
Hydra
Hydrachna
Hydrachnellae
  (Hydracarina)
Hydrachnidae
Hydraecia
Hydraena
Hydranencephaly
Hydrangea
  X-Ref: Hortensia
Hydrases *(Hydrase)*
Hydrastine
Hydrastis
Hydrates *(Hydrate,*
  Hydrated, Hydrous)
Hydration *(Rehydratability,*
  Rehydrated, Rehydrating)
  X-Ref: Rehydration
Hydraulics *(Hydraulic,*
  Hydraulical,
  Hydraulically)
Hydrazides *(Hydrazide)*
Hydrazine
Hydrazinium
Hydrellia
Hydria
Hydrilla
Hydroallantois
Hydroanthracene
Hydrobaenus
Hydrobiology
  (Hydrobiologic,
  Hydrobiological)
Hydrobiotite
Hydroboration
Hydrobromide
Hydrocarbons
  (Hydrocarbon)
Hydrocellulose
Hydrocephalus
Hydrocera
Hydrocharis
Hydrocharitaceae
Hydrocharitetum
Hydrochemistry
  (Hydrochemical)
Hydrochloric
Hydrochloride
Hydrocinnamic
Hydrocleis
Hydrocolloids

(Hydrocolloid)
Hydrocooling
Hydrocortisone
  X-Ref: Cortisol
Hydrocotyle
Hydrocyanic
Hydrodictyon
Hydrodynamics
  (Hydrodynamic)
Hydroecia
Hydroelectric
Hydrofluoric
Hydrogels
Hydrogen
Hydrogenases
  (Hydrogenase)
Hydrogenation
  (Hydrogenated,
  Hydrogenates)
  X-Ref: Unhydrogenated
Hydrogenion
Hydrogenolysis
  (Hydrogenolyses)
Hydrogenomonas
Hydrogeology
  (Hydrogeologic,
  Hydrogeological)
Hydrogeothermal
Hydrography *(Hydrograph,*
  Hydrographic,
  Hydrographically,
  Hydrographs)
Hydrolases *(Hydrolase)*
Hydrology *(Hydrologic,*
  Hydrological,
  Hydrologically)
Hydrolysis *(Hydrolyse,*
  Hydrolysed, Hydrolyses,
  Hydrolysing, Hydrolytic,
  Hydrolyzable, Hydrolyze,
  Hydrolyzed, Hydrolyzing)
  X-Ref: Autohydrolysis
Hydrolyzates *(Hydrolysate,*
  Hydrolysates,
  Hydrolyzate)
Hydromechanics
  (Hydromechanical,
  Hydromechanization)
Hydromelioration
  (Hydromeliorative)
Hydrometeorology
  (Hydrometeorologic,
  Hydrometeorological)
Hydrometers *(Hydrometer,*
  Hydrometric)
Hydrometra
Hydrometridae
Hydromorphic
  (Hydromorph,
  Hydromorphous,
  Hydromorphy)
Hydronephrosis
  (Hydronephrotic)
Hydroperoxidases
Hydroperoxides
  (Hydroperoxide)

Hydroperoxy
Hydroperoxylinoleic
Hydrophilic *(Hydrophile,*
Hydrophilous)
Hydrophilidae
(Hydrophilid)
Hydrophilinae
Hydrophilini
Hydrophiloidea
Hydrophilomyces
Hydrophobic *(Hydrophobe)*
Hydrophorus
Hydrophyllaceae
Hydrophysical
Hydrophytes *(Hydrophytic)*
Hydroponics *(Hydroponic,*
Hydroponically, Soilless)
X-Ref: Nutriculture,
Waterculture
Hydroporinae
Hydroporus
Hydropotes
Hydroprene
Hydrops *see* Edema
Hydropsyche
Hydropsychidae
Hydropsychinae
Hydroptila
Hydroptilidae
Hydropyrus
Hydroquinones
(Hydroquinone)
X-Ref: Quinol
Hydrosalpinx
Hydrostatics *(Hydrostatic,*
Hydrostatically)
Hydrotaea
Hydrothermal
(Hydrothermic)
Hydrotropic
Hydroxamate
Hydroxamic
Hydroxides *(Hydroxide)*
Hydroxyanil
Hydroxyanisole
Hydroxyanthranilic
Hydroxybenzaldehyde
Hydroxybenzoic
X-Ref:
Parahydroxybenzoic
Hydroxybenzyl
Hydroxybutyric
Hydroxycarboxylic
Hydroxycholecalciferol
Hydroxycinnamate
Hydroxycinnamic
Hydroxycinnamoyl
Hydroxycitrate
Hydroxycorticosteroids
(Hydroxycorticosteroid)
Hydroxycorticosterone
Hydroxycorynoline
Hydroxycoumarin
(Hydroxycoumarins)
Hydroxydehydrogenase
Hydroxydeoxycorticosteron

Hydroxydienic
Hydroxydopamine
Hydroxyecdysone
Hydroxyethyl
(Hydroxyethylated,
Hydroxyethylation)
Hydroxyethylphosphonate
Hydroxyethylpiperazine
Hydroxyflavanones
(Hydroxyflavanone)
Hydroxyflavonol
Hydroxygardnutine
Hydroxyiminomethyl
Hydroxyindole
Hydroxyisoleucine
Hydroxylamine
Hydroxylapatite
Hydroxylase *(Hydroxylases)*
Hydroxylation
(Hydroxylated,
Hydroxylating,
Hydroxylations)
Hydroxyls *(Hydroxyl)*
Hydroxyltryptamine
Hydroxylysine
Hydroxymalonate
Hydroxymethyl
Hydroxymethylenechroman
Hydroxymethylfurfurol
Hydroxymethyltransferase
Hydroxynaphthoate
Hydroxyphenethyl
Hydroxyphenoxyl
Hydroxyphenyl
Hydroxyphenylacetic
Hydroxyphenylalanine
Hydroxyprogesterone
Hydroxyproline
Hydroxypropyl
Hydroxyprostaglandin
Hydroxypyridine
Hydroxyquinoline
Hydroxyradiclonic
Hydroxysteroid
Hydroxystilbene
(Hydroxystilbenes)
Hydroxystyrenes
Hydroxytoluene
Hydroxytropanes
Hydroxytryptamides
Hydroxytryptamine
Hydroxytyramine
Hydroxyurea
Hydroxyvitamin
Hydroxywogonin
Hydryphantes
Hyenanchin
Hygiene *(Hygienic,*
Hygienical, Hygienics)
Hygienists *(Hygienist)*
Hygrine
Hygrocybe
Hygrographs
Hygrohypnum
Hygroma *(Hygromas)*
Hygrometry *(Hygrometer)*

Hygrophila
Hygrophoraceae
Hygrophoropsis
Hygrophorus
Hygrophotography
(Hygrophotographic)
Hygroscopy *(Hygroscopic,*
Hygroscopicity)
Hylaeogena
Hylaeus
Hylastes
Hylastinus
Hylecoetus
Hylemya *(Hylemyia)*
Hyles
Hylesinus
Hylobius
Hylocomium
Hyloicus
Hyloniscus
Hylophilidae *see* Aderidae
Hylotrupes
Hylurgopinus
Hymenachne
Hymenaea
Hymenea
Hymenia *(Hymenial)*
Hymenocallis
Hymenogaster
Hymenolepididae
Hymenolepis
Hymenolepsis
Hymenomyces
(Hymenomycete,
Hymenomycetous)
Hymenopappus
Hymenophyllaceae
Hymenophyllum
Hymenophysa
Hymenoptera
(Hymenopteran,
Hymenopterous)
Hymenoxys
Hymexazol
Hyoid
X-Ref: Infrahyoid
Hyoscyamus
Hyostrongylus
Hypagyrtis
Hyparrhenia
Hypeninae
Hypenodes
Hypera
Hyperactivity *(Hyperactive)*
Hyperadrenocorticism
(Hyperadrenocorticalism)
Hyperalimentation *see*
Alimentation
Hyperaspis
Hyperauxiny
Hyperbaric
Hypercalcemia
Hypercapnia *(Hypercapnic)*
Hypercholesterolemia
(Hypercholesteremia,
Hypercholesterolemic)

Hyperemia *(Hyperaemia,*
Hyperaemic)
Hyperestrogenism
Hyperfiltration
Hyperglycemia
(Hyperglycaemia,
Hyperglycaemic,
Hyperglycemic)
Hypericaceae
Hypericum
Hyperimmunization *see*
Immunization
Hyperin
Hyperinsulinism
Hyperkalemia
Hyperkeratosis
Hyperketonemia
Hyperkinesia *(Hyperkinetic)*
Hyperlactacidemia
Hyperlipemia
(Hyperlipaemia,
Hyperlipemic,
Hyperlipidemia,
Hyperlipidemic,
Hyperlipoproteinaemia,
Hyperlipoproteinemia)
Hyperodes
Hyperomyzus
Hyperosmotic *see* Osmosis
Hyperostosis
Hyperoxia
Hyperparasites
(Hyperparasite,
Hyperparasitic,
Hyperparasitism)
Hyperparathyroidism
(Hyperparathyroid)
Hyperphagia
Hyperplasia *(Hyperplastic)*
Hyperploids *(Hyperploid,*
Hyperploidy)
Hypersensibility *see*
Sensitivity
Hypersexuality *see* Sex
Hypertension *(Hypertensive)*
X-Ref: Antihypertensive
Hyperthermia *see* Fever
Hyperthyroidism
(Hyperthyroid,
Hyperthyrosis)
Hypertonicity *(Hypertonic)*
Hypertrophy *(Hypertrophic,*
Hypertrophied,
Hypertrophies)
Hyperuricemia
(Hyperuricemic)
Hyperventilation
Hyperviscosity *see*
Viscosity
Hypervitaminosis
Hyphae *(Hypha, Hyphal)*
Hyphantria
Hyphochytridiales
Hyphomycetes
(Hyphomycete,
Hyphomycetic,

Hyphomycetous)
Hyphopodia
Hyphaceae
Hypnea
Hypnodendraceae
Hypnodendron
Hypnodil
Hypnoidus
Hypnosis *(Hypnotic)*
Hypnum
Hypoadrenocorticism
Hypoalbuminemia
Hypoaspis
Hypobarism *(Hypobaric)*
Hypobetalipoproteinemia
Hypoblast
Hypobletus
Hypocacculus
Hypocalcemia
   (Hypocalcaemia,
   Hypocalcemic)
Hypochaeris *(Hypochoeris)*
Hypochilomorphae
Hypochlorinators
Hypochlorite
Hypocholesterolemia
   (Hypocholesterolemic)
Hypochrysops
Hypocotyls *(Hypocotyl)*
Hypocrea
Hypocreaceae
Hypocreales
Hypocrita
Hypocupremia *(Hypocuprosis)*
Hypoderma
Hypodermatidae
Hypodermis *(Hypodermal)*
Hypodermosis
Hypodiploids *(Hypodiploid,*
   Hypodiploidy)
Hypodiscus
Hypogammaglobulinemia
   (Hypogammaglobulinaemia)
Hypogastrium *(Hypogastric)*
Hypogastrura
Hypogastruridae
Hypogeous
Hypoglossal
Hypoglycemia
   (Hypoglycaemia,
   Hypoglycaemic,
   Hypoglycemic)
Hypoglycin
Hypogonadism
Hypogymnia
Hypohaploids *(Hypohaploid,*
   Hypohaploidy)
Hypokalemia
Hypokinesia
Hypolimnas
Hypolipidemia
   (Hypolipidaemic,
   Hypolipidemic)
Hypomagnesemia
   (Hypomagnesaemia,
   Hypomagnesaemic,

Hypomagnesemic)
Hypomyces
Hyponomeuta
Hyponomeutidae
Hypoparathyroidism
Hypopharynx
   (Hypopharyngeal)
Hypophloeus
Hypophosphatemia
   (Hypophosphatemic)
Hypophosphite
Hypophysectomy
   (Hypophysectomised,
   Hypophysectomized)
Hypophysis *(Hypophyseal,*
   Hypophysial)
Hypopi *(Hypopus)*
Hypoplasia
Hypoponera
Hypoproteinemia
Hyporiboflavinosis
Hyposoter
Hypostome
Hypotension *(Hypotensive)*
Hypotermes
Hypothalamus
   (Hypothalamic,
   Hypothalamically,
   Hypothalamo)
Hypothenemus
Hypothermia *(Hypothermal,*
   Hypothermic)
Hypothyroidism
   (Hypothyroid)
Hypotrachyna
Hypotrigona
Hypotrophy
Hypoventilation
Hypovitaminosis
   (Hypovitaminotic)
Hypovolemia *(Hypovolemic)*
Hypoxanthine
Hypoxia *(Hypoxaemia,*
   Hypoxic)
   X-Ref: Anoxia
Hypoxidaceae
Hypoxis
Hypoxylon
Hypselonotus
Hypsipyla
Hypsometry *(Hypsometric)*
Hyptis
Hyrtanella
Hyssopus
Hysterectomy
   (Hysterectomized)
Hysteresis *(Hysteretic)*
Hysteria
Hysteriales
Hysterionica
Hysterium
Hysterothecia
Hysterotomy
Hystrichopsylla
Hystrichopsyllidae
Hystriomyia

IAA
   X-Ref: Heteroauxin,
      Rhizopin
Iatrogenic
Ibaliidae
Ibidoecus
Iboga
Icacina
Icacinaceae
Ice *(Iced)*
Icerya
Ichneumon
Ichneumonidae
   (Ichneumonid,
   Ichneumonids)
Ichneumoninae
Ichneumonini
Ichneumonoidea
Ictalurus
Icterus see Jaundice
Idaea
Idiasta
Idioblasts *(Idioblast)*
Idiocerinae
Idiocerus
Idioderma
Idiogram
   X-Ref: Karyogram
Idiomelissodes
Idiopathy *(Idiopathic)*
Idiospermaceae
Idiospermum
Idiostatus
Idiostolidae
Iditol
Idotea
Idris
Ifosfamide
Igneous
Ignition *(Ignitability)*
Iguanidae
Igurdin
Ileitis
Ileocystoplasty
Ileum *(Ileal, Ileo)*
Ilex *(Hollies)*
   X-Ref: Holly, Yaupon
Illipe
Illiteracy *(Illiterate)*
Illites *(Illite)*
Illudins *(Illudin)*
Illumination *(Illuminance,*
   Illuminated,
   Illuminations,
   Preilluminated)
   X-Ref: Preillumination
Illuvial
Illuviation *(Illuviated)*
Ilnacora
Imaginal *(Imago, Imagoes)*
   X-Ref: Postimaginal,
      Preimaginal
Imbibition *(Imbibed,*
   Imbibing, Imbibitional)
Imbrasia
Imbrication

Imidazoles *(Imidazole)*
Imidazolidine
Imidazoline
Imide
Imidocarb
Imidoesters
Imine
Imino
Iminoimidazolidine
Iminoxyl
Imipramine
Imitation *(Imitated)*
Immaturity *(Immature,*
   Immatures)
Immersion *(Immerse,*
   Immerses, Immersing)
Immigrants *(Immigrant)*
Immigration *(Immigrations)*
Immobility
Immobilization
   (Immobilisation,
   Immobilized,
   Immobilizing)
Immunity *(Autoimmune,*
   Immune, Immuno)
   X-Ref: Autoimmunity,
      Nonimmune
Immunization
   (Hyperimmune,
   Hyperimmunized,
   Immunisation, Immunise,
   Immunised,
   Immunizations,
   Immunize, Immunized,
   Immunizing)
   X-Ref:
      Hyperimmunization,
      Preimmunized
Immunoabsorption
   (Immunoabsorbent)
Immunoadjuvant
Immunoadsorption
   (Immunoadsorbents)
Immunoassay
   X-Ref:
      Electroimmunoassay
Immunobiology
   (Immunobiologic,
   Immunobiological)
Immunochemistry
   (Immunochemical,
   Immunochemically)
Immunocompetence
Immunocytochemistry
   (Immunocytochemical)
Immunocytology
   (Immunocytological)
Immunodeficiency
Immunodepression
   (Immunodepressant,
   Immunodepressive,
   Immunodepressor,
   Immunodepressors)
Immunodiagnosis
   (Immunodiagnostic)
Immunodiffusion

Immunoelectrophoresis
(Immunoelectrophoretic)
Immunofluorescence
(Immunofluorescent)
Immunogenetics
(Immunogenetic)
Immunogens *(Immunogen,*
Immunogenesis,
Immunogenic,
Immunogenicity)
Immunoglobin
Immunoglobulins
(Immunoglobulin,
Immunoglobuline)
Immunohistochemistry
(Immunohistochemical)
Immunohistology
(Immunohistological)
Immunology *(Immunologic,*
Immunological,
Immunologically)
Immunopathology
(Immunopathologic)
Immunoperoxidase
Immunoprophylaxis
Immunoreactivity
(Immunoreactive)
Immunosuppression
(Immunosuppressive)
Immunosurgery
Immunosurveillance
Immunotherapy
Impala
Impalaia
Impatiens
Impedance
Imperata
Imperforate
Imperialine
Imperialism
Impermeable
(Impermeability)
Impietratura
Implants *(Implant,*
Implantation,
Implantations, Implanted,
Implanting,
Preimplanted)
X-Ref: Postimplantation,
Preimplantation
Importers *(Importer)*
Imports *(Import,*
Importation, Importations,
Imported, Importing)
Impregnation *(Impregnated,*
Impregnating)
Imprinting *(Imprints)*
Impulses *(Impulse,*
Impulsive)
Impurities *(Impure,*
Impurity)
Imugan *see*
Chloraniformethan
Inachis
Inactivation *(Inactivate,*
Inactivated, Inactivates,

Inactivating, Inactivator)
Inanition
Inazuma
Inbreeding *(Inbred,*
Inbreds)
Incandescence
(Incandescent)
Incendiarism *(Incendiary)*
Incentives *(Incentive)*
X-Ref: Disincentives
Inceptisols *(Inceptisol)*
Incineration *(Incinerating,*
Incinerator, Incinerators)
Incisalia
Incision
Incisors *(Incisor)*
Incombustibility *see*
Combustion
Income *(Earning, Incomes)*
X-Ref: Earnings
Incompatibility
(Incompatibilities,
Incompatible)
Incontinence
Increments *(Increment)*
Incubation *(Incubated,*
Incubates, Incubating,
Incubations)
X-Ref: Preincubation
Incubators *(Incubator)*
Incurvariidae
Indarbela
Indehiscence *see* Dehiscence
Indemnity *(Indemnities)*
Indene
Index *(Indexed, Indexes,*
Indexing, Indices)
Indiangrass
Indians
Indicators *(Indicator)*
Indicine
Indigenous
Indigestion *(Indigestible,*
Indigestions)
X-Ref: Dyspepsia
Indigo
Indigofera
Indium
Indole *(Indol, Indoles,*
Indolic, Indolyl)
X-Ref: Bisindole
Indoleacetaldehyde
Indoleacetate
Indoleacetic *(Indolylacetic)*
Indoleacrylic
(Indolylacrylic)
Indolealkylamine
Indoleamine *(Indoleamines)*
Indolebutyric
(Indolylbutyric)
Indolecarboxylic
Indolepropionic
Indolepyruvic
Indolizidine
Indolmycin
Indolopyridoquinazoline

Indomethacin
Indophenol
Indophenyl
Indospicine
Indotristicha
Inducers *(Inducer)*
Inductors *(Inductor)*
Indusia
Industrialists *(Industrialist)*
Inedible
Inelastic *see* Elasticity
Inermoparmena
Inertia *(Inertial)*
Infancy *(Infant, Infantile,*
Infants)
Infanticide
Infarction *(Infarct, Infarcts)*
Infection *(Infect, Infectants,*
Infected, Infectibility,
Infecting, Infectional,
Infections, Infectious,
Infective, Infectivity,
Noninfectious, Reinfected)
X-Ref: Autoinfection,
Noninfected,
Postinfectional,
Pseudoinfection,
Reinfection,
Superinfection,
Uninfected,
Viruliferous
Infertility *(Infertile)*
Infestation *(Infest,*
Infestability, Infestations,
Infested, Infesting,
Infestiveness, Infests,
Reinfested)
X-Ref: Reinfestation,
Superinfestation
Infiltration *(Infiltrate,*
Infiltrates, Infiltrating)
Infiltrometer
Inflammability *see*
Flammability
Inflammation
(Inflammatory)
X-Ref: Antiinflammatory
Inflation *(Inflatable,*
Inflated, Inflationary,
Inflations)
Inflorescence
(Inflorescences,
Inflorescent)
Influenza
X-Ref: Parainfluenza
Informosomes
(Informosomal)
Infrageneric *see* Generic
Infrahyoid *see* Hyoid
Infrared
X-Ref: Ultrared
Infrasound
Infrastructure
(Infrastructural,
Infrastructures)
Infundibulum

(Infundibular)
Infurcitinea
Infusate
Infusion *(Infused, Infusing,*
Infusions)
Infusoria
Ingenol
Ingestion *(Ingest, Ingesta,*
Ingested, Ingesting,
Ingestive)
X-Ref: Postingestional
Ingluviotomy
Ingramia
Ingredients *(Ingredient)*
Inguinal *see* Groin
Inhalation *(Inhalant,*
Inhaled)
Inheritance *(Inherit,*
Inheritability, Inheritable,
Inherited, Inheriting)
Inhibition *(Autoinhibiting,*
Inhibit, Inhibitants,
Inhibited, Inhibiting,
Inhibitions, Inhibitive,
Inhibits)
X-Ref: Autoinhibition
Inhibitors *(Inhibitor,*
Inhibitory)
Inhomogeneity
(Inhomogenities)
Injection *(Inject, Injectable,*
Injected, Injecting,
Injections)
X-Ref: Microinjection
Injectors *(Injector)*
Injury *(Injure, Injured,*
Injures, Injuries, Injuring,
Injurious, Injuriousness)
X-Ref: Uninjured
Innervation *(Innervate)*
Innovar
Innovation *(Innovating,*
Innovations, Innovative,
Innovativeness, Innovator,
Innovators)
Inoculation *(Inocula,*
Inoculant, Inoculants,
Inoculate, Inoculated,
Inoculating, Inoculations,
Inoculative, Inoculum)
X-Ref: Preinoculation,
Reinoculated
Inocybe
Inoperculate *see* Operculum
Inopus
Inosine *(Inosinic)*
Inositol *(Inosite)*
Inotropism *(Inotropic)*
Input *(Inputs)*
Inquilines
Insectaries *(Insectarium,*
Insectary)
Insecticides *(Insecticidal,*
Insecticidally, Insecticide)
Insectivora *(Insectivores,*
Insectivorous)

Insectproofing
Insects *(Bug, Insect,*
Insecta)
X-Ref: Bugs,
Entomofauna
Insemination *(Inseminate,*
Inseminated,
Inseminating,
Inseminations,
Inseminative,
Inseminator)
Insensitivity *(Insensitive)*
Insolation *see* **Sun**
Inspection *(Inspected,*
Inspections, Inspective,
Inspectorial)
X-Ref: Uninspected
Inspectors *(Inspector)*
Instability *(Instabilities)*
Instant *(Instantized,*
Instantizing)
Instars *(Instar)*
Instinct *(Instincts,*
Instinctual)
Institutions *(Institution,*
Institutional)
Instruction *(Instruct,*
Instructional,
Instructions)
Instructors *(Instructor)*
Instruments *(Instrument,*
Instrumentation)
Insufficiency *(Insufficient)*
Insufflation
Insulation *(Insulated,*
Insulating, Insulations)
X-Ref: Uninsulated
Insulin
X-Ref: Proinsulin
Insurance *(Insurances,*
Insure, Insured, Insures,
Insuring)
Intake *(Intakes)*
Integration *(Integrate,*
Integrated, Integrates,
Integrating, Integrational,
Integrative)
Integrators *(Integrator)*
Integuments *(Integument,*
Integumentary,
Tegumental,
Tegumentary, Teguments)
X-Ref: Tegument
Intelligence *(Intelligent,*
Intelligentsia)
Intensity *(Intense,*
Intensification,
Intensified, Intensify,
Intensifying, Intensities,
Intensive, Intensively)
Interallelic *see* **Allele**
Interbasin *see* **Basins**
Interbreed *see* **Breeding**
Intercalation *(Intercalary,*
Intercalated)
Intercapillary *see* **Capillary**

Intercellular *see* **Cells**
Intercerebral *see* **Brain**
Interchromosomal *see*
Chromosomes
Intercisternal *see* **Cisternae**
Intercollective *see*
Collectives
Intercondylar *see* **Condyle**
Interconversion
(Interconversions)
Intercooperative *see*
Cooperatives
Intercostal *see* **Costal**
Intercotyledonary *see*
Cotyledons
Intercropping *(Intercrop,*
Intercroppings, Intercrops)
Intercross *see*
Crossbreeding
Intercultivar *see* **Cultivars**
Interendothelial *see*
Endothelium
Interesterification *see*
Esterification
Interfarm *see* **Farms**
Interferometry
(Interferometric)
Interferons *(Interferon)*
Interfiber *see* **Fibers**
Interfluve *(Interfluvial)*
Interfoliar *see* **Foliage**
Intergeneration
(Intergenerational)
Intergeneric *see* **Generic**
Intergenome *see* **Genomes**
Intergenotypic *see*
Genotypes
Interglacial *see* **Glacial**
Intergovernmental *see*
Government
Intergranular *see*
Granulation
Interlamellar *see* **Lamella**
Interlaminar *see* **Lamina**
Interlayers *(Interlayer)*
Interlinear
Interlocular *see* **Loculus**
Intermittent *(Intermittently)*
Intermolting *see* **Molt**
Interneurons *see* **Neurons**
Internodes *(Internodal,*
Internode)
Interphases *(Interphase)*
Interplanting *(Interplant,*
Interplanted, Interplants)
Interprovenance *see*
Provenance
Interracial *see* **Race**
Interrelation *(Interrelated,*
Interrelations,
Interrelationship,
Interrelationships)
Interrenal *see* **Kidneys**
Interrows *(Interrow)*
Interscapular *see* **Scapula**
Intersection *(Intersect,*

Intersectional)
Interseeding *(Interseed,*
Interseedings)
Intersex *(Intersexes,*
Intersexual,
Intersexuality,
Intersexuals)
Interstem *see* **Stem**
Interstitial
Interstock *(Interstocks)*
Interstratification *see*
Stratification
Intertarsal *see* **Tarsus**
Intertidal *see* **Tide**
Intertilled *see* **Tillage**
Intertrappean *(Intertrap)*
Intertropical *see* **Tropics**
Intervarietal *see* **Varieties**
Intervector *see* **Vectors**
Interventricular *see*
Ventricles
Intervertebral *see* **Vertebrae**
Interviewing *(Interview)*
Intestines *(Intestinal,*
Intestine)
Intine
Intolerances *see* **Tolerance**
Intoxication *(Intoxicated,*
Intoxicating,
Intoxications)
X-Ref: Autointoxication
Intraabdominal *see*
Abdomen
Intraadenohypophysial *see*
Adenohypophysis
Intraarticular *see* **Articulus**
Intraaxonal *see* **Axons**
Intrabreeding *(Intrabred,*
Intrabreed)
Intrabursal *see* **Bursa**
Intracarotid *see* **Carotid**
Intracellular *see* **Cells**
Intracerebral *see* **Brain**
Intrachromosomal *see*
Chromosomes
Intraclonal *see* **Clone**
Intracoelomal *see* **Coelom**
Intracranial *see* **Cranium**
Intracrystalline *see* **Crystals**
Intracutaneous *see* **Skin**
Intracuticular *see* **Cuticles**
Intracytoplasmic *see*
Cytoplasm
Intradermal *see* **Skin**
Intraductal *see* **Ducts**
Intraduodenal *see*
Duodenum
Intraepidermal *see*
Epidermis
Intraerythrocytic *see*
Erythrocytes
Intrafarm *see* **Farms**
Intrafollicular *see* **Follicles**
Intragastric *see* **Stomach**
Intrageneric *see* **Generic**
Intragenic *see* **Genes**

Intraglandular *see* **Glands**
Intrahepatic *see* **Hepatic**
Intralipid *see* **Lipids**
Intramammary *see*
Mammary
Intramedullary *see* **Medulla**
Intramembranous *see*
Membranes
Intramitochondrial *see*
Mitochondria
Intramuscular *see* **Muscle**
Intramyocardial *see*
Myocardia
Intranasal *see* **Nasal**
Intranatal *see* **Birth**
Intraneuronal *see* **Neurons**
Intranuclear *see* **Nucleus**
Intranucleolar *see*
Nucleolus
Intraocular *see* **Eyes**
Intrapancreatically *see*
Pancreas
Intrapelvic *see* **Pelvis**
Intraperitoneal *see*
Peritoneum
Intraplastidial *see* **Plastids**
Intraprovenance *see*
Provenance
Intrapulmonary *see*
Pulmonary
Intrarectally *see* **Rectum**
Intrarenal *see* **Kidneys**
Intraruminal *see* **Rumen**
Intrascrotal *see* **Scrotum**
Intratracheal *see* **Tracheal**
Intratubal *see* **Tube**
Intratumor *see* **Tumor**
Intrauterine *see* **Uterus**
Intravaginally *see* **Vagina**
Intravarietal *see* **Varieties**
Intravascular *see* **Vascular**
Intravenous *(Intravenal,*
Intravenously)
Intraventricular *see*
Ventricles
Intravitelline *see* **Yolks**
Intrazonal
Intubation
Intussusception
X-Ref: Invaginated
Inula
Inuleae
Inulin
Inumakilactone
Inundation *see* **Floods**
Inurois
Invaginated *see*
Intussusception
Invariants *(Invariant)*
Inventories *(Inventory,*
Inventorying)
Inversion *(Inverse,*
Inversions, Invert,
Inverted)
Invertase *(Invertases)*
X-Ref: Sucrase

Invertebrates *(Invertebrata,*
Invertebrate)
Investments *(Investing,*
Investment, Investor,
Investors)
Inviability *see* Viability
Invoice
Involution *(Involuting)*
Iodate
Iodic
Iodide *(Iodides)*
Iodine *(Iodinate, Iodinated,*
Iodinating, Iodination,
Iodization, Iodized, Iodo,
Radioiodinated)
  X-Ref: Radioiodine
Iodoacetate
Iodoacetic
Iodocasein
Iododeoxyuridine
Iodofenphos
Iodometry *(Iodometric)*
Iodonium
Iodophors *(Iodophor,*
Iodophoric)
Iodoprotein
Iodouridine
Iolinidae
Ionites *(Ionite)*
Ionization *(Ionisation,*
Ionising, Ionizable,
Ionized, Ionizing)
Ionogen
Ionol
Ionones *(Ionone)*
Ionophores *(Ionophore)*
Ioxynil
Ipecac *(Ipecacuanha)*
Ipheion
Iphiclides
Iphita
Ipidae
Ipobracon
Ipofos
Ipomeamarone
Ipomoea *(Ipomaea, Ipomea)*
Ipronidazole
Ips
Ipsenol
Iranoraphidia
Iridaceae *(Iridaceous)*
Iridaea
Irides *(Iridial)*
Iridescent *(Iridescence)*
Iridium
Iridoids *(Iridoid)*
Iridomyrmex
Iridovirus
Iriediol
Irieol
Iris *(Irises)*
Iron
  X-Ref: Ferruginous
Ironweed
Irpex
Irradiation *(Irradiance,*

Irradiated, Irradiating,
Irradiations, Irradiators,
Irriadiated, Preirradiated,
Ray)
  X-Ref: Nonirradiated,
  Postirradiation,
  Preirradiation, Rays,
  Unirradiated
Irrigation *(Irrigability,*
Irrigable, Irrigate,
Irrigated, Irrigating,
Irrigational, Irrigations)
  X-Ref: Nonirrigated,
  Subirrigation,
  Unirrigated
Irrigators *(Irrigator)*
Iruana
Irvingia
Iryanthera
Isabelin
Isanthus
Isaria
Isariopsis
Isatin
Iscariotes
Ischaemum
Ischalia
Ischemia *(Ischaemia,*
Ischaemic, Ischemic)
Ischnocera
Ischnoclopius
Ischnodemus
Ischnura
Ischyropsalidae
(Ischyropsalididae)
Iseilema
Iskur
Isoacceptor
Isoallelic *see* Allele
Isoalloxazine
Isoamyl
Isoantibodies *see* Antibodies
Isoantigens *see* Antigen
Isobenzan *(Telodrine)*
  X-Ref: Telodrin
Isoboldine
Isobutane
Isobutanol
Isobutyl
Isobutylidene
Isobutyloctadeca
Isobutylphenyl
Isobutylthiazole
Isobutyrate
Isobutyric
Isocaespitol
Isocaloric *see* Calories
Isocapnia
Isochilus
Isochlorogenic
Isochromazonarol
Isochromosome
(Isochromosomal,
Isochromosomes)
Isochrysis
Isocil

Isocitratase *(Isocitritase)*
Isocitrates *(Isocitrate)*
Isocitric
Isocoma
Isocoumarins *(Isocoumarin)*
Isocyanates *(Isocyanate)*
Isocyanide
Isocynaroside
Isodihydrocadambine
Isodon
Isodontia
Isodrin
Isodromus
Isodur
Isoelectric *(Isoelectrical)*
Isoelectrofocusing *see*
  Electrofocusing
Isoelectronic
Isoentomon
Isoenzymes *(Isoenzymatic,*
Isoenzyme, Isozyme,
Isozymic)
  X-Ref: Isozymes
Isoerythrolysis
Isoesterases
Isoetes
Isoeugenol
Isoflavan
Isoflavones *(Isoflavone)*
Isoflavonoids *(Isoflavonoid)*
Isoflurane
Isofoliol
Isoformosanine
Isogenic *(Isogenicity)*
Isogosferol
Isohalkendin *see* Halkendin
Isohemigossypol
Isohumulate
Isohumulones
(Isohumulone)
Isohydric
Isoimmunization
(Isoimmune)
Isoindol
Isolaureline
Isoleucine *(Isoleucin,*
Isoleucines)
Isoleucyl
Isolines
Isolysergic
Isomaltose
Isomerases *(Isomerase)*
Isomers *(Isomer, Isomeric,*
Isomerisation, Isomerised,
Isomerism, Isomerization,
Isomerized, Isomerizing)
  X-Ref:
  Photoisomerization
Isometopinae
Isometrus
Isometry *(Isometric)*
Isomira
Isomitraphylline
Isomorph *(Isomorphic,*
Isomorphous)
Isoniazid *(Isoniazide)*

Isonicotinic
Isoorientin
Isoosmotic *see* Osmosis
Isoparaffinic
Isopentenyladenine
Isopentenyloxy
Isopentyl *(Isopentenyl)*
Isopeptide
Isoperoxidases
(Isoperoxidase)
Isophya
Isopleths
Isoplexis
Isopoda *(Isopod, Isopodal,*
Isopode, Isopods)
Isoprenaline
Isoprene
Isoprenoid *(Isoprenoids)*
Isoprenyl
Isoprenylation
(Isoprenylated)
Isopropalin
Isopropanol
Isopropenyl
Isopropoxycarbonyl
Isopropyl
Isopropylamine
(Isopropylamino)
Isopropylidene
Isopropylmalate
Isopropylmethanesulfonate
Isoprotein *see* Proteins
Isoproterenol
Isoptera
Isopyrin
Isoquinoline *(Isoquinolines)*
Isorhamnetin
Isorhynchophylline *see*
  Rhynchophylline
Isosamarcandin
Isosmotic *see* Osmosis
Isospora
Isosteres *(Isostere, Isosteric)*
Isotachophoresis
(Isotachophoretic)
Isothea
Isotherapy
Isotherm *(Isothermal,*
Isothermic, Isotherms)
  X-Ref: Nonisothermal
Isothiazolone
Isothiocyanates
(Isothiocyanate)
Isothiocyanic
Isotima
Isotoma
Isotomidae
Isotomurus
Isotopes *(Isotope, Isotopic,*
Isotopically)
Isotypes *(Isotype)*
Isourea
Isovalerate
Isovitexin
Isoxanthopterin
Isoxathion

Isoxazoles *(Isoxazole)*
Isoxazolyl
Isozymes *see* Isoenzymes
Issidae
Itaconic
Itame
Itersonilia
Ithomiidae *(Ithomiinae,*
Ithomiine)
Itonididae *see*
Cecidomyiidae
Itoplectis
Iuloidea
Iva
Ivy
Ixodes
Ixodicides *(Ixodicide)*
Ixodidae *(Ixodid, Ixodiidae)*
Ixodides
Ixodoidea
Ixora
Ixoreae
Iziphya
Jaagziekte *(Jaagsiekte)*
Jacaranda
Jackdaws
Jackets *(Jacket)*
Jackrabbits *(Jackrabbit)*
Jacksonia
Jacobinia
Jaggery
Jaisalmeri
Jalan *see* Molinate
Jaliscodesmus
Jalmenus
Jams *(Jam)*
Jania
Janus
Jaquinia
Jasminum *(Jasmine,*
Jasmines)
Jassidae *(Jassid,*
Jassidomorphic, Jassids)
Jasus
Jatamansi
Jatropha
Jaundice
X-Ref: Icterus
Javesella
Jaw
Jejunoileal
Jejunum *(Jejunal)*
Jelly *(Jellied, Jellies)*
Jellyfish
Jepsonia
Joannesia
Jobs *see* Employment
Johnsongrass
Joinery
Joints *(Joint)*
Jointworms *(Jointworm)*
Joinvetch
Joinvillea
Joists
Jojoba
Jonespeltis

Joosia
Journalism *(Journalist)*
Jovetia
Jovibarba
Jowar *see* Durra
Jowl
Jugalpada
Juglandaceae
(Juglandaceous)
Juglans
Juglone
Jugular
Juices *(Juice, Juicy)*
Jujube
Julbernardia
Julidae
Juloidea
Julostylis
Julus
Jumping
Juncaceae
Juncales
Juncus
Juneberries *(Juneberry)*
X-Ref: Serviceberry
Jungermannia
Jungermannioideae
Jungle
Juniperus *(Juniper,*
Junipers)
Jurassic
Jurinea
Jurinella
Jurisprudence *see* Laws
Jussiaea
Justicia
Jute
Juvabione
Juvenile *(Juveniles,*
Juvenility, Juvenilizing)
Juvenoids *(Juvenoid,*
Juvenoidal)
Juxtanuclear *see* Nucleus
Kabatiella
Kabatina
Kabicidin
Kadang *see* Cadang
Kaempferol
Kafir *(Kaffir)*
Kairomones *(Kairomone)*
Kakothrips
Kalanchoe
Kale *(Collards, Kales)*
X-Ref: Collard
Kallikreins *(Kallikrein)*
X-Ref: Prekallikrein
Kallstroemia
Kallymenia
Kallymeniaceae
Kalmia
Kalopanax
Kalotermes
Kalotermitidae
Kamaboko
Kamatia
Kanamycin *(Kanamycine)*

Kanechlor
Kangaroos *(Kangaroo)*
Kankrej
Kaolinite *(Kaolinites,*
Kaolinitic)
Kaolins *(Kaolin)*
Kapok
Karakul
Karatavigenin
Karavayevo *see* Kostroma
Karbutilate
Karlingia
Karlingiomyces
Karmex *see* Diuron
Karumiidae
Karyogamy
Karyogram *see* Idiogram
Karyokinesis *see* Mitosis
Karyology *(Karyologic,*
Karyological)
Karyometry *(Karyometric)*
X-Ref: Antiketogenic
Karyomorphological
Karyosystematics
(Karyosystematic)
Karyotypes *(Karyotype,*
Karyotypic)
Kaskaval
Kasugamycin
Katschkawalj
Katydids *(Katydid)*
Kauranes *(Kaurane)*
Kaurene
Kaurenoic
Kauri
Kayea
Kefir
Keiferia
Keithia
Kellerin
Kelp
Kelthane *see* Dicofol
Kenaf
X-Ref: Mesta
Kennedia
Kennels *(Kennel)*
Kepone *see* Chlordecone
Keramzit
Keratan
Keratectomy
Keratinization
(Keratinisation,
Keratinized, Keratinizing)
Keratinolytic
Keratinomyces
Keratinophilic
Keratins *(Keratein,*
Kerateine, Keratin,
Keratinic, Keratinous)
Keratitis
Keratoacanthoma
Keratoconjunctivitis
Keratoplasty
Keratose
Kerb *see* Pronamide
Kerigomnia
Kermes

Kernels *(Kernel)*
Kernicterus
Kerogens *(Kerogen)*
Kerosene *(Kerosine)*
Kerria
Kerrichiella
Kerry
Kestrels *(Kestrel)*
Ketalar
Ketals
Ketamine *(Ketamin)*
Ketchup
X-Ref: Catsup
Ketene
Keto
Ketoacidosis
Ketoaciduria
Ketoacyl
Ketocholesterol
Ketogenesis *(Ketogenic)*
X-Ref: Antiketogenic
Ketoglucose
Ketoglucoside
Ketoglutarate
Ketoglutaric
Ketoheptanoic
Ketols *(Ketol)*
Ketonemia
Ketones *(Ketone, Ketonic)*
Ketonuria
X-Ref: Acetonuria
Ketoreductase
Ketose
Ketosis *(Ketoses, Ketotic)*
Ketosteroids *(Ketosteroid,*
Ketosteroidal)
Kettles *(Kettle)*
Keystone
Khapra
Khaya
Kheper
Khoa
Kibbutz
Kickxellales
Kidneys *(Kidney)*
X-Ref: Extrarenal,
Interrenal, Intrarenal,
Renal, Suprarenal
Kids *(Kid)*
Kieselguhr *see*
Diatomaceous
Kieserite
Kievitone
Kilifia
Kiln *(Kilned, Kilning,*
Kilns)
Kinases *(Kinase)*
Kindergarten
Kinematics *(Kinematic)*
Kinesis
Kinetics *(Holokinetic,*
Kinetic)
Kinetins *(Kinetin, Kinetine)*
Kinetochores *(Kinetochore)*
Kinetoplast
Kininases *(Kininase)*

Kininogenase
Kininogens *(Kininogen)*
Kinins *(Kinin)*
  X-Ref: Phytokinins
Kisanthobia
Kitchens *(Kitchen,*
  Kitchenware)
Kittens *see* Cats
Kiwi
Klebsiella
Kloeckera
Kluyveromyces
Knapweed
  X-Ref: Knobweed
Knautia
Kneading *(Kneaded)*
Knees *(Knee)*
Knema
Knemidokoptes
Knightia
Knit *(Knits, Knitted,*
  Knitting)
• Knives *(Knife)*
Knobs *(Knob)*
Knobweed *see* Knapweed
Knots *(Knot, Knotted,*
  Knotter, Knottiness)
Knotweed
Kobresia
Kochia
Koeleria
Koelreuteria
Kohlrabi
Kojic
Kolanut *see* Cola
Konjac
Kopeodin
Kopetdaghin
Kopsia
Kordofan
Korglykon
Korlan
Kostroma
  X-Ref: Karavayevo
Kotoran
Kraft
Krainzia
Krameria
Krasnozem *(Krasnozems)*
Krenite
Krill
Krillium
Krypton
Kudzu
Kumiss
Kuwanaspis
Kwashiorkor
Kyboasca
Kynureninase
Kynurenine
Laban
Labdane
Labdanum
Labedura
Labeling *(Label, Labeled,*
  Labelled, Labelling,

Labels)
Labellum *(Labellar)*
Labia *(Labial, Labium)*
Labiatae *(Labiate)*
Labidocarpidae
Labidognatha
Labidomera
Labidura
Labiduridae
Labidus
Labiidae
Lablab
Labops
Labor *(Labour)*
Laborers *(Laborer,*
  Labourer, Labourers)
Laboulbenia
Laboulbeniales
Laboulbeniomycetes
Laboulbeniopsis
Labrinae
Labrini
Labrum *(Labral)*
Laburnum
Labyrinth
Labyrinthula
Lac
Laccaria
Laccase *(Laccases)*
Laccifer
Lacciferidae
Laccophilus
Lace
Lacerations *(Laceration)*
Lacewing *(Lacewings)*
Lachenalia
Lachesilla
Lachesillidae
Lachnanthes
Lachnanthocarpone
Lachnidae
Lachnocladiaceae
Lachnosterna
Lacinipolia
Lacombe
Lacquer *(Lacquerability,*
  Lacquers)
Lacrimation *(Lachrymal,*
  Lachrymatory, Lacrimal)
Lactalbumins *(Lactalbumin)*
Lactamases *(Lactamase)*
Lactams *(Lactam)*
Lactaria
Lactarius
Lactarorufin
Lactases *(Lactase)*
Lactates *(Lactate)*
Lactation *(Lactated,*
  Lactating, Lactational,
  Lactations, Lactogenesis)
  X-Ref: Nonlactating
Lacteal
Lacteus
Lactic
Lactiferous
Lactim

Lactobacillaceae
Lactobacillus *(Lactobacilli,*
  Lactobacteria,
  Lactobacterium)
Lactodehydrogenase
Lactodynamograms
Lactoferrin
Lactogen *(Lactogenic)*
Lactoglobulins
  (Lactoglobulin)
Lactones *(Lactone,*
  Lactonic)
Lactoperoxidase
Lactose
Lactoserum
Lactuca
Lacuna
Ladybugs
Ladyginia
Ladyginoside
  (Ladyginosides)
Ladyslippers
Laelapidae *(Laelapid,*
  Laelapinae, Laelapine)
Laelaps
Laelaptidae
Laelia *(Laelias)*
Laeliocattleya
Laemobothriidae
Laemobothrion
Laemophloeus
Laetia
Laetinaevia
Lafoensia
Lagarosiphon
Lagenaria
Lagenidiaceae
Lagenidium
Lagenisma
Lagerstroemia
Lagochilin
Lagochilus
Lagomorpha *(Lagomorphs)*
Lagoons *(Lagoon, Lagoonal,*
  Lagooning, Lagune)
Lagriidae
Laguncularia
Lama
Lambdina
Lambing *(Lambed,*
  Lambings)
Lambs *(Lamb)*
Lambsquarters
  (Lambsquarter)
Lamella *(Lamellae,*
  Lamellar)
  X-Ref: Interlamellar,
    Multilamellar,
    Prolamellar
Lamellibranchia
Lamellicornia
Lameness *(Lame,*
  Lamenesses)
Lamiaceae
Lamiastrum
Lamidae

Lamiidae
Lamiinae
Lamina *(Laminae,*
  Laminar)
  X-Ref: Interlaminar
Laminaria
Laminariales
Laminarinases
Lamination *(Laminate,*
  Laminated, Laminates,
  Laminating)
Laminectomy
Laminitis
Lamium
Lampreys *(Lamprey)*
Lamprochernes
Lamproderma
Lampronia
Lampros
Lamproscatella
Lamps *(Lamp)*
Lampteromyces
Lampyridae
Lampyris
Land *(Landless, Lands)*
Landed *see* Landowners
Landfill *(Landfills)*
Landforms
Landholding *see*
  Landowners
Landlords *(Landlord,*
  Landlordism)
Landowners *(Landholders,*
  Landowner,
  Landownership)
  X-Ref: Landed,
    Landholding
Landrace
Landrin
Landscaping *(Landscape,*
  Landscaped, Landscapes)
Landslide *(Landslides,*
  Landslip, Landslips)
Landuse
Langermannia
Languriidae
Lankesteria
Lanosterol
Lantana
Lantanilic
Lantaninilic
Lanthanides *(Lanthanide)*
Lanthanum
Lanthionine
Lanzia
Laodelphax
Lapageria
Laparotomy
Laphria
Laphygma
Lapidaria
Lapinized
Lappa
Lappula
Larches *see* Larix
Lard *(Lards)*

Lardoglyphus
Larentiinae
Largidae
Largus
Lariophagus
Larix *(Larch)*
  X-Ref: Larches
Larkspurs *(Larkspur)*
Larrea
Larridae
Larrinae
Larrini
Larvae *(Larva, Larval)*
Larvaevoridae *see*
  Tachinidae
Larvicides *(Larvicidal,*
  Larvicide, Larviciding)
Larviposition
Larvivorous
Laryngitis
Laryngoplasty
Laryngospasm
Laryngotracheitis
  (Laringotracheitis)
Larynx *(Laryngeal)*
Laserpitium
Lasers *(Laser)*
Lasiagrostis
Lasiobolus
Lasiocampidae
Lasiocarpine
Lasiocephala
Lasiocephalin
Lasiochalcidia
Lasioderma
Lasiodiploidia
Lasiodora
Lasioglossum
Lasiopogon
Lasiopsylla
Lasioptera
Lasiorhynchus
Lasiosiphon
Lasiosphaeria
Lasiotrechus
Lasius
Lasix *see* Furosemide
Laspeyresia
Laspeyresiini
Lasso *see* Alachlor
Lasthenia
Lastrea
Latency *(Latent)*
Laterites *(Laterite,*
  Lateritic, Lateritized,
  Laterization)
Latex *(Latexes)*
Lathe
Latheticus
Lathraea
Lathridiidae
Lathrolestes
Lathyrus *(Lathyritic,*
  Lathyrogenic,
  Lathyrogens)
  X-Ref: Osteolathyrismus

Lathyrus
Laticifers
Latifundia
Latitude *(Latitudes,*
  Latitudinal)
Latoia
Latosols *(Latosol, Latosolic)*
Latrodectus
Lattice *(Latticed)*
Lauan
Launaea
Laundry *(Laundered,*
  Laundering)
Lauraceae
Laureatin
Laurels *(Laurel)*
Laurencia
Laurenobiolide
Lauric *(Lauryl)*
Laurocerasus
Lauroyl
Laurus
Lauterborniella
Lauxania
Lauxaniidae
Lava
Lavandula
Lavender
Laver
Lavinia
Lawn *(Lawngrass,*
  Lawngrasses, Lawns)
Laws *(Law, Lawful,*
  Lawyer, Legally,
  Ordinance, Statute,
  Statutory)
  X-Ref: Jurisprudence,
  Legal, Ordinances,
  Statutes
Lawsonia
Layering *(Layered)*
  X-Ref: Multilayered
Layers *(Layer)*
Laying *(Lay)*
Layouts *(Layout)*
Leaching *(Leach,*
  Leachability, Leachable,
  Leachate, Leachates,
  Leached, Leachings)
Lead *(Leaden)*
Leafblotch
Leafcurl
Leafcutters *(Leafcutter,*
  Leafcutting)
Leafhoppers *(Leafhopper)*
Leafiness *(Leafless)*
Leafminers *(Leafminer)*
Leafroll
Leafrollers *(Leafroller)*
Leafspot
Leafworms *(Leafworm)*
Leakage *(Leaks, Leaky)*
Lean *(Leaner, Leanness)*
Leasing *(Lease, Leaseback,*
  Leased, Leasehold,
  Leasement, Leases)

Leather *(Leathers)*
Leatherjackets
Leatherleaf
Leaven *(Leavened,*
  Leavening)
Leavenworthia
Leaves *(Leaf, Leafing,*
  Leaflets)
Lebaycid *see* Fenthion
Lebeckia
Lebedin
Lebetanthus
Lecanium
Lecanora
Lecanoraceae
Lecanoric
Leccinum
Lecidea
Lecithin *(Lecithins)*
Lectins *(Lectin)*
Lectotypes *(Lectotype,*
  Lectotypification,
  Lectotyping)
Lecythidaceae
Ledermuelleriopsis
Leeaceae
Leeches *(Leech)*
Leeks *(Leek)*
Leeuwenhoekia
Leeuwenhoekiidae
Leftovers *(Leftover)*
Legal *see* Laws
Legendrella
Leghemoglobins
  (Leghaemoglobin,
  Leghaemoglobins,
  Leghemoglobin)
Leghorn *(Leghorns)*
Legislation *(Legislative,*
  Legislator, Legislature,
  Legislatures)
Legs *(Leg)*
  X-Ref: Foreleg, Hindlegs
Legumes *(Legume,*
  Leguminaceous,
  Leguminous)
Legumin
Leguminosae
Leguminous
  X-Ref: Nonleguminous
Leidynema
Leiobunidae
Leiobunum
Leiodidae *(Liodidae)*
Leiodinychus
Leiomyoma
Leiomyosarcoma
Leiopelta
Leiophron
Leioproctus
Leiostomus
Leiothecium
Leishmania *(Leishmanias)*
Leishmaniasis
Leistus
Leitneria

Leiurus
Lejeunea
Lejeuneaceae
  (Lejeuneoideae)
Lelya
Lema
Lemaireocereus
Lemanea
Lembeja
Lemmaphyllum
Lemmings *(Lemming)*
Lemna
Lemnaceae
Lemonades *(Lemonade)*
Lemongrass
Lemonia
Lemons *(Lemon)*
Lemphus
Lemurs
Lenacil *(Lenacyl)*
  X-Ref: Venzar
Lending *(Lend, Lender,*
  Lenders, Lends, Loan,
  Loaned, Loaning)
  X-Ref: Loans
Lenses *(Lens)*
Lentibulariaceae
Lenticels *(Lenticel,*
  Lenticellar)
Lentils *(Lentil)*
Lentinic
Lentinus
Lentogenic
Lenzites
Leonotis
Leontice
Leontodon
Leonurus
Leopoldia
Leotia
Leperisinus
Lepidiota
Lepidium
Lepidochora
Lepidochrysops
Lepidocyrtus
Lepidodermella
Lepidoptera *(Lepidopteran,*
  Lepidopterans,
  Lepidopterous)
Lepidopterists
  (Lepidopterist)
Lepidosaphes
Lepidospartum
Lepinotus
Lepiota
Lepisma
Lepismachilis
Lepismatidae
Lepismodes
Lepisorus
Lepista
Leporidae
Lepra
Lepraria
Leprosy

Leptacis
Leptadenia
Lepthyphantes
Leptidea
Leptinotarsa
Leptinus
Leptobatopsis
Leptobryum
Leptocera
Leptoceridae
Leptochilus
Leptochloa
Leptoconops
Leptocoris
Leptocorisa
Leptocycas
Leptodemus
Leptoderris
Leptodes
Leptodiridae
Leptodon
Leptodontiella
Leptodontium
Leptogamasus
Leptogastridae
Leptogenys
Leptogium
Leptoglossus
Leptohylemyia
Leptolaena
Leptolejeunea
Leptomastix
Leptomeningitis
    (Leptomeningeal)
Leptomonas
Leptomyrina
Leptonia
Leptopharsa
Leptophelbia
Leptophlebiidae
Leptophos
Leptoporus
Leptopsylla
Leptopsyllidae
Leptopteromyia
Leptorhabdos
Leptorhabine
Leptorrhabdium
Leptosiropsis
Leptospermum
Leptosphaeria
Leptosphaerulina
Leptospira *(Leptospirae,*
    Leptospiral, Leptospires)
Leptospirosis
    (Antileptospiral,
    Leptospiroses)
    X-Ref: Antileptospira
Leptostroma
Leptotarsus
Leptothorax
Leptotrichus
Leptotrombidium
Leptura
Lepus
Lerps

Lesions *(Lesion)*
Lespedeza
Lespesia
Lesquerella
Lessonia
Lessoniopsis
Lestes
Lestidae
Lestodiplosus
Lethaeini
Lethal *(Lethality, Lethals,*
    Semilethals, Sublethally)
    X-Ref: Nonlethal,
    Semilethal, Sublethal,
    Supralethal
Lethocerus
Lethrus
Lethus
Lettuce *(Lettuces)*
Leucadendron
Leucaena
Leucania
Leucanthemum
Leucheria
Leucinodes
Leuciscus
Leucoanthocyanidins
    (Leucoanthocyanidin)
Leucoanthocyanins
    (Leucoanthocyanin)
Leucochloridium
Leucocrinum
Leucocytes *see* Leukocytes
Leucocytozoon
Leucocytozoonosis
Leucojum
Leucoma
Leuconostoc
Leuconotis
Leucopenia
    (Panleucopaenia,
    Panleucopenia)
    X-Ref: Panleukopenia
Leucophaea
Leucopholis
Leucoplasts *(Leucoplast)*
Leucoplema
Leucopogon
Leucoptera
Leucosis *see* Leukosis
Leucospermum
Leucostoma
Leucothoe
Leucothrix
Leuctra
Leucyl
Leucylglycyl
Leukemia *(Leukaemia,*
    Leukaemic,
    Leukaemogenesis,
    Leukemias, Leukemic)
    X-Ref: Antileukemic
Leukocidin
Leukocytes *(Leucocyte,*
    Leucocytic, Leukocyte,
    Leukocytic)

X-Ref: Leucocytes
Leukodystrophy
    (Leukodystrophic)
Leukosis *(Leucoses,*
    Leucotic, Leukoses,
    Leukotic)
    X-Ref: Leucosis
Levamisole *see* Tetramisole
Levee *(Levees)*
Leveillula
Levisticum
Levoglucosan
Levoglucosenone
Levopimaric
Levulinic
Levulose *see* Fructose
Lewisia
Liagoropsis
Liana *(Lianas)*
Liatris
Libanotis
Libellula
Libellulidae
Libertia
Libido
Libocedrus
Libraries *(Library)*
Libytheana
Libytheidae
Licania
Licaria
Lice
    X-Ref: Louse
Licea
Licensing *(License,*
    Licensed, Licenser)
Lichenin
Lichenization *(Lichenized)*
Lichenologists
    (Lichenologist)
Lichens *(Lichen, Lichenes,*
    Lichenic, Lichenologic,
    Lichenological,
    Lichenous)
    X-Ref: Macrolichen
Lichina
Licodione
Licorice
Lidbeckia
Lidophia
Lifetime *(Lifespan,*
    Lifetimes)
Lifters *(Lifted, Lifter,*
    Lifting)
Ligaments *(Ligament,*
    Ligamentous,
    Ligamentum)
Ligands *(Ligand)*
Ligases *(Ligase)*
    X-Ref: Carboligase
Ligation *(Ligated,*
    Ligations, Ligature,
    Ligatured, Ligatures,
    Ligaturing)
Light *(Lights)*
Lightning

Ligia
Ligidium
Lignan *(Lignans)*
    X-Ref: Neolignans
Lignicolous
Lignins *(Ligneous,*
    Lignification, Lignified,
    Lignifying, Lignin)
Lignites *(Lignite)*
Lignituber
Lignocarbohydrate
Lignocellulose
Lignosulfonates
    (Lignosulfonate,
    Lignosulphonate,
    Lignosulphonates)
Lignosulfonic
    (Lignosulphonic)
Ligula
Ligularia
Ligules *(Ligule, Liguleless)*
Ligustrum
Likubin
Lilac *(Lilacs)*
Liliaceae *(Liliaceous)*
Liliales
Lilies *(Lillies, Lily)*
Liliifloris
Lilium *(Liliums)*
Lima *(Limas)*
Limabeans *(Limabean)*
Limacodidae
Liman
Limax
Limbing
Limbs *(Forelimb, Limb)*
    X-Ref: Antibrachium,
    Forearm, Forelimbs,
    Hindlimb
Lime *(Limed, Limes,*
    Liming)
    X-Ref: Unlimed
Limenitis
Limestone *(Limestones)*
Limnanthaceae
Limnanthes
Limnephilidae *(Limnephilid,*
    Limnephilids)
Limnephilus
Limnesia
Limnesiidae
Limnia
Limnichidae
Limnobium
Limnology *(Limnological)*
Limnophila
Limnophyes
Limnoria
Limnoxenus
Limobius
Limonene
Limonia
Limoniidae
Limonin
Limonium
Limonius

Limonoate
Limonoids *(Limonoid)*
Limothrips
Limousin *(Limousine)*
Limpets *(Limpet)*
Linaceae
Linalool *(Linalol)*
Linalyl
Linamarase
Linaria
Linaridial
Linarius
Linaroside
Lincocin
Lincomycin
Lindane *see*
  Hexachlorocyclohexane
Lindelofia
Lindenius
Lindens *see* Tilia
Lindera
Linderina
Lindernia
Lindingaspis
Lindleyin
Lineage
Linear *(Linearity, Linearly)*
Linebreeding *(Linebred)*
Linecrosses *(Linecross)*
Linen *(Linens)*
Linerboards *(Linerboard)*
Liners *(Liner)*
Lingual
Linguatula
Linings *(Lining)*
Linkage *(Linkages)*
Linnaemyia
Linognathidae
Linognathus
Linoleic *(Linoleate)*
Linolenic *(Linolenate)*
Linospora
Linseed *(Linseeds)*
Linshcosteus
Lint *(Lintered, Linters,*
  Lints)
Linum
Linuron
  X-Ref: Afalon
Linyphiidae *(Linyphiid)*
Lioadalia
Liochthonius
Liorchis
Liothrips
Lipaphis
Liparis
Lipases *(Lipase)*
Lipemia *(Lipaemia)*
Lipeurus
Liphistius
Lipidosis
Lipids *(Lipid, Lipidic,*
  Lipoid, Lipoidic, Lipoids)
  X-Ref: Alkoxylipides,
  Intralipid
Lipitsa

Lipoamide
Lipoate
Lipocarpha
Lipofuscin *(Lipofuscinosis)*
Lipogenesis *(Lipogenetic,*
  Lipogenic)
Lipoic
Lipolysis *(Lipolytic)*
Lipoma
Lipomyces
Liponeura
Liponucleotides
Lipoperoxides
Lipophages
Lipophilic *(Lipophil)*
Lipopolypeptides
Lipopolysaccharides
  (Lipopolysaccharide)
Lipoproteins
  (Apolipoprotein,
  Lipoprotein)
  X-Ref: Apolipoproteins
Lipoptena
Lipoquinones
Liposarcoma
Liposcelidae
Liposcelis
Liposomes
Lipostatic
Lipotropin
Lipotropy *(Lipotropes,*
  Lipotropic)
Lipovitellins
Lipoxidases *(Lipoxidase)*
Lipoxygenases
  (Lipoxygenase)
Lippia
Lips *(Lip)*
Liqcoumarin
Liquefaction *(Liquefied)*
Liqueurs *(Liqueur)*
Liquidambar
Liquidation *(Liquidating)*
Liquorice
Liquors
Lirinidine
Liriodendron
Liriodenine
Liriomyza
Liriope
Lirioresinol
Liris
Lirula
Lisianthius
Lissamine
Lissencephaly
Lissonota
Lissorhoptrus
Listera
Listeria *(Listeric)*
Listeriosis *(Listerellosis)*
Listroderes
Listronotus
Listrophoridae
Listrophorus
Litchi

X-Ref: Lychee
Lithiasis
  X-Ref: Calculosis
Lithium
Lithobiidae
Lithobiomorpha
Lithobius
Lithocarpus
Lithocharis
Lithocolletis
Lithology *(Lithologic)*
Lithophane
Lithops
Lithosia
Lithospermic
Lithospermum
Lithothamnion
Lithothamnium
Litoblatta
Litomastix
Litomosoides
Litotetothrips
Litsea
Litterfall
Litters *(Litter, Litterless,*
  Littermates)
Littonia
Littoral
  X-Ref: Sublittoral,
  Supralittoral
Littorine
Liver *(Livers)*
Liverwort *see* Hepaticae
Livestock
Liveweights *(Liveweight)*
Livistona
Lixophaga
Lizards *(Lizard)*
Llamas *(Llama)*
Loaders *(Loader)*
Loading *(Loaded)*
  X-Ref: Unloading
Loams *(Loam, Loamed,*
  Loamless, Loamy)
Loans *see* Lending
Loasaceae
Loaves *(Loaf)*
Lobaria
Lobaspis
Lobbying *(Lobbies)*
Lobectomy
Lobeliaceae
Lobelias *(Lobelia)*
Lobeline
Lobelioidae
Lobella
Lobesia
Lobiopa
Lobivia
Loblolly
Lobodirina
Lobomonas
Loboplusia
Lobospira
Lobsters *(Lobster)*
Locomotion *(Locomotive,*

  Locomotor)
Locoweed
Loculus *(Loculi)*
  X-Ref: Interlocular
Locusta
Locustacurus
Locustana
Locustol
Locusts *(Locust)*
Lodgepole
Lodging *(Lodged)*
Lodicules
Lodoicea
Lodophor
Loess *(Loesses, Loessial,*
  Loessic, Loessive, Loessy)
Loganberries *(Loganberry)*
Loganiaceae
Loganic
Loganin
Loggers *(Logger)*
Logging *(Felled, Fellings,*
  Logged)
  X-Ref: Felling
Logs *(Log)*
Loin *(Loins)*
Lolium
Lomandra
Lomatium
Lomopneus
Lonatura
Lonchaea
Lonchaeidae
Lonchocarpus
Lonchoptera
Lonchopteridae
Longevity
Longicorn
Longidoridae
Longidorus
Longifolene
Longipeditermes
Longistyly
Longitarsus
Longitude *(Longitudinal)*
Longiunguis
Lonicera
Loofah *see* Luffa
Looms *(Handloom, Loom)*
  X-Ref: Handlooms
Loopers *(Looper)*
Lopezieae
Lophiostomataceae
Lophocateres
Lophocereus
Lophocerine
Lophochloa
Lophocolea
Lophodermella
Lophodermium
Lophodesmus
Lophophora
Lophopteryx
Lophopyxidaceae
Lophopyxis
Lophosoria

Lophozia
Lophoziaceae
Lophuromys
Lophyroplectus
Lopidea
Lopping
Loquats *(Loquat)*
Loranthaceae
(Loranthaceous)
Loranthus
Lordello
Loroglossine
Lorry *see* Trucks
Lorryia
Lorsban *see* Chlorpyrifos
Lotoideae
Lotononis
Lotus
Louping
Louse *see* Lice
Lovage
Lovebug
Lowlands *(Lowland)*
Loxagrotis
Loxodellic
Loxodonta
Loxosceles
Loxostege
Loxsoma
Lubimin
Lubricants *(Lubricant)*
Lubrication
Lucanidae *(Lucanid)*
Lucassenia
Lucernes *see* Alfalfa
Lucienola
Luciferases *(Luciferase)*
Luciferin
Luciferyl
Lucilia
Lucilin
Luciola
Ludwigia
Luffa
    X-Ref: Loofah
Lulworthia
Lumbar
Lumber *(Lumbering.*
    Lumbers)
Lumbermen *(Lumberman)*
Lumbosacral
Lumbricidae
Lumbricus
Lumichrome
Luminescence
    (Luminescent. Luminous)
Lumisterol
Lunar
    X-Ref: Moon, Moonlight
Lunaria
Lunarine
Lunatipula
Lunch *(Luncheon. Lunches.*
    Lunchroom)
Lungs *(Lung)*
Lungworms *(Lungworm)*

Lunularia
Lunularic
Lupanine *(Lupanin)*
Luperus
Lupine *(Lupin, Lupines.*
    Lupins)
Lupinine
Lupinosis
Lupinus
Lupus
Lures *(Lure)*
Luster
Lutein
Luteinization *(Luteinizing)*
Luteolin
Luteolysis *(Luteolytic)*
Luteoskyrin
Luteotrophy *(Luteotrophic.*
    Luteotropic)
Luteotropin
Lutoids
Lutzomyia
Luxations *(Luxation.*
    Subluxated, Subluxation)
    X-Ref: Subluxations
Luzula
Lyases *(Lyase)*
Lycaeides
Lycaena
Lycaenidae *(Lycaenid)*
Lycaste *(Lycastes)*
Lychee *see* Litchi
Lychnis
Lycia
Lycidae
Lycium
Lycogala
Lycomarasmin
Lycopene *(Lycopenic)*
Lycoperdon
Lycopersicon
    (Lycopersicum)
Lycopodiaceae *(Lycopod)*
Lycopodiophyta
Lycopodium
Lycoptis
Lycoriella
Lycorine
Lycoris
Lycosa
Lycosidae *(Lycosid)*
Lyctidae
Lyctus
Lyda
Lye *(Lyes)*
Lygaeidae *(Lygaeid)*
Lygaeus
Lygistorrhina
Lygocoris
Lygodium
Lygris
Lygromma
Lygus
Lymantria
Lymantriidae
Lymexylonidae

Lymnaea
Lymnaeidae
Lymph *(Lymphatic,*
    Lymphatics, Lympho,
    Lymphology)
Lymphadenitis
Lymphadenosis
Lymphangitis
Lymphoblastoid
Lymphoblastoma
Lymphoblasts
    (Lymphoblast,
    Lymphoblastic)
Lymphocytes
    (Antilymphocytic,
    Lymphocyte,
    Lymphocytic)
    X-Ref: Antilymphocyte
Lymphocytosis
Lymphogranuloma
Lymphoid
    X-Ref: Nonlymphoid
Lympholeucosis
Lymphoma *(Lymphomas)*
Lymphomatosis
Lymphoproliferative
Lymphosarcoma
    (Lymphosarcomas)
Lymphosarcomatosis
Lyngbya
Lynx
Lyofolic
Lyonetiidae
Lyonia
Lyophilization
    (Lyophilisation.
    Lyophilised, Lyophilized.
    Lyophilizing)
Lyophyllum
Lyphia
Lypolytic
Lysandra
Lysergic
Lysiloma
Lysimachia
Lysimetry *(Lysimeter,*
    Lysimeters, Lysimetric)
Lysine *(Lysin)*
Lysinoalanine
Lysiphlebus
Lysis *(Lysates, Lyse, Lysing,*
    Lytic)
Lysogeny *(Lysogenic)*
Lysolecithin
Lysophospholipases
    (Lysophospholipase)
Lysosomes *(Lysosomal.*
    Lysosome)
Lysozymes *(Lysozyme)*
    X-Ref: Muramidase
Lysyl
Lythraceae
Lythrum
Lytta
Lyxose
Macaca

Macadamia *(Macadamias)*
Macaranga
Macaroni
Macchia
Macdermotti
Mace
Maceration *(Macerate,*
    Macerated, Macerates,
    Macerating)
Macgillivraya
Machaeranthera
Machilidae
Machilinus
Machilis
Machilus
Machinery
Mackerel
Macleaya
Maclura
Macoubea
Macracanthorhynchus
Macrobenthos *see* Benthos
Macrobiology
Macrobiotics *(Macrobiotic)*
Macrocentrus
Macrocera
Macrocheles
Macrochelidae
Macroclimatic
Macroclinidium
Macrocoma
Macroconidia
    (Macroconidial,
    Macroconidium)
Macrocystis
Macrocytosis
Macrodactylini
Macroeconomics
    (Macroeconometric,
    Macroeconomic,
    Macroeconomical,
    Macroeconomy)
Macroelements
    (Macroelement)
Macrofauna *see* Fauna
Macrofossils *see* Fossils
Macrogalea
Macrogametogenesis *see*
    Gametogenesis
Macroglobulinemia
    (Macroglobulinaemia)
Macroglobulins
    (Macroglobulin)
Macroglossa
Macroglossum
Macroinvertebrates
    (Macroinvertebrate)
Macrolepidoptera
Macrolichen *see* Lichens
Macrolide
Macromalocera
Macromolecules
    (Macromolecular.
    Macromolecule)
Macromutants
Macromycetes

(Macromycete)
Macronoctua
Macronutrients
 (Macronutrient)
Macronychus
Macronyssidae
Macronyssus
Macrophages *(Macrophage)*
 X-Ref: Antimacrophage
Macrophoma
Macrophomina
Macrophya
Macrophytes *(Macrophyta,*
 Macrophyte,
 Macrophytical)
Macropiper
Macropoliana
Macroposthonia
Macropsidius
Macroptilium
Macroreticular
Macroscopy *(Macroscopic,*
 Macroscopical)
Macroscytus
Macrosiphoniella
Macrosiphum
Macrosolen
Macrosporium
Macrosposthonia
Macrosteles
Macrotermes
Macrotermitinae
Macrotyloma
Macrourus
Macrovalsaria
Macrovegetation *see*
 Vegetation
Macrozamia
Maculate
Madamyobia
Madecacesta
Madecastalia
Madhuca
Madotrogus
Madrone
Maduromycosis
Maggots *(Maggot)*
Magicicada
Magnaporthe
Magnesemia *(Magnesaemia)*
Magnesite
Magnesium
Magnetic *(Magnet,*
 Magnetically, Magnetised,
 Magnetism, Magnets)
 X-Ref: Paleomagnetic
Magnetochemistry
Magnolia *(Magnolias)*
Magnoliaceae
 (Magnoliaceous)
Magnoliales
Magnolidin
Magonia
Maguey
Mahanarva
Mahogany *(Mahoganies)*

Mahonia
Maianthemum
Maingayic
Maize *(Cornfields, Maizes)*
 X-Ref: Corn, Cornfield,
 Zea
Majorana
Makisterone
Malabsorption
 (Malabsorptive)
Malaceae
 X-Ref: Pomaceae
Malachiidae
Malachius
Malachra
Malacosoma
Malacothrix
Maladera
Malameba
Malamoeba
Malaoxon
Malaraeus
Malaria *(Malarial,*
 Malariogenic,
 Malariology, Malarious)
 X-Ref: Antimalarial
Malate
Malathion
 X-Ref: Carbophos
Malaxodes
Malbranchea
Malcolmia
Male *(Males)*
Maleate
Maleic
Maleopimaric
Malformation
 (Malformations,
 Malformed)
Malformins *(Malformin)*
Malic
Malignancy *(Malignant)*
Mallada
Mallard *(Mallards)*
Mallomonas
Mallophaga
Mallotin
Mallotus
Mallow *see* Malvaceae
Malnutrition
 (Malnourished,
 Malnourishment,
 Underfed,
 Undernourished)
 X-Ref: Underfeeding,
 Undernourishment,
 Undernutrition
Malocclusions
 (Malocclusion)
Malolactic
Malonaldehyde
Malonanilides
Malonate
Malonic
Malonyl
Malopolski *(Malopolska)*

Malouetia
Malpighamoeba
Malpighiaceae
 (Malpighiaceous)
Malpighian
Malpractice
Malsecco
Malt *(Malted, Malthouse,*
 Malting, Malts)
 X-Ref: Unmalted
Maltase
Malthinus
Maltitol
Maltlage
Maltodextrin
Maltol
Maltooligosaccharides
Maltose
Maltosides *(Maltoside)*
Maltotriose
Malus
Malva
Malvaceae *(Malvaceous)*
 X-Ref: Mallow
Malvastrum
Malvaviscus
Malvidin
Mamegakinone
Mamestra
Mammalia *(Mammal,*
 Mammalian, Mammals)
Mammary
 X-Ref: Intramammary,
 Transmammary
Mammea
Mammilla *see* Teats
Mammillaria
 (Mammillarias)
Mammillitis
Man *(Men)*
Mandarins *(Mandarin)*
Mandelic
Mandevilla
Mandibles *(Mandible,*
 Mandibular)
 X-Ref: Submandibular
Mandioca *see* Cassava
Mandola
Manduca
Maneb
Manganate
Manganese *(Mangan,*
 Manganic, Manganous)
Mange
Mangels *see* Beets
Mangers *(Manger)*
Mangifera
Mangiferin
Mangoes *(Mango, Mangos)*
Mangohoppers
 (Mangohopper)
Mangolds *(Mangold)*
Mangosteen
Mangroves *(Mangrove)*
Manihot
Manilkara

Manioc *see* Cassava
Maniola
Mannagettaea
Mannanase
Mannans *(Mannan)*
Mannitol
Mannolipids
Mannoproteins
Mannoses *(Mannose)*
Mannosidase
Mannosidosis
Mannosyl *(Mannosylation)*
Mannosyltransferases
Manometry *(Manometer,*
 Manometric)
Manpower
Mansonella
Manteidae *(Mantid,*
 Mantidae, Mantids,
 Mantoidea)
 X-Ref: Mantodea
Mantis
Mantispa
Mantispidae *(Mantispid,*
 Mantispids)
Mantodea *see* Manteidae
Manual *(Manually)*
 X-Ref: Nonmanual
Manufacture
 (Manufactured,
 Manufacturer,
 Manufacturers,
 Manufactures,
 Manufacturing)
Manure *(Manured,*
 Manures, Manurial,
 Manuring)
Manzanitas
Maples *see* Acer
Mappia
Mapping *(Chart, Map,*
 Mapped, Maps)
 X-Ref: Charts
Maquis
Maracanthus
Marah
Marantaceae
Marasmiellus
Marasmius
Marasmus
Marattia
Marattiaceae
Marattiales
Marbling
Marcgravia
Marcgraviaceae
Marchantia
Marchantiales
Marcipa
Marcs
Marcuzziella
Marek *(Mareks)*
Maremma
Mares *(Mare)*
Maretin
Margaric

Margarine *(Margarines)*
Margarodidae
  (Margarodids)
Margaronia
Margosa
Marigolds *(Marigold)*
Marijuana *(Marihuana)*
Marinating *(Marinated)*
Marine
Marjoram
Marketing *(Marketability,*
  Marketable, Marketed,
  Marketer, Marketers,
  Marketings)
Markets *(Market,*
  Marketplace)
Markings *(Marking)*
Markov *(Markoff,*
  Markovian)
Marl
Marlattiella
Marlin
Marmalade
Marmosets *(Marmoset)*
Marmota
Marmots *(Marmot)*
Maro
Marpesia
Marrow
Marrubium
Marsdenia
Marshes *(Marsh,*
  Marshland, Marshlands,
  Marshy, Saltmarsh,
  Wetland)
  X-Ref: Saltmarshes,
    Wetlands
Marsilea
Marsileaceae
Marssonina
Marsupialia *(Marsupial,*
  Marsupials)
Martarega
Martinapis
Martynia
Martyniaceae
Maruca
Marzipan
Masaridae *(Masarid)*
Mascaroside
Masculinity
  (Masculinization)
Masdevallia *(Masdevallias)*
Mashes *(Mash, Mashed,*
  Mashing)
Masonaphis
Masoreidae
Massage
Massalongia
Massarina
Massecuites *(Massecuite)*
Massoilactone
Massospora
Mast
Mastectomy
Mastic

Mastication
Mastigocladus
Mastitis *(Mastitic)*
Mastocytoma
Mastogloia
Mastomys
Mastotermes
Matacil
Maternal *(Maternalism,*
  Maternally, Maternity)
Matgrass
Mathematics
  (Mathematical,
  Mathematically)
Mathiola *(Matthiola)*
Mating *(Courting, Mate,*
  Mated, Mates, Matings)
  X-Ref: Courtship,
    Postmating, Premating,
    Remating
Matric
Matricaria
Matrices *(Matrix)*
Matsucoccus
Matsumuraeses
Mattesia
Matteuccia
Mattock
Matudea
Maturity *(Maturation,*
  Maturational, Mature,
  Matured, Maturing,
  Maturities, Premature,
  Prematurely)
  X-Ref: Prematurity
Mauritin
Maxillaria
Maxillary *(Maxilla,*
  Maxillae)
  X-Ref: Submaxillary
Mayetiola
Mayflies *(Mayfly)*
Mayonnaise
Maytenus
Mazindol
MBC *see* Carbendazim
MCPA
MCPB
Meadowgrass
Meadows *(Meadow,*
  Meadowlands)
Meals *(Meal, Mealtime)*
Mealworms *(Mealworm)*
Mealy
Mealybugs *(Mealybug)*
Measles
Meat *(Meatiness, Meats,*
  Meaty)
Meatpacking
Mebendazole
Mecadox
  X-Ref: Carbadox
Mecarbam
Mechanics *(Mechanic,*
  Mechanical,
  Mechanically)

Mechanization
  (Mechanisation,
  Mechanised, Mechanising,
  Mechanize, Mechanized,
  Mechanizing)
  X-Ref: Nonmechanized,
    Telemechanization
Mechanoreceptors
  (Mechanoreceptive,
  Mechanoreceptor)
Mechanosensitive
Mecistocirrus
Meclofenamate
Mecodium
Meconella
Meconematinae
Meconium
Meconopsis
Mecopodinae
Mecoptera
  X-Ref: Scorpionflies
Mecostibus
Medetera
Mediastinum *(Mediastinal)*
Medic *(Medics)*
Medicago
Medicarpin
Medicine *(Medical,*
  Medically, Medicaments,
  Medicated, Medication,
  Medications, Medicinal,
  Medicines)
  X-Ref: Premedicated
Medroxyprogesterone
Medulla *(Medullar,*
  Medullary, Medullated)
  X-Ref: Extramedullary,
    Intramedullary
Medulloblastoma
Medusahead
Megabothris
Megaceras
Megachile
Megachilidae
Megacolon
Megacyllene
Megaesophagus *see*
  Esophagus
Megafloras *see* Flora
Megafossil *see* Fossils
Megagametogenesis *see*
  Gametogenesis
Megagametophytes
  (Megagametophyte)
Megakaryocytes
  (Megakaryocyte)
Megaloceroea
Megaloptera
Megalopyge
Megalopygidae
Megamerina
Megaselia
Megasennius
Megasoma
Megaspores *(Megasporal,*
  Megaspore,

Megasporocyte,
  Megasporogenesis)
Megastigmus
Megastylus
Megasyrphus
Megathymidae
Megavitamin *see* Vitamin
Megestrol
Megisthanidae
Megisthanus
Megninia
Megodontus
Megophthalmine
Megoura
Meigenia
Meimuna
Meineckia
Meiocytes *see* Meiosis
Meiosis *(Meiocyte, Meioses,*
  Meiotic)
  X-Ref: Meiocytes,
    Premeiotic
Meiosporangia
Melaleuca
Melamine *(Melamin)*
Melampodium
Melampolide
Melampsora
Melampsoraceae
Melampsoridium
Melampyrum
Melanagromyza
Melanaphis
Melanargia
Melanaspis
Melanconiaceae
Melanconiales
Melandrium
Melandryidae
Melanins *(Melanin)*
Melanism *(Melanic,*
  Melanistic, Melanization,
  Melanized)
Melanitis
Melanocytes *(Melanocyte,*
  Melanocytic)
Melanogenesis
Melanoidins *(Melanoidin)*
Melanolophia
Melanoma
Melanonotus
Melanophila
Melanoplus
Melanorrhoea
Melanose
Melanosis *(Melanotic)*
Melanotus
Melanthera
Melanthesa
Melasoma
Melastomaceae
  (Melastomataceae,
  Melastomataceaes)
Melatonin
Melchiora
Meleageria

Meleagris *see* **Turkey**
Melengestrol
Meleoma
Melezitase
Melezitose
Melia
Meliaceae *(Meliaceous)*
Meliacins
Melianthus
Melibiose
Melica
Meligethes
Melilotus
Melinda
Melinis
Meliodogyne
Melioidosis
Meliola
Meliolaceae
Melioration
  (Agromeliorative,
  Ameliorant, Ameliorating,
  Ameliorations,
  Ameliorative, Meliorated,
  Meliorating, Meliorations,
  Meliorative)
  X-Ref: Agromelioration,
  Amelioration
Melipona
Meliponinae
Meliponini
Melipotis
Melissa
Melissopus
Melitaea
Melittin *(Melittins)*
Melittobia
Melittomma
Melitturga
Melliferous *see* **Honey**
Mellinus
Melobaline
Melocactus
Melochia
Melochinone
Melodinus
Meloidae *(Meloid)*
Meloidogyne
Melolontha
Melolonthidae
Melolonthinae
Melolonthini
Melons *(Melon)*
Melophagus
Melosira
Melothria
Meltwater *see* **Water**
Melyridae
Membracidae *(Membracid,*
  Membracids)
Membracis
Membranes *(Membrana,*
  Membranal, Membrane,
  Membraneous,
  Membranous)
  X-Ref: Intramembranous,

Submembranous,
  Transmembrane
Memnoniella
Memory *see* **Biography**
Menacanthus
Menadione *(Menadion)*
Menazon
  X-Ref: Sayphos
Mendelian
Mendoncia
Menemerus
Mengovirus
Menhaden
Menichlopholan *see* **Bilevon**
Meninges *(Meningeal,*
  Meningo)
Meningioma
Meningitis
Meningoencephalitis
Meningoencephalomyelitis
Meniscus
Menispermaceae
  (Menispermaceous)
Menispermum
Menochilus
Menodora
Menopon
Menoponidae
Mensuration
Mental
Mentha
Menthol
Menthyl
Mentzelia
Menus *(Menu)*
Meotica
Mephitis
Mepyramine *see* **Pyrilamine**
Meraporus
Mercantilism *(Mercantile)*
Mercaptan *(Mercaptans)*
Mercapto
Mercaptoethanol
Mercaptopurine
Mercerization *(Mercerized)*
Merchandise *(Merchandiser,*
  Merchandising)
Merchants *(Merchant)*
Mercurialis
Mercury *(Mercurated,*
  Mercuration, Mercurial,
  Mercurials, Mercuric)
Merendera
Mergers *(Merger, Merging)*
Meria
Meridiocolax
Merino *(Merinos)*
Merismopedia
Meristems *(Meristem,*
  Meristematic, Meristemic)
Meristotropis
Mermis
Mermithidae *(Mermithid,*
  Mermithids)
Merodon
Meroglossa

Meromyosin
Meromyza
Merosporangia
  (Merosporangial)
Merothrips
Merozoites
Merpelan
Merremia
Mertect
Mertilanidea
Merulius
Merxmuellera
Mesaphorura
Mescaline
Mesembrine
Mesembryanthemaceae
Mesembryanthemum
  (Mesembryanthemums)
Mesencephalon
  (Mesencephalic)
  X-Ref: Midbrain
Mesenchyme
  (Mesenchymal)
Mesenchymoma
Mesenteron
Mesentery *(Mesenteric)*
Mesh
Mesic
Mesidia
Mesitius
Mesocarp
Mesocestoides
Mesocotyls *(Mesocotyl)*
Mesocricetus
Mesoderm *(Mesodermal)*
Mesofauna
Mesophiles *(Mesophilic,*
  Mesophilous)
Mesophylax
Mesophyll *(Mesophyllous,*
  Mesophylls)
Mesophytes *(Mesophyte,*
  Mesophytic)
Mesopolobus
Mesosa
Mesosemia
Mesosomes *(Mesosome)*
Mesostenini
Mesostigmata
Mesotaenium
Mesotheliomas
  (Mesothelioma)
Mesothorax *(Mesothoracic)*
Mesotrichia
Mesotrophic
Mesovelia
Mesoveliidae
Mesozoic
Mesquite
Messerschmidia
Messor
Mesta *see* **Kenaf**
Mestome
Mestranol
Mesua
Mesurol

Mesylate
Metabisulfite *see* **Bisulfite**
Metabolism *(Metabolic,*
  Metabolically,
  Metabolisable,
  Metabolising,
  Metabolisms,
  Metabolizability,
  Metabolizable,
  Metabolize, Metabolized,
  Metabolizing)
  X-Ref: Antimetabolic,
  Cometabolism,
  Nonmetabolic
Metabolites *(Antimetabolite,*
  Metabolite, Metabolities)
  X-Ref: Antimetabolites
Metabus
Metacarpal *(Metacarpus)*
Metacarpophalangeal
Metacercaria
  (Metacercariae)
Metacestode *see* **Cestoda**
Metacnephia
Metacysostethum
Metadenopus
Metagagrella
Metagame *see* **Game**
Metagonimus
Metagonistylum
Metaldehyde
Metallibure *see*
  Methallibure
Metallized
Metallocarboxypeptidases
Metallogenium
Metalloporphyrins
  (Metalloporphyrin)
Metalloproteins
  (Metalloproteic)
Metallothionein
Metals *(Metal, Metallic,*
  Metalliferous)
  X-Ref: Nonmetallic
Metamasius
Metameric
Metamorphosis
  (Metamorphic,
  Metamorphosed,
  Metamorphosing)
Metapenaeus
Metaphase
  X-Ref: Prometaphase
Metaphidippus
Metaphosphates *see*
  Phosphates
Metaphycus
Metaplasia
Metarhodopsin *see*
  Rhodopsin
Metarrhizium
Metaseiulus
Metasequoia
Metasilicate
Metastasis *(Metastases,*
  Metastatic)

Metastigmata
Metastrongylidae
Metastrongylus
Metasyrphus
Metatarsus *(Metatarsal)*
Metathion *see* Fenitrothion
Metathorax *(Metathoracic)*
Metation *see* Fenitrothion
Metatropis
Metaxenia
Metaxya
Metazoa *(Metazoal)*
Meteloidine
Meteorologists
  (Meteorologist)
Meteorology *(Meteorologic,*
  Meteorological,
  Meteorologically,
  Micrometeorological)
  X-Ref: Micrometeorology
Metepa
Meters *(Meter, Metering)*
Methacrylates
  (Methacrylate)
Methacrylic
Methallibure *(Metallibur,*
  Methallibur)
  X-Ref: Metallibure
Methallyl
Methane
Methanearsonate
  (Methanearsonates,
  Methanearsonic)
Methanesulfonate
  (Methanesulfonic,
  Methanesulphonate)
Methanethiol
Methanobacterium
Methanol *(Methanolic)*
Methanolysis
Methazole
Methbenzthiazuron
  (Methabenzthiazuron)
  X-Ref: Tribunil
Methdilazine
Methemoglobin
  (Methaemoglobin,
  Methaemoglobins)
Methemoglobinemia
  (Methaemoglobinemia)
Methidathion
Methide
Methiocarb
Methiodide
Methional
Methionine
Methionyl
Methiotepa
Methohexitone
Methomidate
Methomyl
Methoprene
Methotrexate
Methoxsalen *see*
  Xanthotoxin
Methoxyallybenzene

Methoxycamptothecin
Methoxychalkone
Methoxychlor
Methoxychlorine
Methoxychrysazin
Methoxycinnamate
Methoxycoumarin
Methoxyerythraline
Methoxyethoxy
Methoxyethylmercury
Methoxyflurane
Methoxyharmalan
Methoxyharman
Methoxyl
Methoxylation
  (Methoxylated)
Methoxyphenol
Methoxyphenyl
Methoxypyrazines
Methoxyvincadifformine
Methylacetophenone
Methylacridinium
Methylacridone
Methyladrenaline
Methylalkanes
Methylamines
  (Methylamine)
Methylammonium
Methylanthraquinone
Methylarginine
Methylarsonate
Methylarsonic
Methylases *(Methylase)*
Methylation *(Methylated,*
  Methylating)
Methylbayin
Methylbenzene
Methylcarbamate
  (Methylcarbamates)
Methylcarbamoyl
Methylcellulose
  X-Ref: Tylose
Methylcholanthrene
Methylcholine
Methylchromone
Methylcimigenol
Methylcitrate
Methylcurine
Methylcyclohex
Methylcyclohexylamine
Methylcysteine
  (Methylselenocysteine)
Methylcytosine
Methyldecinine
Methylellagic
Methylene
Methylenecholesterol
Methylenedammarenol
Methylenedioxy
Methylenedioxyphenyl
Methyleneoxindole
Methylenetetrahydrofolate
Methylester
Methyleugenol
Methylformamidine
Methylglutaryl

Methylglyoxal
Methylgrandifloroside
Methylguanosine
Methylherbacetin
Methylhistidine
Methylhydroxylamine
Methylindole
Methylisoflavones
Methylisoxazole
Methyljuglone
Methylmalonic
Methylmercapto
Methylmercuric
  (Methylmercury)
Methylmethacrylate
Methylmethionine
Methylnitrophos
Methylnitrosamine
Methylnonadecanes
Methyloctadecanes
Methylol *(Methylols)*
Methylolamide
Methylparathion *see*
  Parathion
Methylphenazinium
Methylphenoxyacetic
Methylpropionamide
Methylpyridinium
Methylquinoline
Methylsalicylic
Methylsterol
Methyltestosterone *see*
  Testosterone
Methyltetrahydrofolate
  (Methyltetrahydrofolates)
Methylthioadenosine
Methylthiobutane
Methylthiohydantoin
Methylthiouracil
Methyltransferases
  (Methyltransferase,
  Transmethylase)
  X-Ref: Transmethylases
Methyluracil *see* Thymine
Methylurea *(Methylureas)*
Methylurethane
  (Methylurethan)
Methylviologen
Methylxanthines
  (Methylxanthine)
Methylxanthomegnin
Methylxanthone
Methyridine
  X-Ref: Mintic, Promintic
Metiamide
Metiazinic
Meticlorpindol
  (Metichlorpindol)
  X-Ref: Clopidol
Metmyoglobin
Metobromuron
Metopiinae
Metopium
Metopolophium
Metorchis
Metoxuron

Metresura
Metribuzin
Metric *(Metrical,*
  Metrication, Metrics,
  Metrification)
Metridia
Metriocampa
Metrioptera
Metritis
Metronidazole
  X-Ref: Flagyl
Metrosideros
Meturin
Metyrapone
Metzgeria
Metzgeriaceae
Metzgeriales
Metzgeriopsis
Mevalonate
Mevalonic *(Mevalonoid)*
Mevinphos
Mexacarbate
  X-Ref: Zectran
Mexichthonius
Miastor
Mica *(Micas)*
Micarea
Mice *(Mouse)*
Micelle *(Micella, Micellar,*
  Micelles)
Michelia
Miconazole
Micracanthia
Micraedes
Micrantheum
Micrasterias
Micrencephaly
Microaerophile
  (Microaerophilic)
Microagglutination *see*
  Agglutination
Microalgae *see* Algae
Microanalyser *see*
  Analyzers
Microanalysis
Microanatomy *see* Anatomy
Microangiography
  (Microangiographic)
Microangiopathy
  (Microangiopathic)
Microappendages *see*
  Appendages
Microapplicator *see*
  Applicators
Microarthropods *see*
  Arthropoda
Microautoradiography *see*
  Autoradiography
Microbembex
Microbes *(Microbe,*
  Microbial, Microbially,
  Microbic)
  X-Ref: Microbiota
Microbicidal *see* Antibiotic
Microbioassay *see* Assay
Microbiologists

(Microbiologist)
Microbiology
(Microbiologic,
Microbiological,
Microbiologically)
Microbiota *see* **Microbes**
Microbracon
Microcaeculus
Microcalculi *see* **Calculi**
Microcalorimetry *see*
Calorimetry
Microcautery *see*
Cauterization
Microcerotermes
Microchemical
Microcirculation *see*
Circulation
Microclava
Microclimate
(Microclimates,
Microclimatic,
Microclimatical,
Microclimatology)
Micrococcaceae
(Micrococcal, Micrococci)
Micrococcus
Microcoenoses *see*
Microecology
Microcoleus
Microcolony *see* **Colonies**
Microconidiating *see*
Conidium
Microcordylomyia
Microcoulometric *see*
Coulometry
Microcreagris
Microcrypticus
Microcrystalline *see*
Crystals
Microcrystals *see* **Crystals**
Microctonus
Microcyclus
Microcysis
Microcystis
Microcysts *(Microcyst)*
Microdiffusion *see*
Diffusion
Microdigestion *see*
Digestion
Microdiptera *see* **Diptera**
Microdispus
Microdissection *see*
Dissection
Microdistillation *see*
Distillation
Microdus
Microecology
(Microecosystem)
X-Ref: Microcoenoses,
Microecosystems
Microeconomics
(Microeconomic,
Microeconomical)
Microecosystems *see*
Microecology
Microelectrodes

(Microelectrode)
Microelectrophoresis
(Microelectrophoretic)
Microelements
(Microelement)
Microencapsulation *see*
Encapsulation
Microenvironment
(Microenvironmental,
Microenvironments)
Microepidemiology *see*
Epidemiology
Microeuzercon
Microevolution
(Microevolutionary)
Microfauna
Microfertilizers
(Microfertilization,
Microfertilizing)
Microfibril *(Microfibrillar,*
Microfibrils)
Microfilament *see*
Filaments
Microfilaria *(Microfilariae,*
Microfilarial)
Microfilaricide
Microflora *(Microfloral,*
Microfloras)
Microfossils *see* **Fossils**
Microfungi *see* **Fungi**
Microgametes *see* **Gametes**
Microgametogenesis *see*
Gametogenesis
Microgametophytes
(Microgametophyte)
Microglobulin
Microgramma
Microgranulated *see*
Granulation
Micrography *(Micrographic,*
Micrographs)
Microhabitats *see* **Habitats**
Microinjection *see* **Injection**
Microlarinus
Microlejeunea
Microlenin
Microlepidoptera
(Microlepidopteran,
Microlepidopterous)
Microlophium
Micromalthus
Micromanipulators
(Micromanipulating,
Micromanipulation,
Micromanipulator)
Micromelia *(Micromelic)*
Micromelum
Micrometeorology *see*
Meteorology
Micrometer *(Micrometry)*
Microminerals *see* **Minerals**
Micromonoliths *see*
Monolith
Micromonospora
Micromorphological *see*
Morphology

Micromorphometric *see*
Morphometric
Micromutations *see*
Mutation
Micromycetes *see* **Fungi**
Micronectriella
Micronization *(Micronized)*
Micronodules *see* **Nodules**
Micronuclear *see* **Nucleus**
Micronutrients *see*
Nutrients
Microorganisms
(Microorganism)
Micropalaeontology *see*
Paleontology
Micropedology *see*
Pedology
Micropeplus
Micropezidae
Microphenologic *see*
Phenology
Microphotography
(Microphotographic,
Microphotographing)
Microphotometry *see*
Photometry
Microphthalmia
Microphysidae
Microplanktons *see*
Plankton
Microplaque *see* **Plaque**
Microplasmosis
Microplax
Microplitis
Microplots *see* **Plots**
Micropollutants *see*
Pollutants
Micropolyspora
Microponds *see* **Pond**
Micropopulations *see*
Population
Micropterus
Micropyles *(Micropylar,*
Micropyle)
Microquedius
Microradiographic *see*
Radiography
Microreactor
Microrelief
Microrespirometry *see*
Respirometry
Microscale
Microsclerotia
Microscopy *(Microscope,*
Microscopes, Microscopic,
Microscopical)
X-Ref: Submicroscopic
Microsites *see* **Site**
Microsome *(Microsomal,*
Microsomes)
X-Ref: Submicrosomal
Microspecies
Microspectrofluorimetric
see **Spectrofluorometry**
Microspectrophotometry *see*
Spectrophotometry

Microsphaera
Microsphaeropsis
Microspongium
Microsporangium
Microspores *(Microspore,*
Microsporocyte,
Microsporocytes,
Microsporogenesis)
Microsporidia
(Microsporida,
Microsporidan,
Microsporidian,
Microsporidiosis)
Microsporon
Microsporum
Microstigmus
Microstroma
Microtendipes
Microtermes
Microterritorial *see*
Territory
Microthamnion
Microtomy *(Microtoming)*
X-Ref: Ultramicrotomy
Microtopography
(Microtopographic,
Microtopographical)
Microtractors *see* **Tractors**
Microtrombicula
Microtropis
Microtubules *see* **Tubules**
Microtus *(Microtine)*
Microvascular *see* **Vascular**
Microvelia
Microvilli *see* **Villi**
Microviscosity *see* **Viscosity**
Microwave *(Microwaves)*
Micrurus
Micryphantidae
Midbrain *see*
Mesencephalon
Middlemen *(Middleman)*
Middlings
Midge *(Midges)*
Midgut
Midparent *see* **Parent**
Midrib *(Midribs)*
Midsummer *see* **Summer**
Mielichhoferia
Miersia
Migrants *(Migrant)*
Migration *(Migrate,*
Migrated, Migrating,
Migrational, Migrations,
Migratory)
X-Ref: Nonmigration
Mikania
Mildbraedeodendron
Mildew *(Mildews)*
Milesiini
Milfoil *(Milfoils)*
Milichiidae
Milk *(Milch, Milkability,*
Milks)
X-Ref: Foremilk
Milkfats *(Milkfat)*

Milkhouse
Milking *(Milked, Milker.*
 Milkers, Milkings)
 X-Ref: Overmilking,
 Postmilking
Milkshed
Milkvetch *(Milkvetches)*
Milkweed
Milkwort *(Milkworts)*
Milky
Millers *(Miller)*
Millet *(Millets)*
Millettia *(Milletia)*
Millfeed *(Millfeeds)*
Milling *(Millability,*
 Millable, Milled)
 X-Ref: Upmilling
Millingtonia
Millipede *(Milliped,*
 Millipedes, Millipeds)
Mills *(Millyard)*
Millsonia
Milo
Milorganite
Miltonia *(Miltonias)*
Miltonidium
Mimastra
Mimetidae
Mimicry *(Mimetic, Mimic.*
 Mimicking, Mimics)
 X-Ref: Biomimetic
Mimomyia
Mimosa
Mimosaceae
Mimosine
Mimosoideae
Mimulus
Mimusops
Mincing *(Minced, Minces)*
Mindaridae
Mindarus
Mineralogy *(Mineralogic.*
 Mineralogical)
Minerals *(Mineral.*
 Mineralisation.
 Mineralised.
 Mineralization.
 Mineralized)
 X-Ref: Microminerals,
 Remineralization
Miners *(Miner)*
Miniature
Minicomputers *see*
 Computers
Minimization *(Minimise.*
 Minimising. Minimize.
 Minimized, Minimizing)
Mining *(Mine, Mined.*
 Mines)
Minipoinsettias *see*
 Poinsettia
Minirotation *see* Rotation
Minks *(Mink)*
Minnow
Minority
Mint *(Mints)*

Minthea
Mintic *see* Methyridine
Minuartia
Minytus
Miocene
Miomantis
Miospores *(Miospore)*
Mirabilis
Miracidia
Mire
Mirex
Mirgorod
Miridae *(Mirid, Mirids)*
Miscanthus
Miscarriages *see* Abortion
Miscella
Miserotoxin
Mish
Miso
Misolampidius
Mistblower *(Mistblowers)*
Mistletoe *(Mistletoes)*
Mists *(Mist, Misting. Misty)*
Mite *(Mites)*
Miticide *(Miticidal,*
 Miticides)
Mitochondria
 (Mitochondrial,
 Mitochondrially,
 Mitochondrion)
 X-Ref:
 Anisomitochondrial,
 Extramitochondrial,
 Intramitochondrial,
 Submitochondrial
Mitogens *(Mitogen,*
 Mitogenetic, Mitogenic)
Mitomycin
Mitopus
Mitoribosomes *see*
 Ribosomes
Mitosis *(Mitoses, Mitotic.*
 Mitotically)
 X-Ref: Antimitotic,
 Caryokinetic,
 Endomitosis,
 Karyokinesis,
 Postmitotic, Premitotic
Mitospores *see* Spores
Mitragyna
Mitragyninae
Mitral
Mitraphylline
Mitrula
Mixing *(Mix, Mixed,*
 Mixer, Mixers, Mixes)
Mixoploidy *see* Ploidy
Miyagawanella
Miyagawanellosis
Mnais
Mniaceae
Mnium
Mocap
Model *(Modeling.*
 Modelling, Models)
Modena

Modiolus
Moduli *(Modulus)*
Moehnia
Moellerodiscus
Moghania
Mogoltadin
Mogoltadone
Mogoltavin
Mogoltavinin
Mogoltin
Mohair
Moisture *(Moist, Moistener,*
 Moistening, Moistures,
 Moisturization,
 Overmoistened)
 X-Ref: Overmoistening
Moko
Molasses
Mold *(Molded, Molding,*
 Molds, Moldy, Moulding,
 Moulds, Mouldy)
 X-Ref: Mould
Moldboard *(Moldboards,*
 Mouldboard)
 X-Ref: Mouldboards
Moles *(Mole)*
Molinate
 X-Ref: Jalan, Ordram,
 Yalan
Molinia
Mollicutes
Mollinedia
Mollisols
Molluginaceae
Mollugo
Molluscicides
 (Molluscicidal,
 Molluscicide)
Mollusks *(Mollusc,*
 Mollusca, Molluscan,
 Molluscs, Mollusk)
Molophilus
Molt *(Molting. Molts.*
 Moulting, Moults)
 X-Ref: Antimoulting,
 Intermolting, Moult
Molucella
Molybdates *(Molybdate)*
Molybdenosis
Molybdenum
 (Molybdenized)
Molybdophosphoric
Momilactone
Momordica
Momphidae
Monacosanone
Monacrosporium
Monallantus
Monarda
Monarthrum
Monascus
Monellia
Monellin
Monelliopsis
Monensin
Money *(Monetary)*

Mongooses *(Mongoose)*
Moniezia
Monieziasis *(Monieziosis)*
Monilia
Moniliaceae
Moniliales
Moniliasis
Moniliella
Moniliformis
Monilinia
Monilochaetes
Monimiaceae .
Monitoring *(Monitor,*
 Monitored, Monitors)
Monkeys *(Monkey)*
Monkshood *see* Aconitum
Monnina
Monoacetate *see* Acetates
Monoacetoxyscirpenol
Monoacyl
Monoamines *(Monoamine,*
 Monoamino)
Monoammonium
Monoarylglycol
Monoblepharidales
Monocalcium
Monocarbonyl
Monocarboxycellulose
Monocarboxylic
Monochaetia
Monochamus
Monochoria
Monochromatic
Monochrysis
Monocontamination *see*
 Contamination
Monocotyledons *(Monocot,*
 Monocotyledon,
 Monocotyledonae,
 Monocotyledoneae,
 Monocotyledonism,
 Monocotyledonous)
Monocotyledous *(Monocots)*
Monocrotophos
Monocrystals *see* Crystals
Monocultures
 (Monocultural,
 Monoculture)
Monocytes *(Monocyte,*
 Monocytic. Monocytoid)
Monodictys
Monodiet *see* Diets
Monodontomerus
Monodora
Monodus
Monoecious *(Monoecism)*
Monoepoxides *see* Epoxides
Monoethanolamine
Monoethylenic
Monoferric *see* Ferric
Monofluoroacetate
Monogalactosyls
 (Monogalactosyl)
Monogastrics *(Monogastric)*
Monogenetic *(Monogenic)*
Monogerm

Monoglucosides
(Monoglucoside)
Monoglycerides
(Monoglyceride)
Monogyny (Monogynous)
Monohaploids see Haploids
Monohelea
Monohybrid see
Hybridization
Monohydrate
Monohydroxybenzoic
Monohydroxydienoic
Monoiodoacetate
Monokaryotic
Monolayer (Monolayers)
Monolepta
Monolinuron (Arezin,
Arezine)
X-Ref: Aresin
Monolith (Monolithic,
Moliliths)
X-Ref: Micromonoliths
Monomer (Monomeric,
Monomers)
Monomorium
Monomycin
Mononucleated see Nucleus
Mononucleotides see
Nucleotides
Monooxygenase see
Oxygenase
Monooxygenated see
Oxygenation
Monophosphates
(Monophosphate)
Monophyllaea
Monoploids (Monoploid,
Monoploidy)
Monopodial
Monopoly (Monopolistic)
Monopsyllus
Monopteryx
Monorosiella
Monosaccharide
(Monosaccharides)
Monosan
Monoseed see Seeds
Monosilicic
Monosodium
Monosome (Monosomes,
Monosomic)
Monospermy
Monosporium
Monostearate
Monostroma
Monosugars see Sugar
Monosulfates see Sulfate
Monosulphide see Sulfide
Monotarsobius
Monoterpenes see Terpenes
Monoterpenoids see
Terpenes
Monothioquinol
Monotropa
Monoxide
Monozygotic see Zygote

Monsoon (Monsoonal)
X-Ref: Premonsoon
Monstera
Monstereae
Montane
Montbretia see Tritonia
Montia
Montmorillonite
(Montmorillonites,
Montmorillonitic)
Monuron
Monvillea
Moon see Lunar
Moonflower
Moonia
Moonlight see Lunar
Moor (Moorland,
Moorlands, Moors)
Moose
Moraba
Moraceae
Moraea
Morain (Morainal, Moraine,
Morainic)
Morantel
Morawitzia
Moraxella
Morchella
Morchellaceae
Mordant (Mordanting)
Mordellidae
Morellia
Morels (Morel)
Morestan
Moretenol
Morgania
Moribund see Death
Moricandia
Moriculture see Morus
Morimus
Morinda
Moringa
Mormodes
Mormolyca
Mormoniella
Morningglory (Morning)
Morphactin (Morphactines,
Morphactins)
Morphinandienone
Morphine
X-Ref: Apomorphine
Morphochemical
Morphogenetic
(Morphogenesis,
Morphogenetics,
Morphogenic)
Morpholine
Morpholinium
Morphology
(Biomorphology,
Micromorphologic,
Micromorphology,
Morpho,
Morphoanatomical,
Morphobiological,
Morphodynamic,

Morphoecological,
Morphogenetical,
Morphologic,
Morphological,
Morphologically,
Morphologies,
Morphologo,
Morphostructure)
X-Ref: Biomorphological,
Micromorphological
Morphometric
(Micromorphometrical,
Morphometrical,
Morphometrics,
Morphometry)
X-Ref:
Micromorphometric
Morphophysiological
Morphoregulators see
Regulators
Morphosis (Morphoses,
Morphotic)
Morphotypes (Morph,
Morphotype, Morphs)
Morrenia
Mortality (Mortal,
Mortalities)
Mortar (Mortars)
Mortem see Death
Mortgage (Mortgages)
Mortierella
Morula
Morus
X-Ref: Moriculture,
Mulberry
Mosaic (Mosaicism,
Mosaics)
Mosquitoes (Mosquito,
Mosquitos)
Mosquitofish
Moss see Musci
Motacilla
Mothers (Mother,
Mothercraft, Mothering,
Motherless)
Mothproofing (Mothproof,
Mothproofer,
Mothproofers)
Moths (Moth)
Motility (Motile)
Motivation (Motivated,
Motivating, Motivations,
Motive, Motives)
Motoneurons
(Motoneuronal,
Motoneurones)
Motors (Motor,
Motorization, Motorized)
Motsugo
Mottle (Mottled, Mottling)
Moufflon
X-Ref: Mufflon
Mougeotia
Mould see Mold
Mouldboards see
Moldboard

Moult see Molt
Mounds (Mound)
Mountains (Mountain,
Mountainous)
Mounting (Mounted,
Mounts)
Mouth (Mouthpart,
Mouthparts)
Movement (Movable, Move,
Movements, Moves,
Moving)
Mower (Mow, Mowed,
Mowers, Mowing,
Mowings, Mown)
Mozzarella
mRNAsee RNA
MSMA
Mucidin
Mucilage (Mucilages,
Mucilaginous)
Mucin (Mucinous)
Muck (Mucked, Muckland,
Mucks)
Mucocele
Mucocutaneous
Mucoid (Mucoids)
Mucolysis (Mucolytic)
Mucopolysaccharide
(Mucopolysaccharides)
Mucopolysaccharidosis
Mucoproteins (Mucoprotein)
Mucor
Mucoraceae
Mucorales
Mucormycosis
Mucosa (Mucosae, Mucosal)
Mucosubstances
Mucuna
Mucus (Mucous)
Mud (Mudded, Muddiness,
Muddy, Mudflow, Muds)
Mudstone (Mudstones)
Muelleriella
Mufflon see Moufflon
Mugwort
Muhlenbergia
Muilla
Mukulol
Mulberry see Morus
Mulch (Mulched, Mulches,
Mulching)
Mules (Mule)
Mulesing
Mullein
Mulleripollis
Mullet
Multibank see Banks
Multichannel see Channels
Multicolor see Color
Multienzyme see Enzymes
Multigeneration see
Generation
Multihusked see Husks
Multijuga
Multijugin
Multijuginol

Multilamellar *see* Lamella
Multilayered *see* Layering
Multinucleate *see* Nucleus
Multioxygenated *see*
  Oxygenation
Multiplasms *see* Plasm
Multisample *see* Sampling
Multiseptate *see* Septum
Multispored *see* Spores
Multistaminate *see* Stamen
Multistriatin
Mulfisuckling *see* Suckling
Multivariate
Multivarietal *see* Varieties
Multivitamin *see* Vitamin
Mumps
Mung *(Mungbean)*
  X-Ref: Urd
Mungo
Munira
Muramidase *see* Lysozymes
Murdannia *(Murdania)*
Murexide
Muridae
Murine
Murmurs *(Murmur)*
Murrah
Murraya
Murrayacine
Musa
Musaceae
Musanga
Musca
Muscadinia
Muscardine
Muscari
Muscarine *(Muscarinic)*
Musci *(Mosses)*
  X-Ref: Moss
Muscidae *(Muscid,*
  Muscineae)
Muscidifurax
Muscle *(Intramuscularly,*
  Muscled, Muscles,
  Muscling, Muscular,
  Muscularity, Musculation,
  Musculature, Musculo)
  X-Ref: Intramuscular
Muscoidea *(Muscoid)*
Muscovy
Museums *(Museum)*
Mushrooms *(Mushroom)*
  X-Ref: Champignon,
  Toadstools
Musk
Muskeg *(Muskegs)*
Muskmelons *(Muskmelon)*
Muskox
Muskrat *(Muskrats)*
Musoda
Mussaenda
Mussel *(Mussels)*
Mussidia
Mustard *(Mustards)*
Mustela
Mustelidae *(Mustelids)*

Musts
Mutagenesis *(Mutagen,*
  Mutagenic, Mutagenicity,
  Mutagenous, Mutagens)
  X-Ref: Antimutagenic,
  Automutagens,
  Radiomutagenesis
Mutants *(Mutant,*
  Radiomutant)
  X-Ref: Photomutants,
  Radiomutants
Mutase
Mutation *(Mutabilities,*
  Mutability, Mutable,
  Mutated, Mutational,
  Mutationally, Mutations,
  Mutator)
  X-Ref: Micromutations,
  Paramutation,
  Premutation
Mutilation
Mutilla
Mutillidae *(Mutillid)*
Mutinus
Mutisia
Mutton
Mutualism *see* Symbiosis
Muzzle
Mya
Myasthenia
Mycalesis
Mycangia
Mycelia *(Mycelial,*
  Mycelium)
  X-Ref: Amycelial
Mycena
Mycetocytes
Mycetoma
Mycetome
Mycetophagous
Mycetophila
Mycetophilidae
Mycetozoans *see*
  Myxomycetes
Mycobacteriaceae
  (Mycobacteria,
  Mycobacterial)
  X-Ref: Antimycobacterial
Mycobacteriosis
Mycobacterium
Mycobiont
Mycococcus
Mycoderma
Mycodiplosis
Mycoferritin
Mycoflora *see* Fungi
Mycogone
Mycolaminarans
Mycoleptodiscus
Mycologists *(Mycologist)*
Mycology *(Mycologic,*
  Mycological)
Mycolysis
Mycoorganisms *see* Fungi
Mycoparasites
  (Mycoparasite,

Mycoparasitic,
  Mycoparasitism)
Mycophagy *(Mycophagous)*
Mycophenolic
Mycoplasm *(Mycoplasms)*
Mycoplasma *(Mycoplasmal,*
  Mycoplasmalike, ·
  Mycoplasmas,
  Mycoplasmata,
  Mycoplasmic)
Mycoplasmacides
Mycoplasmataceae
Mycoplasmatales
Mycoplasmosis
  (Mycoplasmoses,
  Mycoplasmotic)
Mycorrhizae *(Mycorrhiza,*
  Mycorrhizal, Mycorrhizas,
  Mycorrhizogenic,
  Mycorrhyzal)
  X-Ref: Nonmycorrhizal
Mycosis *(Mycoses, Mycotic)*
  X-Ref: Antimycotic,
  Otomycosis
Mycosphaerella
Mycosphaerellaceae
Mycosporine
Mycotoxicosis *(Mycotoxic,*
  Mycotoxicoses)
Mycotoxins
  (Mycotoxicological,
  Mycotoxin)
Mycotrophic *(Mycotrophy)*
Mycoviruses *(Mycovirus)*
Mycteristes
Mycterodus
Myelin *(Myelinated,*
  Myelinic)
Myelitis
Myeloblastosis
  (Myeloblasts)
Myeloencephalitis
Myelography *(Myelogram,*
  Myelograms)
Myeloid
Myeloliposarcoma
Myeloma
Myelomalacia
Myelomonocytic
Myelopathy
Myeloperoxidase
Myelophilus
Myelophthisic
Myeloproliferative
Mygalomorphae
  (Mygalomorph)
Myiasis
Myiomma
Mylia
Mylio
Myllocerus
Mylone *see* Dazomet
Mymaridae *(Mymarid)*
Myobia
Myobiidae *(Myobiid)*
Myobiosis

Myoblast *(Myoblasts)*
Myocardia *(Myocardial,*
  Myocardiopathy,
  Myocardium)
  X-Ref: Intramyocardial,
  Postmyocardial
Myocastor *see* Coypus
Myocoptes
Myocoptidae
Myodystrophy
Myoepithelial
Myofibril *(Myofibrillar,*
  Myofibrillary, Myofibrils)
Myogenesis *(Myogenic)*
Myoglobin *(Myoglobins)*
Myoglobinuria
Myoinositol
Myokinase
Myoma
Myometrium *(Myometrial)*
Myopathy *(Myopathia,*
  Myopathies)
Myopia
Myoporaceae
Myoporum
Myorelaxants *see* Relaxants
Myosin *(Myosins)*
Myositis
Myosotis
Myotonia *(Myotonic)*
  X-Ref: Nonmyotonic
Myotubes
Myrcenes
Myrceugenia
Myrcia
Myriapoda *(Myriapod,*
  Myriapods)
Myrica
Myricaceae
Myricetin
Myricetol
Myrimica
Myriococcum
Myrionemataceae
Myriophyllum
Myristic
Myristica
Myristicaceae
Myristicin
Myrmecia
Myrmecidae
Myrmecoblatta
Myrmecochorous
Myrmecocides
Myrmecocystus
Myrmecofauna
Myrmecology
Myrmecophila
Myrmecophilous
  (Myrmecophilic,
  Myrmecophilism,
  Myrmecophily)
Myrmecormiris
Myrmeleontidae
  (Myrmeleon,
  Myrmeleontid)

Myrmeleotettix
Myrmica
Myrmicinae *(Myrmicine)*
Myrmosidae
Myrosinases *(Myrosinase)*
Myrothamnus
Myrothecium
Myrsidea
Myrtaceae
Myrtopsine
Myrtopsis
Myrtus
Mythimna
Mytilaspis
Myxamoebae
Myxarium
Myxin
Myxobacteriales
Myxococcus
Myxoma *(Myxomatous)*
Myxomatosis
Myxomphalia
Myxomycetes
    (Myxomycete)
    X-Ref: Mycetozoans
Myxophyceae
Myxovirus *(Myxoviruses)*
Myzocallis
Myzocytium
Myzodes
Myzodium
Myzus
NAA
    X-Ref: Planofix
Nabam
Nabidae *(Nabids)*
Nabis
Nacoleia
NAD
NADase
NADH
NADP
NADPH
Naegleria
Naftalofos
Naiadacarus
Naiads
Naias *see* Najas
Nail *(Nailed, Nailing, Nails)*
Nais
Najadaceae
Najas
    X-Ref: Naias
Nakataea
Naled
    X-Ref: Dibrom
Nalepella
Nali *see* Bikaneri
Nalidixate
Nalidixic
Naloxone
Naltrexone
Nandina
Nannochloris
Nannotrigona
Nanomutilla

Nanophyetus
Nanophyton
Nanorchestes
Naphthaleneacetic
Naphthalenes *(Naphthalene,*
    Naphthaline)
Naphthalenesulfonate
Naphthalenone
Naphthalic
Naphthalophos
Naphthaquinone *see*
    Naphthoquinones
Naphthenate *(Naphthenates)*
Naphthenic
Naphthindan
Naphthionic
Naphthofuran
Naphthoic
Naphthol
Naphthoquinones
    (Naphthoquinone)
    X-Ref: Naphthaquinone
Naphthoxyacetic
Naphthylacetamide
Naphthylacetic
Naphthylamine
Naphthylphthalamic
Naphthyridine
Napiergrass
Napomyza
Napropamide
Napthenate
Naranga
Narathura
Narcissin
Narcissus *(Narcissi)*
Narcolepsy
Narcosis *(Narcotized)*
Narcotics *(Narcotic)*
Nardostachys
Nardus
Naringenin
Naringin
Narthecium
Nasal *(Intranasally)*
    X-Ref: Intranasal,
    Paranasal
Nasolabial
Nasolacrimal
Nasonia
Nasopharynx
    (Nasopharyngeal)
Nasturtium *(Nasturtiums)*
Nasutapis
Nasutitermes
Nasutitermitinae *(Nasute)*
Natal *see* Birth
Natalenone
Nathalis
Natto
Nauclea
Naucledal
Nauclederine
Naucleeae
Nauclefine
Naucleonidine

Naucoria
Naucoriaeae
Naucoridae
Naucoris
Naude
Naufraga
Nauphoeta
Naval
Navel *(Navels)*
Navia
Navicula
Neanura
Neaphis
Nearctaphis
Nebria
Neburon
Neck *(Necked)*
Neckera
Necrobacillosis
Necrobia
Necrobiosis *(Necrobiotic)*
Necrodes
Necrophorus *(Necrophoric)*
Necropsy
Necrosis *(Necroses,*
    Necrotic, Necrotizing)
    X-Ref: Subnecrotic
Nectandra
Nectar *(Nectareous,*
    Nectaries, Nectariferous,
    Nectariless, Nectars,
    Nectary)
    X-Ref: Prenectar
Nectarines *(Nectarine)*
Nectria
Nectriaceae
Nedcosmospora
Neduba
Needle *(Needles)*
Needlecast
Needlegrass
Neelides
Neem
Neftin *see* Furazolidone
Neguvon *see* Trichlorfon
Neisseria
Neivamyrmex
Nelima
Nellite
Nelumbium
Nelumbo
Nemacur
Nemagon
Nemalion
Nemasoma
Nemastoma
Nemastomataceae
Nemastomatidae
Nematinae *(Nematine)*
Nematocera
Nematocide *(Nematicidal,*
    Nematicide, Nematicides,
    Nematocidal,
    Nematocides)
Nematode *(Nematoda,*
    Nematodes, Nematofauna,

Nematological,
Nematology,
Nematophagous)
Nematodiasis *(Nematodosis)*
Nematodirus
Nematomorpha
Nematophagus
Nematospiroides
Nematospora
Nematothecium
Nemeritis
Nemesia
Nemobius
Nemognatha
Nemophila
Nemopteridae
Nemotelus
Nemoura
Nemouridae
Nenteria
Neoagigenin
Neoalsomitra
Neoaplectana
Neoaplectanidae
Neoascaris
Neobavaisoflavone
Neobisiidae
Neobisium
Neobulgaria
Neocallimastix
Neocene
Neochetina
Neochilenia
Neocoelidiinae
Neoconocephalus
Neocortex *see* Cortex
Neocosmospora
Neocteniza
Neocypholaelaps
Neodecanoic
Neodiprion
Neodusmetia
Neoechinulin
Neoeucheyla
Neofabraea
Neoflavanoid
    (Neoflavanoids,
    Neoflavonoids)
Neogene
Neogregarinida
    (Neogregarines)
Neohopane
Neohydronomus
Neokoehleria
Neolaccogrypota
Neolecta
Neolignans *see* Lignan
Neolitsea
Neolloydia
Neomaskellia
Neomenthol
Neomesomermis
Neomoorea
Neomusca
Neomycin
Neomyobia

Neomyzus
Neonatal see Birth
Neopelomyia
Neopetalonema
Neopetasitenine
Neophasia
Neophylax
Neophyllaphis
Neophyllobius
Neopinamine
Neoplasia
Neoplasm *(Neoplasms,*
   Neoplastic)
   X-Ref: Antineoplastic
Neoplea
Neopodismopsis
Neopsylla
Neorautanenia
Neoregelia
Neorickettsia
   (Neorickettsial)
Neoschongastia
Neoseiulus
Neospongiococcum
Neostrearia
Neoteny *(Neotenic)*
Neotermes
Neothiobinupharidine
Neotigogenin
Neotorreyol
Neotrombicula
Neotropics see Tropics
Neottia
Neotylenchidae
Neovitamin see Vitamin
Neovossia
Neowerdermannia
Neoxanthins *(Neoxanthin)*
Neozimiris
Nepa
Nepenthaceae
Nepenthes
Nepeta
Nepetaefolin
Nephantis
Nephelea
Nephelium
Nephelodes
Nephelometry
   (Nephelometer,
   Nephelometric)
Nephila
Nephilengys
Nephotettix
Nephrectomy
   (Nephrectomized)
Nephridia *(Nephridial)*
Nephritis
Nephroarctin
Nephroblastoma
Nephrocalcinosis
Nephrolepis
Nephrolithiasis
   (Nephroliths)
Nephrolithotomy
Nephroma

Nephron
Nephropathy
Nephrosis *(Nephrotic)*
Nephrotoma
Nephrotoxicity
   (Nephrotoxic)
Nepidae
Nepoviruses *(Nepo)*
Nepticula
Nepticulidae
Nereocystis
Nerine
Nerisyrenia
Nerium
Nerolidol
Nerosine
Nerozin
Nerves *(Nerve, Neural)*
Nervous *(Nervousness)*
Neryl
Nesidiocoris
Nesogordonia
Nesomylacris
Nesothrips
Nessus
Nest *(Nesting, Nestling,*
   Nestlings, Nests)
   X-Ref: Nidicolous
Netocia
Netomocera
Netropsin
Nettle *(Nettles)*
Nettlehead
Network *(Networks)*
Neuraminic
Neuraminidases
   (Neuraminidase)
Neurectomy
Neurigona
Neuritis
Neuroanatomy
Neuroblast *(Neuroblasts)*
Neuroblastoma
Neurochemistry
   (Neurochemical)
Neurocolpus
Neurocranium see Cranium
Neurocytes
Neuroendocrine
Neuroepithelial
Neurogenesis *(Neurogenic)*
   X-Ref: Nonneurogenic
Neurogenia
Neuroglia see Glia
Neurohemal *(Neurohaemal)*
Neurohistology
   (Neurohistological)
Neurohormone
   (Neurohormones)
Neurohumoral
   (Neurohumor)
Neurohypophysial
   (Neurohypophysis)
Neuroleptanalgesia
   (Neuroleptanalgesic)
Neuroleptic

Neurology *(Neuro,*
   Neurologic, Neurological)
Neuroma
Neuromuscular
Neurons *(Interneuron,*
   Interneurone,
   Interneurones, Neuron,
   Neuronal, Neurones)
   X-Ref: Extraneuronal,
   Interneurons,
   Intraneuronal
Neuropathology
   (Neuropathies,
   Neuropathologic,
   Neuropathological,
   Neuropathy)
Neurophysins *(Neurophysin)*
Neurophysiology
   (Neurophysiological)
Neuroptera *(Neuropteran)*
Neuropteroidea
Neurosecretion
   (Neurosecretions,
   Neurosecretory)
Neurospora
Neurosurgery
   (Neurosurgical)
Neurotensin
Neurotomy
Neurotoxins *(Neurotoxic,*
   Neurotoxicity,
   Neurotoxicoses,
   Neurotoxicosis,
   Neurotoxin)
Neurotransmitters
   (Neurotransmitter)
Neurotropy *(Neurotropic,*
   Neurotropism)
Neurovirulence
   (Neurovirulent)
Neurulation
Neutron *(Neutrons)*
Neutropenia
Neutrophil *(Neutrophils)*
Neviusia
Newborn see Birth
Newcastle
Newsprint
Newsteadia
Newt
Nexagan
Nexion
Nexton see Bromophos
Nezara
Niacin
Niacinamide
Nicandra
Nicandrenone
Nickel
Niclofolan see Bilevon
Niclosamide *(Mansonil)*
Nicoletia
Nicotiana
Nicotinamide
Nicotinate

Nicotine
Nicotinic
Nidicolous see Nest
Nidulariales
Nierembergia
Nifurprazine
Nifursol
Nigakihemiacetals
Nigella
Night *(Nights, Nocturnally)*
   X-Ref: Nocturnal
Nightflowering see
   Flowering
Nightshade
Nightsoil
Nigrosine
Nigrospora
Nilaparvata
Nilverm see Tetramisole
Nimetazepam
Ninhydrin
Niobium
Nipa see NYPA
Nipaecoccus
Nipple see Teats
Nipponorhynchus
Nippostrongylus
Niptus
Niridazole
   X-Ref: Ambilhar
Nisin
Nitela
Nitella
Nitellopsis
Nitidulidae
Nitragin
Nitralin
Nitramines *(Nitramine)*
Nitrapyrin
Nitraria
Nitrates *(Nitrate, Nitrated,*
   Nitration)
Nitrazepam
Nitric
Nitricides *(Nitricide)*
Nitrification *(Nitrified,*
   Nitrifiers, Nitrifying)
Nitrile *(Nitriles)*
Nitrilotriacetic
Nitrilotrimethyl
Nitrites *(Nitrite)*
Nitroalkenylation
Nitroamines
Nitroammophos
Nitroammophoska
Nitroanilide *(Nitroanilid)*
Nitroaniline
Nitroaromatic
Nitrobacter
Nitrobenzene
Nitrobenzoic
Nitrobenzol
Nitrobenzyl
Nitrobenzylbromide
Nitrocellulose
Nitrofen

Nitrofuran
Nitrofurazone
 (Nitrofurazon)
Nitrofuryl
Nitrofurylacrylic
Nitrogen (Nitro,
 Nitrogenized, Nitrogenous,
 Nitrogens)
 X-Ref: Nonnitrogenous
Nitrogenase (Nitrogenases)
Nitroglycerin
Nitrohumic
Nitroimidazole
Nitromethane
Nitrophenol (Nitrophenols)
Nitrophenylglycosidases
Nitrophile (Nitrophilous)
Nitrophosphate
Nitropropane
Nitropropionic
Nitropyridine
Nitroquinoline
Nitrosalicylanilide
Nitrosamides
Nitrosamidines
Nitrosamine (Nitrosamines)
Nitrosamino
Nitrosation (Nitrosating)
Nitrose
Nitroso
Nitrosoalkyl
Nitrosoalkylureas
Nitrosoamino
Nitrosodimethylamine
Nitrosoethylurea
Nitrosoethylurethane
Nitrosoguanidines
 (Nitrosoguanidine)
Nitrosohemoglobin
Nitrosoiminodiacetate
Nitrosomethyl
Nitrosomethylurea
Nitrosomethylurease
Nitrosomonas
Nitrosophenol
Nitrosopyrrolidine
Nitrososarcosine
Nitrosothiol
Nitrosourea
Nitrosylhaems
Nitrosylsulphuric
Nitrourethane
Nitrous
Nitrovin
Nitrovine
Nitroxynil
 X-Ref: Dovenix
Nitzschia (Nitzchia)
Nivalenol
Nivation
NMR
Nocardia (Nocardial)
Nocardiosis
Nocobactins
Noctua
Noctuidae (Noctuid,

Noctuids, Noctuinae)
Noctuiseius
Nocturnal see Night
Node (Nodal, Noded,
 Nodes)
Nodita
Nodules (Nodular,
 Nodulated, Nodulates,
 Nodulating, Nodulation,
 Nodule, Nonnodulated)
 X-Ref: Micronodules,
 Nonnodulating
Nodulosphaeria
Noelucanus
Noemacheilus
Nogos
Nogra
Noise (Noisy)
Nomada
Nomadacris
Nomadinae (Nomadine)
Nomadism (Nomad,
 Nomadic, Nomadization,
 Nomads)
Nomaphila
Nomenclature
 (Nomenclatorial,
 Nomenclatural)
Nomia
Nomography (Nomogram,
 Nomograms, Nomograph,
 Nomographic,
 Nomographs)
Nomophila
Nomuraea
Nonachlor
Nonacosanone
Nonadecenoic
Nonadienal
Nonadienol
Nonalcoholic see Alcohol
Nonallelic see Allele
Nonaqueous see Aqueous
Nonbiological see Biology
Nonbloating see Bloat
Noncalcareous see
 Calcareous
Noncancerous see Cancer
Noncarcinogen see
 Carcinogens
Noncatalytic see Catalysis
Noncellular see Cells
Nonchemical
Nonchernozem see
 Chernozem
Noncohesive see Cohesion
Noncombustible see
 Combustion
Noncrop see Crops
Noncrystallizing see
 Crystallization
Noncutting see Cuttings
Noncytotoxic see
 Cytotoxicity
Nondairy see Dairy
Nondegenerate see

Degeneration
Nondestructive
 (Nondestructively)
Nondiapausing see Diapause
Nondiel see Diels
Nondisjunction
Nondomesticated see
 Domestication
Nondormant see Dormancy
Nonenal
Nonendosperm see
 Endosperm
Nonenzymatic see Enzymes
Nonerodible see Erosion
Nonerosive see Erosion
Nonesterified see
 Esterification
Nonfarm see Farms
Nonfat see Fats
Nonfertile see Fertility
Nonfilterable see Filtration
Nonflowering see Flowers
Nonfluorescent see
 Fluorescence
Nonforest see Forest
Nonfumigant see Fumigants
Nongerminating see
 Germination
Nongrazed see Grazing
Nonhardy see Hardiness
Nonheme see Heme
Nonhemin see Hemin
Nonheterosexual see
 Heterosexual
Nonhomologous see
 Homology
Nonhost see Hosts
Nonimmune see Immunity
Noninfected see Infection
Nonirradiated see
 Irradiation
Nonirrigated see Irrigation
Nonisothermal see Isotherm
Nonlactating see Lactation
Nonleguminous see
 Leguminous
Nonlethal see Lethal
Nonlymphoid see Lymphoid
Nonmanual see Manual
Nonmechanized see
 Mechanization
Nonmetabolic see
 Metabolism
Nonmetallic see Metals
Nonmetropolitan
Nonmigration see Migration
Nonmycorrhizal see
 Mycorrhizae
Nonmyotonic see Myotonia
Nonneurogenic see
 Neurogenesis
Nonnitrogenous see
 Nitrogen
Nonnodulating see Nodules
Nonnuclear see Nucleus
Nonnutrition see Nutrition

Nonobese see Obesity
Nonoccluded see Occlusion
Nononcogenic see Tumor
Nonoxidative see Oxidation
Nonparasitic see Parasites
Nonpathogenic see
 Pathogenesis
Nonpathogical see
 Pathology
Nonpenetrating see
 Penetration
Nonphosphorylated see
 Phosphorylation
Nonphotosynthetic see
 Photosynthesis
Nonpigment see Pigments
Nonplaquing see Plaque
Nonplowing see Plowing
Nonpolyadenylated see
 Polyadenylate
Nonprecipitating see
 Precipitation
Nonprofessionals see
 Occupations
Nonprotein see Proteins
Nonrefrigerated see
 Refrigeration
Nonrhizosphere see
 Rhizosphere
Nonribosomal see
 Ribosomes
Nonripening see Ripening
Nonruminants
 (Nonruminant)
Nonseed see Seeds
Nonsprouting see Sprout
Nonsterile see Sterile
Nonsterol see Sterol
Nonstressed see Stress
Nonsucrose see Sucrose
Nonsugar see Sugar
Nonsurgical see Surgery
Nonsusceptible see
 Susceptibility
Nonsymbiotic see Symbiosis
Nonsystemic see Systemic
Nontarget see Target
Nonteratogenic see
 Teratogenesis
Nonterpenoid see Terpenoid
Nontidal see Tide
Nontillage see Tillage
Nontillering see Tillers
Nontoxic see Toxicity
Nontransformed see
 Transformation
Nontransmissible see
 Transmission
Nontuber see Tuber
Nontumorigenic see Tumor
Nonurban see Rural
Nonvaccinated see
 Vaccination
Nonvolatile see Volatile
Nonwaxy see Wax
Nonwettability see Wet

Nonwood
Nonwoven see Weaving
Noodle (Noodles)
Nootkatone
Nopalea
Nopaline
Noradrenaline see
    Norepinephrine
Norbornane
Norbornene
Norcapillene
Nordihydrocapsaicin
Nordihydroguaiaretic
Norditerpene
    (Norditerpenoids)
Norea
Norepinephrine
    (Noradrenalin)
    X-Ref: Noradrenaline
Norethandrolone
Norethisterone
Norflurazon
Norgestrel
Norleucine
Nornicotine
Norpyridoxol
Norsesquiterpenes
Norseychelanone
Norspirostanol
Norsqualene
Nortestosterone
Nortron
Noryangonin
Nosculation
Nose (Nosed, Noses)
Nosema
Nosematidae
Nosematosis
Nosopsyllus
Nostoc
Nostoxanthin
Noteridae
Nothofagus
Notholaena
Notholopus
Nothomorpha
Nothoscordum
Notiophilus
Notiophygus
Notocactus
Notocolletes
Notocotylus
Notodonta
Notodontidae
Notonecta
Notonectidae
Notopthalmus
Notoscyphus
Notozus
Nourishment (Nourish,
    Nourished, Nourishes,
    Nourishing)
Novaraenol
Novius
Novobiocin
Novocaine see Procaine

Novomessor
Nowellia
Nozzles (Nozzle)
Nubian
Nucellangium
Nucellus (Nucellar)
Nuciferine
Nuclear see Nucleus
Nucleases (Nuclease)
Nucleate see Nucleus
Nucleic
Nucleides (Radionucleides)
Nucleolus (Nucleolar,
    Nucleoli, Nucleologenesis)
    X-Ref: Intranucleolar,
    Uninucleolate
Nucleolytic
Nucleophilic (Nucleophiles)
Nucleopolyhedrosis
Nucleoproteids
Nucleoprotein
    (Nucleoproteic,
    Nucleoproteins)
Nucleosidase
Nucleoside (Nucleosides)
Nucleosome
Nucleotidases
    (Nucleotidase)
Nucleotides
    (Mononucleotide,
    Nucleotide)
    X-Ref: Mononucleotides,
    Polynucleotide
Nucleotidyl
Nucleotidyltransferase
Nucleus (Binucleated,
    Mononuclear,
    Multinuclear,
    Multinuclearity,
    Multinucleated,
    Nucleated, Nucleating,
    Nucleation, Nuclei,
    Nucleoid)
    X-Ref: Anucleate,
    Binucleate,
    Extranuclear,
    Intranuclear,
    Juxtanuclear,
    Micronuclear,
    Mononucleated,
    Multinucleate,
    Nonnuclear, Nuclear,
    Nucleate, Polynuclear,
    Pronuclear,
    Pronucleus,
    Trinucleate,
    Uninucleate
Nuclides (Radionuclide)
    X-Ref: Radionuclides
Nudaurelia
Nuia
Nulliparous
Nullisomics
Numicia
Numida
Nummularine

Nuphar
Nupharolutine
Nupserha
Nurseries (Nursery)
Nurserymen (Nurseryman)
Nursing (Nurse, Nursed)
Nutation
Nutgrass
Nutlet
Nutmeg
Nutria see Coypus
Nutriculture see
    Hydroponics
Nutrients (Micronutrient,
    Nutrient, Nutriment,
    Nutriments)
    X-Ref: Antinutrients,
    Micronutrients
Nutrition (Antinutritive,
    Nonnutritive, Nutrimeter,
    Nutritional, Nutritionally,
    Nutritions, Nutritious,
    Nutritive, Nutritives)
    X-Ref: Antinutritional,
    Nonnutrition,
    Overnutrition
Nutritionists (Nutritionist)
Nuts (Nut, Nutfall,
    Nutmeats)
Nutsedge (Nutsedges)
Nuttalia
Nuttalliella
Nuttalliosis
Nuvan see Dichlorvos
Nuvanol
Nychthemeron
    (Nychthemeral)
    X-Ref: Nycthemeron
Nyctaginaceae
Nyctanthes
Nyctegretis
Nycteola
Nycteribia
Nycteribiidae
Nycterosea
Nycthemeron see
    Nychthemeron
Nyctinasty (Nyctinastic)
Nyctiphantus
Nygmia
Nylon (Nylons)
Nymphaea
Nymphaeaceae
Nymphalidae (Nymphalids,
    Nymphalid, Nymphalinae)
Nymphalis
Nymphoides
Nymphomyiidae
Nymphs (Nymph, Nymphal)
Nymphula
NYPA
    X-Ref: Nipa
Nysius
Nyssa
Nyssaceae
Nystatin

Oak see Quercus
Oakwood (Oakwoods)
Oasis (Oases)
Oatfields
Oatgrass
Oatmeal
Oats (Oat)
Obeliscoides
Oberea
Oberonia
Obesity (Obese)
    X-Ref: Nonobese,
    Overweight,
    Superobesity
Obesumbacterium
Obidoxime
Obituary see Biography
Obsolescence (Obsolete)
Obstetrics (Obstetric,
    Obstetrical)
Obstruction (Obstructing,
    Obstructive)
Obtusilactones
    (Obtusilactone)
Occipital
    X-Ref: Suboccipital
Occlusion (Occluded)
    X-Ref: Nonoccluded
Occupations (Career,
    Occupation, Occupational,
    Occupationally)
    X-Ref: Careers,
    Nonprofessionals,
    Professions
Oceanography
Ocelli (Ocellar, Ocellus)
Ochna
Ochnaceae
Ochodaeinae
Ochotensanes
Ochratoxin (Ochratoxicosis,
    Ochratoxins)
Ochrocarpus
Ochroconis
Ochrolechia
Ochroma
Ochromonas
Ochrops
Ochrosia
Ochrosiineae
Ochrotrichia
Ochthephilus
Ochtodes
Ochyroceratidae
Ocimum
Ocnogyna
Ocotea
Ocotillol
Octaacetate
Octacarbonyl
Octadecadienoic
Octadecenoates
    (Octadecenoate)
Octadecenoic
Octadiene
Octalactone

Octanal
Octane *(Octanes)*
Octanoic
Octanol
Octapeptide *see* Peptides
Octenoic
Octoglutamates
Octomeles
Octopamine
Octopine
Octoploid
Octospora
Octosporea
Octulose
Octulosonic
Octyl
Octylamine
Octylbicycloheptene
Ocular
Oculomotor
Ocymum
Ocypus
Odagmia
Odinia
Odites
Odocoileus
Odoiporus
Odonata *(Odonate)*
Odontacarus
Odontestra
Odontites
Odontogenesis *see* Teeth
Odontoglossum
   (Odontoglossums)
Odontomachus
Odontomantis
Odontophorus
Odontopus
Odontopygidae
Odontoschisma
Odontotermes
Odor *(Odoriferous, Odorous,*
   Odors, Odour, Odourless,
   Odours, Scented,
   Scentless, Scents)
   X-Ref: Pungent, Scent
Oebalus
Oecanthidae *(Oecanthinae)*
Oecanthus
Oecetis
Oecia
Oecobiidae
Oecophoridae *(Oecophorid)*
Oecophylla
Oedaleus
Oedema *see* Edema
Oedemagena
Oedemeridae *(Oedemerid)*
Oedipoda
Oedipodinae
Oedipodius
Oedocladium
Oedogonium
Oedothorax
Oenanthe
Oeneis

Oenocytes
Oenothera
Oeobia
Oerskovia
Oesophagitis *see*
   Esophagitis
Oesophagostomum
Oesophagus *see* Esophagus
Oestradiol *see* Estradiol
Oestridae
Oestrogen *see* Estrogen
Oestromyia
Oestrone
Oestrus *see* Estrus
Offal *(Offals)*
Offspring *see* Progeny
Ogovia
Ohia
Oidiodendron
Oidiopsis
Oidium *(Oidia)*
Oiketicus
Oil *(Oiliness, Oiling, Oils,*
   Oily)
   X-Ref: Oleaginous
Oilcrop
Oilmeal *(Oilcake, Oilcakes,*
   Oilmeals)
Oilpalm *(Oilpalms)*
Oilseed *(Oilseeds)*
Ointment *(Ointments)*
Okanagana
Okra
   X-Ref: Bhindi, Gumbo
Olacaceae
Oldenlandia
Oldfield
Oldsters *see* Elderly
Olea
Oleaceae
Oleaginous *see* Oil
Oleander
Oleandomycin
Oleandra
Oleanolic
Olearia
Oleate
Olecranon
Olefin *(Olefinic, Olefins)*
Oleic
Oleodipalmitin
Oleogesaprim
Oleoresins *(Oleoresin)*
Olethreutidae
   X-Ref: Eucosmidae
Olethreutinae
Oleyl
Olfactometry *(Olfactometer,*
   Olfactometers)
Olfactory *(Olfaction,*
   Olfactive, Smell, Smelled,
   Smells)
   X-Ref: Smelling
Oliarus
Olibanum *see* Frankincense
Oligia

Oligocene
Oligochetes *(Oligochaeta,*
   Oligochaete, Oligochaetes)
Oligoelements
Oligoglycoside
Oligomeric *(Oligomers)*
Oligomycin
Oligonichus
Oligonitrophyllic
   (Oligonitrophylls)
Oligonucleotides
   (Oligonucleotide)
Oligonychus
Oligopyrene
Oligosaccharidases
Oligosaccharide
   (Oligosaccharides)
Oligota
Oligotrichum
Oligotrophic
Oligotrophidi
Oligotrophus
Oligouronides
Oliniaceae
Olivacine
Olives *(Olive)*
Olivine
Olneya
Olophoeus
Olpidium
Omasum *(Omasal)*
Omelet
Ommatidia *(Ommatidial,*
   Ommatidium)
Ommatoiulus
Ommochrome
   (Ommochromes)
Omocestus
Omphale
Omphaliaster
Omphalitis
Omphalodiscus
Omphisa
Onagraceae *(Onagraceous)*
Onchocerca
Onchocerciasis
   (Onchocercosis)
Oncidium
Oncoba
Oncogenic *see* Tumor
Oncologic *see* Tumor
Oncopeltus
Oncopera
Oncopsis
Oncorhynchus
Oncornaviruses
   (Oncornavirus)
Oncotheca
Onega
Onions *(Onion)*
Oniscidae
Oniscoidea
Oniscomorpha
Oniscus
Onobrychis
Onoclea

Ononis
Onopordum
Onthophagini
Onthophagus
Ontogeny *(Ontogenesis,*
   Ontogenetic,
   Ontogenetical, Ontogenic)
Onychine
Onychium
Onychiurus
Onychogomphus
Onychonema
Onychopetalum
Onychophora
Onymacris
Oocampsa
Oocardium
Oocyst *(Oocysts)*
Oocystis
Oocyte *(Oocytes)*
Ooencyrtus
Oogenesis *(Oogenetic)*
   X-Ref: Ovogenesis
Oogonia *(Oogonial,*
   Oogonium)
Oomorphus
Oomycetales
Oomycete *(Oomycetes)*
Oonopidae
Oophagous *(Oophagy)*
Ooplasm
Oosorption
Oosphere
Oosponol
Oospora
Oospores *(Oospore)*
Ootheca *(Oothecae,*
   Oothecal)
Opacifrons
Opaque *(Opacity)*
Opegrapha
Operculella
Operculum *(Opercular,*
   Operculate)
   X-Ref: Inoperculate,
   Suboperculate
Operophtera
Ophiobolus
Ophiocarpine
Ophiocytium
Ophioglossaceae
Ophioglossum
Ophiomyia
Ophioninae
Ophionini
Ophiorrhiza
Ophiostoma
Ophonus
Ophrys
Ophthalmia
Ophthalmic
Ophthalmitis
Ophthalmology
   (Ophthalmological)
Ophthalmoscopy
   (Ophthalmoscopic)

Ophthalmotropes
(Ophthalmotropic)
Ophyra
Opifex
Opiinae
Opiliones
Opilionidea *(Opilionid)*
Opilones
Opiptacris
Opisocrostis
Opisthorchiasis
Opisthorchis
Opium
Opius
Oplitis
Opogona
Opomyza
Opomyzidae
Opopaea
Oporinia
Opossum
Oppia
Oppiidae *(Oppioid)*
Opsiphanes
Opsonins *(Opsonic,*
Opsonized)
Optics *(Optic, Optical,*
Optically)
X-Ref: Preoptic
Optimization *(Optimal,*
Optimalization,
Optimally, Optimisation,
Optimised, Optimising,
Optimize, Optimized,
Optimizing, Optimum)
Opuntia
Oral *(Orally)*
Oranges *(Orange)*
Orangeworm
Orchardgrass
X-Ref: Cocksfoot
Orchards *(Orchard,*
Orcharding, Orchardist)
Orchelimum
Orchesella
Orchidaceae *(Orchid,*
Orchidaceous,
Orchidology, Orchids)
Orchidoeca
Orchiectomy
Orchis
Orcinol
Ordinances *see* Laws
Ordram *see* Molinate
Oregano
Oreocarabus
Oreoherzogia
Orexigen
Organelles
Organic *(Organically,*
Organics)
Organoaluminum
(Organoaluminium)
Organoarsenicals
Organoboron
Organochloric

Organochloride
Organochlorines
(Organochlorine)
Organogenesis
(Organogenetic,
Organogenic)
Organohalogen
(Organohalogenated)
Organoids
Organoleptic
(Organoleptical)
Organomercury
(Organomercurial,
Organomercurials,
Organomercuries)
Organometallic
Organophosphates
(Organophosphate)
X-Ref: Phosphororganic
Organophosphorus
(Organophosphoric,
Organophosphorous)
Organotaxic *see* Taxonomy
Organotin *(Organotins)*
Organs *(Organ)*
Orgilus
Orgotein
Orgyia
Oria
Oribata
Oribatei
Oribatella
Oribatellidae
Oribatidae *(Oribatid,*
Oribatids)
Oribatoidea *(Oribatoid)*
Oricia
Orientalidine
Orientation *(Orientated,*
Orientational,
Orientations, Orientative,
Oriented, Orienting)
Orifice *(Orifices)*
Origanum
Oriola
Oritrophium
Orius
Ormia
Ormosia
Ornamental *(Ornamentals,*
Ornamentation)
Ornatoraphidia
Ornithacris
Ornithine
Ornithobilharzia
Ornithodoros
(Ornithodorus)
Ornithogalum
Ornithomyia
Ornithonyssus
Ornithophilic
Ornithopus
Ornithosis
Ornithospermum
Orobanchaceae
Orobanche

Oropetium
Oropodes
Orosius
Orotate
Orotic
Orphninae
Orphrys
Ortene
Orthellia
Orthene
Orthetrotrema
Orthetrum
Orthezia
Ortheziidae
Orthochaetes
Orthocladius
Orthomorpha
Orthomus
Orthonil
Orthopedics *(Orthopaedic,*
Orthopedic)
Orthophosphates *see*
Phosphates
Orthophosphoric *see*
Phosphoris
Orthophotography *see*
Photography
Orthopodomyia
Orthoporus
Orthoptera
(Orthopterological,
Orthopteron,
Orthopterous)
Orthopteroidea
Orthosia
Orthotrichum
Orthotropy *(Orthotropic)*
Orthungini
Orus
Oryctanthus
Oryctes
Oryctolagus
Orystes
Oryza
Oryzaephilus
Oryzalin
Oryzias
Oryzophagus
Oryzopsis
Osazone
Oscillatoria
Oscillography
(Oscillograms,
Oscillographic)
Oscillometric
Oscillopolarography
(Oscillopolarographic)
Oscilloscope *(Oscilloscopic)*
Oscillospira
Oscinella
Osier *(Osiers)*
Osmanthus
Osmaronia
Osmia
Osmium
Osmocote

Osmoderma
Osmolyte
Osmometry
Osmophilic *(Osmophilous)*
Osmoregulation
(Osmoregulatoric,
Osmoregulatory)
Osmosis *(Osmolality,*
Osmolarity, Osmolysis,
Osmose, Osmotic,
Osmotically)
X-Ref: Exosmosis,
Hyperosmotic,
Isoosmotic, Isosmotic
Osmunda
Osmundaceae
(Osmundaceous)
Osmundastrum
Osseous
Ossification
(Hyperossification,
Ossifying)
Ossimi *see* Ausimi
Osteitis
Osteoarthritis
Osteoarthropathy
Osteoarthrosis
Osteoarticular
Osteochondral
Osteochondritis
Osteochondroma
Osteochondromatosis
Osteochondropathy
(Osteochondropathies)
Osteochondrosis
Osteoclast *(Osteoclasts)*
Osteodystrophy
(Osteodystrophia,
Osteodystrophic)
Osteogenesis *(Osteogenetic,*
Osteogenic)
Osteolathyrismus *see*
Lathyrism
Osteology *(Osteo)*
Osteoma
Osteomalacia
Osteomyelitis
Osteopathy *(Osteopathies)*
Osteopetrosis *(Osteopetrose)*
Osteophytes *(Osteophyte)*
Osteoporosis
Osteosarcoma
Osteospermum
Osteotomy
Osternus
Ostertagia
Ostertagiasis
Ostracod
Ostreobium
Ostrinia
Othnonius
Othonna *(Othonnas)*
Othonoius
Othreis
Otiorrhynchus
Otitidae

Otitis
Otoacariasis
Otocyst
Otodectes
Otoliths *(Otolith)*
Otology
Otomycosis *see* Mycosis
Ottelia
Ottonia
Ouabain
Oudemansicheyla
Oudemansiella
Oulema
Outbreak *(Outbreaks)*
Outbred *see* Breeding
Outcrossing *see*
    Crossbreeding
Outfall *(Outfalls)*
Outflows
Outlook *see* Forecast
Outplanting
Ova *(Ovum)*
Ovacaridae
Ovacarus
Ovalbumin
Ovariectomy
    (Ovariectomies,
    Ovariectomised,
    Ovariectomized)
Ovariohysterectomy
Ovarioles *(Ovariole)*
Ovary *(Intraovarian,*
    Ovarial, Ovarian, Ovaries,
    Ovario)
    X-Ref: Transovarial
Oven *(Ovens)*
Ovenbird
Overcutting *see* Cuttings
Overdominance *see*
    Dominance
Overdrilled *see* Drilling
Overevolution *see* Evolution
Overfeeding *see* Feeding
Overfermented *see*
    Fermentation
Overgrazing *see* Grazing
Overgrown *(Overgrowing,*
    Overgrowth, Overgrowths)
Overheated *see* Heat
Overmilking *see* Milking
Overmoistening *see*
    Moisture
Overnutrition *see* Nutrition
Overpopulation *see*
    Population
Overseeding *see* Seeding
Oversnow *see* Snows
Oversowing *see* Sowing
Overstory
Overweight *see* Obesity
Overwinter *see* Winter
Overwrap *see* Packing
Ovex
Ovicide *(Ovicidal, Ovicides)*
Oviduct *(Oviductal,*
    Oviducts)

Ovine
Oviparous *(Oviparity)*
Oviposition *(Oviposited,*
    Ovipositing, Ovipositional,
    Ovipositor, Ovipositors)
    X-Ref: Preoviposition
Ovis
Ovitraps *see* Traps
Ovogen *(Ovogenic)*
Ovogenesis *see* Oogenesis
Ovoinhibitors *(Ovoinhibitor)*
Ovomucin
Ovomucoid
Ovoplanum
Ovotestes *(Ovotestis)*
Ovulation *(Ovulate,*
    Ovulating, Ovulations,
    Ovulator, Ovulatory,
    Superovulate,
    Superovulated,
    Superovules)
    X-Ref: Periovulatory,
    Preovulatory,
    Superovulation
Ovule *(Ovular, Ovules)*
Ovulinia
Owadofos
Owls *(Owl)*
Ownership *(Owned, Owner,*
    Owners, Ownerships,
    Owning)
Oxadiazoles *(Oxadiazole)*
Oxalacetase
Oxalacetate
    X-Ref: Oxaloacetate
Oxalacetic
    X-Ref: Oxaloacetic
Oxalate *(Oxalates)*
Oxalic
Oxalidaceae
Oxalis
Oxaloacetate *see*
    Oxalacetate
Oxaloacetic *see* Oxalacetic
Oxalylamino
Oxalylaminopropionic
Oxamyl
Oxanthiin *(Oxanthin)*
Oxathiin
Oxazolidine
    (Oxazolidinethione)
Oxazolidone
Oxazolines
Oxen *(Ox)*
Oxfendazole
Oxibendazole
Oxidants *(Oxidant)*
    X-Ref: Bioantioxidants
Oxidases *(Oxidase)*
    X-Ref: Oxydase
Oxidate
Oxidation *(Autoxidative,*
    Autoxidized, Oxidated,
    Oxidating, Oxidations,
    Oxidative, Oxidatively,
    Oxidised, Oxidized,

Oxidizers, Oxidizing,
    Oxydation,
    Photooxidative)
    X-Ref: Autooxidation,
    Autoxidation,
    Biooxidation,
    Nonoxidative,
    Photooxidation,
    Reoxidation
Oxide *(Oxides)*
    X-Ref: Oxyde
Oxidoreductases
    (Oxidoreductase)
Oxidus
Oxime *(Oximes)*
Oximecarbamate
Oxindole
Oxine
Oxirane
Oxisol *(Oxisols)*
Oxoaphyllidine
Oxoaporphine
Oxocrinol
Oxoglutarate
Oxokopsinine
Oxoproline
Oxowithanolide
Oxyacids
Oxybelus
Oxybenzoic
Oxycarboxin
Oxycetonia
Oxychloride
Oxyclozanide
Oxycoccus
Oxydase *see* Oxidases
Oxyde *see* Oxide
Oxydemetonmethyl
Oxydendrum
Oxydesmus
Oxydipyridine
Oxydoreductases
Oxyethira
Oxyethyl
Oxyethylcellulose
Oxygen *(Oxygenic)*
Oxygenase
    X-Ref: Monooxygenase
Oxygenation *(Oxygenated)*
    X-Ref: Monooxygenated,
    Multioxygenated,
    Photooxygenation,
    Trioxygenated
Oxyhemoglobin
Oxyhydantoin
Oxylipeurus
Oxylobium
Oxymercuration
Oxymethylaminotriazines
Oxyopes
Oxyopidae
Oxypaeoniflorin *see*
    Paeoniflorin
Oxyproline
Oxypurines
Oxyquinolines

Oxyrrhis
Oxysarcodexia
Oxytelinae
Oxytetracycline
Oxytocin *(Oxytocic,*
    Oxytocins)
Oxytocinase
Oxytropis
Oxyuriasis
Oysters *(Oyster)*
Ozone *(Ozonated,*
    Ozonation, Ozonization)
Ozonolysis
Pachira
Pachodynerus
Pachybasium
Pachybrachys
    (Pachybrachis)
Pachycereus
Pachycormus
Pachycrepoideus
Pachydiplosis
Pachygaster
Pachygasterinae
Pachyiulus
Pachylobius
Pachyman
Pachymetopius
Pachynematus
Pachynomidae
Pachyphragma
Pachyphytum
Pachypleurum
Pachypodanthium
Pachyrhizus
Pachysandra
Pachyschelus
Pachystachys
Pachytene
Pachyzancla
Packaging *(Package,*
    Packaged, Packages,
    Prepackaged)
    X-Ref: Prepackaging
Packers *(Packer)*
Packing *(Overwraps, Pack,*
    Packed, Packet, Packets,
    Packings, Packs,
    Prepacked, Wrapped,
    Wrapper, Wrappers,
    Wrapping, Wrappings,
    Wraps)
    X-Ref: Overwrap,
    Prepacking,
    Unpacking, Wrap
Packinghouse *(Packhouse,*
    Packinghouses)
Paculla
Pacullidae
Padan
Paddock *(Paddocks)*
Paddy *see* Rice
Padina
Pads *(Pad, Padding)*
Padus
Paecilomyces

Paederia
Paederinae
Paederus
Paedogenesis *see*
  Pedogenesis
Paeonia
Paeoniflorin
  X-Ref: Albiflorin,
  Benzoylpaeoniflorin,
  Oxypaeoniflorin
Paeonoside
Pagiantha
Pain
Paints *(Paint. Paintability.*
  Painted. Painter.
  Painting)
Pairing *(Paired)*
Pairs *(Pair)*
Paktia
Palaeacaridae *(Paleacarid)*
Palaearctic *see* Palearctic
Palaemonetes
Palaeobotany *see*
  Paleobotany
Palaeodipteron
Palaeopalynology *see*
  Palynology
Palaeorhiza
Palaeosclerotium
Palamnaeus
Palarus
Palatability *see* Taste
Palate
Paleacrita
Palearctic
  X-Ref: Palaearctic
Palembus
Paleobotanists
  (Palaeobotanists)
Paleobotany
  (Palaeobotanical.
  Paleobotanic.
  Paleobotanical)
  X-Ref: Palaeobotany
Paleocene *(Palaeocene)*
Paleoclimatology
  (Palaeoclimatic.
  Palaeoclimatology)
Paleoecology
  (Palaeoecological.
  Palaeoecology.
  Paleoecologic.
  Paleoecological)
Paleoentomology
Paleogene *(Palaeogene)*
Paleogeography
  (Paleogeographic)
Paleolimnology
  (Paleolimnological)
Paleomagnetic *see* Magnetic
Paleontology
  (Micropaleontology.
  Palaeontology.
  Paleontological)
  X-Ref:
  Micropalaeontology

Paleopedology
  (Paleopedological)
Paleosols *(Palaeosol.*
  Palaeosols. Paleosol)
Paleozoic *(Palaeozoic)*
Paleudults *(Paleudult)*
Palexorista
Palicourea
Palimna
Palisade
Palisota
Palla
Palladium
Pallets *(Pallet. Palletization.*
  Palletizations. Palletized)
Pallidin
Palmae *(Palm. Palms)*
Palmarosa
Palmatine
Palmeria
Palmette *(Palmetta)*
Palmetto
Palmitate
Palmitic
Palmitodiolene
Palmitoleic
Palmitoyl
Palmitoylcarnitine
Palmitoyltransferase
Palmoil
Palmotoxin
Palomena
Palpigrades
Palpimanidae
Palpita
Palpomyia
Palpus *(Palp. Palpi)*
Paltothyreus
Palynology
  (Palaeopalynological.
  Palyno. Palynologic.
  Palynological.
  Palynomorph.
  Palynomorphs.
  Palynostratigraphic)
  X-Ref: Aeropalynology,
  Palaeopalynology
Palynotaxonomy
Palytoxin
Pamoate
Pampa
Pamphagidae
Pamphiliidae
Pamphilius
Panacide
Panacis
Panacur
Panaeolus
Panagrellus
Panax
Panaxadiol
Pancakes *(Pancake)*
Pancratium
Pancreas *(Intrapancreatic.*
  Pancreatic)
  X-Ref:

Intrapancreatically
Pancreatectomy
  (Pancreatectomized)
  X-Ref: Depancreatized
Pancreatitis
Pancreolithiasis
Pancreozymin
Pandaca
Pandaceae
Pandanaceae
Pandanus
Pandemis
Panderia
Pandorina
Panellus
Panels
Panelus
Panencephalitis *see*
  Encephalitis
Pangium
Pangolagrass *(Pangola)*
Panicle *(Panicles)*
Panicoideae
Paniculatan
Paniculatoside
Panicum
Panleukopenia *see*
  Leucopenia
Panmixia
Panniculitis
Panolis
Panonychus
Panorpa
Panstrongylus
Pantala
Panteniphis
Pantorhytes
Pantothenate
Pantothenic
Panurginus
Panus
Panzeria
Papad
Papain
Papaver
Papaveraceae
  (Papaveraceous)
Papaverine
Papaws *(Papaw)*
Papayas *(Papaya)*
Paper *(Papermaking)*
Paperboard *see* Fiberboard
Papermill
Paphiopedilum
  (Paphiopedilums)
Papilio
Papilionaceae
  (Papilionaceous)
Papilionidae *(Papilionid)*
  X-Ref: Swallowtail
Papilledema
Papilloma *(Papillomas.*
  Papillomatosis.
  Papillomatous)
Papio
Papovaviruses *(Papovavirus)*

Pappogeomys
Paprika
Papulaspora
Papulus
Papyriogenin
Parabenzoin
Parabiosis
Parabrachypterus
Parabronchial *see* Bronchi
Parabronema
Paracaryum
Paraceras
Paraclemensia
Paracoccidioides
Paracolon
Paracopium
Paracrystals
  (Paracrystalline)
Paracymus
Paracynoglossum
Paradasyhelea
Parafallia
Parafasciolopsis
Paraffin *(Paraffinic.*
  Paraffins)
Parafilaria
Paragangliomas
Paragonia
Paragonimiasis
Paragonimus
Paragordius
Paragus
Paragymnomerus
Parahermetia
Parahormones
  (Parahormone)
Parahydroxybenzoic *see*
  Hydroxybenzoic
Parahypochaeta
Parainfluenza *see* Influenza
Parajapygidae
Parajapyx
Parakeets *(Parakeet)*
Parakeratosis
Paraleucobryum
Paraliochthonius
Parallergic *see* Allergy
Paralobesia
Paralysis *(Paralytic.*
  Paralyzing)
  X-Ref: Paresis
Paramagnetic
Paramastacinae
Paramecia *(Paramecium)*
Paramenia
Parametriotes
Paramongaia
Paramphistomiasis
  (Paramphistomosis)
Paramphistomidae
Paramphistomum
Paramutation *see* Mutation
Paramyelois
Paramyxoviruses
  (Paramyxovirus)
Paranasal *see* Nasal

Paranevisanus
Paraoesophageal *see*
　Esophagus
Paraoxon
Paraperiglischrus
Paraperlinae
Paraphlepsius
Paraphysomonas
Paraplegia
Paraponera
Paraprotein
Paraproteinaemia *see*
　Proteinemia
Parapudaspis
Paraquat *(Gramoxon)*
　X-Ref: Gramoxone
Parasarcophaga
Parascaris
Paraschizognathus
Parascorpiops
Parasetigena
Parasexual
Parashorea
Parasitaphelenchus
Parasites *(Nonparasitized,*
　Parasite, Parasitic,
　Parasitically, Parasitise,
　Parasitised, Parasitising,
　Parasitism, Parasitisms,
　Parasitization, Parasitize,
　Parasitized, Parasitizes,
　Parasitizing, Parasitologic,
　Parasitological,
　Parasitology)
　X-Ref: Antiparasitic,
　Autoparasitism,
　Nonparasitic,
　Superparasitism,
　Unparasitized
Parasiticol
Parasitidae
Parasitiformes
Parasitoidea
Parasitoids *(Parasitoid)*
Parasitologists
Parasitoses *(Parasitosis)*
Parasitus
Paraspidodera
Parasympathetic
Paratemnus
Paratenic
Paratheresia
Parathion
　X-Ref: Methylparathion,
　Thiophos
Parathyroid
　X-Ref: Proparathyroid
Parathyroidectomy
Paratoxodera
Paratrichobius
Paratrichodorus
Paratrioza
Paratriphleps
Paratuberculosis
Paratydeidae
Paratylenchidae

(Paratylenchinae)
Paratylenchus
Paratyphoid
Paratyphus
Paravespula
Parbendazole
　X-Ref: Helmatac
Parboiling *(Parboil)*
Parcoblatta
Pardosa
Paregle
Parelaphostrongylosis
Parelaphostrongylus
Parena
Parenchyma *(Parenchymal,*
　Parenchymatous)
Parent *(Parentage, Parental,*
　Parents)
　X-Ref: Biparental,
　Midparent
Parenteral *(Parenterally)*
Parentomobrya
Parentucellia
Paresis *see* Paralysis
Parholaspidae
Parholaspis
Paria
Parietal
Parietaria
Paripartal *see* Birth
Pariphyllin
Parishia
Parkerella
Parkeriaceae
Parkia
Parks *(Park)*
Parlatoria
Parmelia
Parmeliaceae
Parmesan
　X-Ref: Parmigiano
Parmigiano *see* Parmesan
Parnalius
Parnassia
Parnassiana
Parnassiidae
Parnassius
Parnkalla
Parodia
Parolinia
Paromalus
Paronellinae
Paronychia
Parotid
Paroxyna
Paroxysms *(Paroxysmal)*
Parrots *(Parrot)*
Parsley
Parsnips *(Parsnip)*
Parsonsia
Partal *see* Birth
Parthenice
Parthenin
Parthenium
Parthenocarpy
　(Parthenocarpic)

Parthenocissus
Parthenogenesis *(Apogamic,*
　Apogamous, Apogamy,
　Apomictic,
　Parthenogenetic,
　Parthenogenetical,
　Parthenogenic)
　X-Ref: Apogamia,
　Apomixis, Uniparental
Parthenolecanium
Particleboards
　(Particleboard)
Particulate *(Particulates)*
Partnuniella
Partridges *(Partridge)*
Parturition *see* Birth
Paruroctonus
Parvine
Parvoraphidia
Parvoviruses *(Parvovirus)*
Pasilobus
Paspalidium
Paspalum
Passalidae
Passaloecus
Passalora
Passalurus
Passeriformes *(Passerine,*
　Passerines)
Passeromyia
Passiflora
Passifloraceae
Passiflorine
Passionfruit
Pasta
Paste *(Pastes, Pasty)*
Pasteurella
Pasteurellosis
Pasteurization
　(Pasteurisation,
　Pasteurised, Pasteuriser,
　Pasteurized, Pasteurizer,
　Pasteurizing)
　X-Ref: Subpasteurization
Pastinaca
Pastoral *(Pastoralism,*
　Pastoralists)
Pastry *(Pastries, Pie)*
　X-Ref: Pies
Pastures *(Pasturage,*
　Pasture, Pastured,
　Pastureland,
　Pasturelands, Pasturing)
Patanga
Patchoulenes
Patchouli
Patella *(Patellar)*
Patellaria
Patents *(Patent, Patented)*
Pathochemical
Pathogenesis
　(Etiopathogeny,
　Pathogenetic,
　Pathogenetical,
　Pathogenic,
　Pathogenically,

Pathogenicity,
　Pathogenous, Pathogeny)
　X-Ref: Apathogenic,
　Etiopathogenesis,
　Nonpathogenic
Pathogens *(Pathogen)*
Pathologists *(Pathologist)*
Pathology
　(Pathohistological,
　Pathologic, Pathological,
　Pathologically)
　X-Ref: Nonpathogical
Pathomorphology
　(Pathomorphological)
Pathophysiology
　(Pathophysiologic,
　Pathophysiological)
Pathotoxin *see* Toxicity
Pathotypes
Patinoa
Patrobus
Patulin
Paulownia
Pawpaw *(Pawpaws)*
Paxillus
Payments *(Payment)*
　X-Ref: Repayment
PBB
PCB *(PCBs)*
PCNB *see*
　Pentachloronitrobenzene
PCP
Peaches *(Peach)*
Peacock *see* Peafowl
Peafowl
　X-Ref: Peacock
Peanuts *(Groundnut,*
　Peanut)
　X-Ref: Groundnuts
Peanutworm
Pearlmillet *(Bajri,*
　Pearlmillets)
　X-Ref: Bajra
Pears *(Pear)*
Peas *(Pea)*
Peasants *(Peasant,*
　Peasantry)
Peat *(Peatmoss, Peatmosses,*
　Peats, Peaty)
Peatlands *(Peatbogs,*
　Peatland)
Peavine
Pebbles *see* Gravel
Pebrine
Pebulate
　X-Ref: Tillam
Pecans *(Pecan)*
Pectate
Pectawamorin
Pectinaria
Pectinases *(Pectinase)*
Pectinesterase
Pectiniseta
Pectinmethylesterase
Pectinophora
Pectins *(Pectic, Pectin,*

Pectinolytic, Pectolytic)
Pectobacterium
Pedagogy *see* Education
Pedal
Pedaria
Pediasia
Pediastrum
Pediatrics *(Pediatric)*
Pedicel *(Pedicels, Pedicle,*
  Pedicles)
Pedicularis
Pediculidae
Pediculosis
Pediculus
Pedigrees *(Pedigree)*
Pedilanthus
Pediobius
Pediocactus
Pediococcus *(Pediococcal,*
  Pediococci)
Pedobotanical
Pedochemical
Pedogenesis *(Paedogenetic,*
  Paedogenetically,
  Pedogenetic, Pedogenic)
  X-Ref: Paedogenesis
Pedologists *(Pedologist)*
Pedology *(Micropedological,*
  Pedo, Pedologic,
  Pedological)
  X-Ref: Agrology,
    Micropedology
Pedomorphology
  (Pedomorphic)
Pedons *(Pedon)*
Pedozoology
  (Pedozoological)
Peduncle *(Peduncles,*
  Pedunculate)
Peels *(Peel, Peelability,*
  Peelable, Peeled, Peeling)
  X-Ref: Unpeeled
Peganum
Pegomya *(Pegomyia)*
Pekin
Pekingese
Pelargonidin
Pelargonin
Pelargonium *(Pelargoniums)*
Pelecitus
Pelecyphora
Pelgigera
Pelicans
Pelidnota
Peliosis
Pellagra
Pelletization *(Pellet,*
  Pelletability, Pelleted,
  Pelleter, Pelleting,
  Pelletized, Pellets,
  Pelletting)
Pellia
Pellicle *(Pellicular)*
Pellicularia
Pellionia
Pelochrista

Pelodera
Peltandra
Peltigera
Peltodytes
Pelts *(Pelt, Pelting)*
Pelvecyan
Pelvetia
Pelvis *(Pelvic)*
  X-Ref: Intrapelvic
Pemoline
Pemphigus *(Pemphigoid)*
Penaeus
Pencilloyl
Pendarus
Pendepon
Penetration *(Penetrability,*
  Penetrable, Penetrant,
  Penetrating)
  X-Ref: Nonpenetrating
Penetrometer
  (Penetrometry)
Penguins *(Penguin)*
Penial *see* Penis
Penicillanic
Penicillic
Penicillidia
Penicillin *(Penicillins)*
Penicillium *(Penicillia)*
Penicilloyl
Peniocereus
Peniophora
Penis *(Penile)*
  X-Ref: Penial
Penitrem
Penitrothion
Pennisetum
Penoxalin
Pens *(Pen, Penned,*
  Penning)
Penstemon
Pentaacetic
Pentac
Pentachaeta
Pentachloride
Pentachloroaniline
Pentachloronitro
Pentachloronitrobenzene
  X-Ref: PCNB,
    Quintozene, Terraclor
Pentachlorophenate
Pentachlorophenols
  (Pentachlorophenol)
Pentaclethra
Pentacme
Pentafluorobenzyl
Pentagastrin
Pentaketide
Pentalobus
Pentalonia
Pentandra
Pentane
Pentanediol
Pentanone
Pentapanax
Pentapeptide *see* Peptides
Pentaphalangium

Pentaploids *(Pentaploid)*
Pentas
Pentasaccharide
Pentatomidae *(Pentatomid,*
  Pentatomids)
Pentatomoidea
Pentenoic
Penthaleus
Penthalodidae
Pentites
Pentitols
Pentobarbital
Pentobarbitone
Pentodontini
Pentopetia
Pentor
Pentosans *(Pentosan,*
  Pentosane)
Pentoses *(Pentose)*
  X-Ref: Anhydropentoses
Pentoxide
Pentzia
Peonin
Peony *(Peonies)*
Peperomia
Peponapis
Peponium
Pepper *(Peppers)*
Peppermint
Pepperoni
Pepro
Pepsin *(Pepsinogen, Pepsins)*
Pepsis
Pepstatin
Peptic
Peptidases *(Peptidase)*
Peptides *(Peptid, Peptide,*
  Peptidic, Peptido)
  X-Ref: Octapeptide,
    Pentapeptide,
    Propeptides
Peptidoglycans
  (Peptidoglycan)
Peptidyl
Peptone *(Peptones)*
Peracarida
Peracetic
Peramphistomum
Peranema
Peraphyllum
Perborate
Percassa
Perchlorates *(Perchlorate)*
Perchloric
Percolation *(Percolate,*
  Percolated, Percolating)
Percursaria
Percutaneous *see* Skin
Perdix
Peregrine
Peregrinus
Perennation *(Perennating)*
Perennial *(Perennials)*
  X-Ref: Plurennial
Pereskia
Pereskiopsis

Perezia
Perfluidone
Perfluorinated
Perforations *(Perforated,*
  Perforating, Perforation)
Perforine
  X-Ref: Anhydroperforine
Performic
Perfumes *(Perfume,*
  Perfumery)
Perfusion *(Perfusable,*
  Perfused)
Perga
Pergularia
Perhydrol
Perianal *see* Anus
Perianth
Periarteritis *see* Arteritis
Peribuccal *see* Buccal
Pericallia
Pericarditis
Pericardium *(Pericardial)*
Pericarp *(Pericarps)*
Periclinal
Periconia
Pericopsis
Pericycle
Perideridia
Periderm
Peridermium
Peridesmia
Peridinin
Peridinium
Peridroma
Perigynium
Perilampidae
Perilampus
Perileptus
Perilitus
Perilla
Perillene
Perillus
Perinatal *see* Birth
Perineum *(Perineal)*
Periodates *(Periodate)*
Periodicity *(Periodicities)*
Periodontal
Periodontopathy
Periosteum *(Periosteal)*
Periovulatory *see* Ovulation
Peripartal *see* Birth
Peripatoides
Periphyllus
Periphyton
Periplaneta
Periplanone
Periploca
Periplogenin
Peripsocidae
Perishables *(Perishable)*
Peristalsis *(Peristaltic)*
  X-Ref: Antiperistaltic
Peristenus
Peristeria
Peristome
Peristrophe

Perithecium *(Perithecia,*
  Perithecial,
  Protoperithecia)
  X-Ref: Protoperithecial
Peritoneum
  (Intraperitoneally,
  Peritoneal)
  X-Ref: Intraperitoneal
Peritonitis
Peritrophic
Perityle
Perivaginal *see* Vagina
Perivascular *see* Vascular
Perivitelline *see* Vitelline
Periwinkle *(Periwinkles)*
Perizoma
Perkinsiella
Perla
Perlidae
Perlinella
Perlite
Perloline
Permafrost
  X-Ref: Subpermafrost
Permanganate
Permeability *(Permeation)*
  X-Ref: Semipermeable
Permeameter
Permease
Perniphora
Perocin *see* Zineb
Peromyscopsylla
Peromyscus
Peroneutypa
Peronospora
Peronosporaceae
Peronosporales
Peronosporosis
Perosis
Peroxidases *(Peroxidase,*
  Peroxydase, Peroxydasic)
  X-Ref: Apoperoxidase
Peroxides *(Peroxidation,*
  Peroxidative, Peroxide,
  Peroxidized)
Peroxisomes *(Peroxisomal,*
  Peroxisome)
  X-Ref: Extraperoxisomal
Peroxobenzoic
Peroxyacetic
Peroxyacetyl
Peroxyferolide
Peroxysomes *(Peroxysomal)*
Perphenazine
Perrieri
Persea
Persectania
Persicargas
Persicinus
Persimmon *(Persimmons)*
Personnel
Persteril
Pertya
Pertyol
Perulifera
Peschiera

Pestalotia *(Pestalozzia)*
Pestalotiopsis
Pesticides *(Pesticidal,*
  Pesticide)
Pests *(Pest)*
Petalostemon
Petals *(Petal, Petaloid)*
Petasitenine
Petasites
Petechia *(Petechial)*
Pethidine
Petioles *(Petiolar, Petiole,*
  Petioled)
Petitgrain
Petiveria
Petrea
Petrobius
Petrocelis
Petrochemical
Petrodavisia
Petroglossum
Petrol *see* Gasoline
Petroleum
Petrology
Petroprotein
Petroselinum
Petunia *(Petunias)*
Peucedanum
Peumus
Peyote
Pezicula
Peziza
Pezizaceae
Pezizales
Pezizella
Phacelia
Phacidium
Phaeanthus
Phaedon
Phaedranassa
Phaenacantha
Phaenicia
Phaenocarpa
Phaenops
Phaenopsectra
Phaeocollybia
Phaeocryptopus
Phaeodactylum
Phaeoisaria
Phaeoisariopsis
Phaeophyceae
Phaeophyta
Phaeoseptoria
Phaeotrichoconis
Phaffia
Phages *(Bacteriophage,*
  Phage)
  X-Ref: Bacteriophages
Phagocarpus
Phagocytes *(Phagocyte,*
  Phagocytic, Phagocytose,
  Phagocytosis,
  Phagocytotic)
Phagodeterrency
Phagostimulation
Phagotrophic

Phaius
Phakopsora
Phalacrocorax
Phalacrotophora
Phalaenopsis *(Phalaenopsid)*
Phalalis
Phalange *(Phalanges,*
  Phalanx)
Phalangida
Phalangiidae
Phalangiinae
Phalaris
Phaleria
Phallacidin
Phallacin
Phallisacin
Phallolysin
Phallotoxins
Phallus *(Phalloid)*
Phalsa
Phanacis
Phaneranthy
Phanerochaete
Phanerogam
  (Phanerogamia,
  Phanerogamic,
  Phanerogams)
Phanerotoma
Phaneta
Phanonychus
Phanuropsis
Phaonia
Pharate
Pharbitis
Pharmaceutics
  (Pharmaceutic,
  Pharmaceutical,
  Pharmaceuticals,
  Pharmacopedia,
  Pharmacy)
Pharmacodynamics
  (Pharmacodynamic,
  Pharmacokinetic)
  X-Ref: Pharmacokinetics
Pharmacognosy
  (Pharmacognostic,
  Pharmacognostical)
Pharmacokinetics *see*
  Pharmacodynamics
Pharmacology *(Pharmaco,*
  Pharmacologic,
  Pharmacological,
  Pharmacologically)
Pharmacotherapy
  (Pharmacotherapeutics)
Pharohylaeus
Pharyngeal *see* Pharynx
Pharyngitis
Pharynx
  X-Ref: Pharyngeal
Phaseic
Phasemy
Phaseolamin
Phaseollin *(Phaseolin)*
Phaseolus
Phaseotoxin

Phases *(Phase, Phasic)*
Phasganophora
Phasianidae
Phasianus
Phasiidae
Phasmatidae *(Phasmatid,*
  Phasmid, Phasmidae)
Phasmatodea
Phatnocader
Phaulacridium
Phauloppia
Pheasants *(Pheasant)*
Phebalium
Pheidole
Phelline
Phellinus
Phellodendron
Phellogen
Phelloids
Phenacaspis
Phenacetin
Phenalenone
Phenamiphos
Phenanthrene
Phenanthroline
Phenarctin
Phenazopyridine
Phenethyl
Phenethylamines
  (Phenethylamine)
Phenetic
Phenformin
Phenmedipham *see* Betanal
Phenobarbital
  (Phenobarbitone)
Phenocopies
Phenogenetics
Phenolases *(Phenolase)*
Phenolcarboxylic
Phenoldisulfonic
Phenology *(Phenological,*
  Phenologically)
  X-Ref: Biophenology,
  Microphenologic
Phenoloxidases
  (Phenoloxidase)
Phenols *(Phenol, Phenolic,*
  Phenolics)
Phenothiazine
Phenothrin
Phenotypes *(Phenotype,*
  Phenotypic, Phenotypical,
  Phenotypically,
  Phenotyping)
Phenoxazine
Phenoxy
Phenoxyacetic
Phenoxyalkanoic
Phenoxybenzamine
Phenoxybenzyl
Phenoxypropionitriles
Phenthion *see* Fenthion
Phenthoate
Phenyl *(Phenylation,*
  Phenylic)
Phenylacetates

(Phenylacetate)
Phenylacetic
Phenylalanine
Phenylalaninol
Phenylalanyl
Phenylamides
 (Phenylamide)
Phenylarsonic
Phenylboronic
Phenylbutazone
Phenylcarbamates
 (Phenylcarbamate)
Phenylcoumarins
 (Phenylcoumarin)
Phenylene
Phenylephrine
Phenylethanol
Phenylethanolamine
Phenylethers
Phenylethyl
Phenylethylamines
 (Phenylethylamine)
Phenylglycine
Phenylhydrazine
Phenylhydrazones
 (Phenylhydrazone)
Phenylketonuria
Phenylmercury
 (Phenylmercuric)
Phenylphenate
Phenylphenol
Phenylphosphatase
Phenylpropanoid
Phenylpropanol
Phenylpyruvate
Phenylquinolizidines
Phenylureas (Phenylurea)
Pheochromocytoma
 (Pheochromocytomas)
Pheophytin
Pheopigments
Pheosiopsis
Pherbellia
Pheretima
Pheromones (Pheromonal,
 Pheromone)
Pherotesia
Phiala
Phialides (Phialide)
Phialophora
Phialospore
Phibalothrips
Phigalia
Philadelphus
Philaenus
Philanthinae
Philanthus
Philately
Phillipsia
Philodendron
 (Philodendrons)
Philodromus
Philoliche
Philolithus
Philonotis
Philophthalmus

Philophylla
Philopteridae
Philosamia
Philoscia
Philoside
Phimosis
Phityogamasus
Phlaeothripidae
Phlebia
Phlebophyllum
Phlebotomidae
Phlebotominae
 (Phlebotomine)
Phlebotomus
Phleomycin
Phleum
Phloem (Phloematic)
 X-Ref: Protophloem
Phloeosinus
Phloeoxena
Phlojodicarpus
Phlomis
Phlorizin (Phloridzin)
Phloroglucinol
Phlox
Phlugiola
Phlyctaena
Phlyctochytrium
Phocanema
Phoebis
Phoenix
Pholcidae
Pholeoixodes
Pholiota
Pholus
Phoma
Phomazarin
Phomopsis
Phoneutria
Phoniomyia
Phonocardiography
Phonotaxis
Phoracantha
Phoradendron
Phorate
 X-Ref: Thimet, Timet
Phorbia
Phorbic
Phorbol
Phoresy (Phoretic)
Phoridae (Phorid)
Phormia
Phormidium
Phormium
Phorodon
Phoroncidia
Phorone
Phosalone
Phosdrin
Phosfolan
 X-Ref: Cyolane
Phosfon
 X-Ref: Phosphon
Phosphagens
Phosphamide see
 Dimethoate

Phosphamidon
Phosphatases (Phosphatase,
 Phosphatasic)
 X-Ref: Bisphosphatase
Phosphates (Bisphosphate,
 Metaphosphate,
 Orthophosphate,
 Phosphate, Phosphatic,
 Superphosphate)
 X-Ref: Biphosphate,
 Metaphosphates,
 Orthophosphates,
 Superphosphates
Phosphatidases
Phosphatidate
Phosphatides (Phosphatide)
Phosphatidyl
Phosphatidylcholines
 (Phosphatidylcholine)
Phosphatidylethanolamine
Phosphatidylglycerol
Phosphatidylinositol
Phosphides (Phosphide)
Phosphine (Phosphin)
Phosphobacterin
Phosphocellulose
Phosphocreatine
Phosphodiesterases
 (Phosphodiesterase)
Phosphoenol
Phosphoenolpyruvate
Phosphoenolpyruvic
Phosphofructokinase
 (Phosphofructokinases)
Phosphoglucomutase
Phosphogluconate
Phosphoglycerate
Phosphoglyceric
Phosphoglycerides
Phosphoglyceromutase
Phosphoglycolate
Phosphohexose
Phosphohistidine
Phosphohomoserine
Phosphohydrolase
Phosphokinases
 (Phosphokinase)
Phospholipases
 (Phospholipase)
Phospholipids
 (Phospholipid)
Phospholipolysis
Phosphomannose
Phosphomolybdic
Phosphomonoesterase
 (Phosphomonoesterases)
Phosphon see Phosfon
Phosphonates (Phosphonate)
Phosphonic
Phosphonium
Phosphonobutyric
Phosphonodithioates
Phosphonomethyl
Phosphonomethylglycine
Phosphopeptide
Phosphoproteins

(Phosphoprotein)
Phosphoramidothioate
Phosphorescence
Phosphoribosyl
Phosphoribosyltransferase
Phosphoribulokinase
Phosphoric
Phosphoris
 X-Ref: Orthophosphoric,
 Polyphosphoric,
 Superphosphoric
Phosphorite
Phosphorodithioate
Phosphorodithiolate
Phosphorolysis
Phosphororganic see
 Organophosphates
Phosphorothiate
Phosphorothioates
 (Phosphorothioate)
Phosphorothionate
Phosphorus (Phospho,
 Phosphorated,
 Phosphorous)
 X-Ref: Radiophosphorus
Phosphoryl
Phosphorylases
 (Phosphorylase)
Phosphorylation
 (Phosphorylated,
 Phosphorylating,
 Phosphorylations,
 Phosphorylative)
 X-Ref:
 Alkylphosphorylation,
 Nonphosphorylated,
 Photophosphorylation,
 Transphosphorylation
Phosphorylcholine
Phosphorylethanolamine
Phosphorylinositol
Phosphosulfate
 (Phosphosulphate)
Phosphotransferases
 (Phosphotransferase)
Phostoxin
Phosvel
Phosvitin
Photic (Photically)
Photinia
Photinopteris
Photinus
Photo
Photoactivation
 (Photoactive,
 Photoinactive)
 X-Ref: Photoinactivation
Photoaffinity
Photoautotrophic see
 Phototrophism
Photobehavior see
 Photobiology
Photobiology
 (Photobiological)
 X-Ref: Photobehavior,
 Photophysiology

Photobleaching *see*
 Bleaching
Photochemiluminescence *see*
 Chemiluminescence
Photochemistry
 (Photochemical,
 Photochemically)
Photoconductivity
 (Photoconductive)
Photocontrol
Photoconversion
Photocyclization
 (Photocyclisation)
Photodecomposition *see*
 Photolysis
Photodegradation
 (Photodegradable)
Photodestruction
Photodieldrin *see* Dieldrin
Photodynamics
 (Photodynamic)
Photoelasticity
 (Photoelastic,
 Photoelastimetric)
Photoelectric
Photoelectron
 (Photoelectronic)
Photoemission
Photoenolization
 (Photoenolisation)
Photoenzymatic *see*
 Enzymes
Photogrammetry *(Airphoto,*
 Photogrammetric)
 X-Ref: Airphotos
Photography *(Photograph,*
 Photographic,
 Photographically,
 Photographing,
 Photographs, Photos)
 X-Ref: Orthophotography
Photoheterotrophy
 (Photoheterotrophic)
Photoinactivation *see*
 Photoactivation
Photoinduction
 (Photoinduced,
 Photoinductive)
Photoinsensitivity *see*
 Photosensitization
Photointerpretation
Photoirradiation
 (Photoirradiated)
Photoisomerization *see*
 Isomers
Photokinesis *see*
 Phototropism
Photolability *(Photolabile)*
Photolysis *(Photolytic)*
 X-Ref:
 Photodecomposition
Photometry
 (Histophotometric,
 Microphotometer,
 Photometer, Photometers,
 Photometric)

X-Ref: Histophotometry,
 Microphotometry
Photomicrography
 (Photomicrographs)
Photomodulation
Photomorphic
Photomorphogenesis
 (Photomorphogenetic,
 Photomorphogenic)
Photomutants *see* Mutants ,
Photon *(Photonic)*
Photoorganotrophically *see*
 Phototrophism
Photooxidation *see*
 Oxidation
Photooxygenation *see*
 Oxygenation
Photoperiodism
 (Photoperiod,
 Photoperiodic,
 Photoperiodical,
 Photoperiodically,
 Photoperiodicity,
 Photoperiods)
Photophase *(Photophases)*
Photophobia *(Photophobic)*
Photophosphorylation *see*
 Phosphorylation
Photophysiology *see*
 Photobiology
Photopigments *see* Pigments
Photopolymerization *see*
 Polymerization
Photoproduction
 (Photoproduced,
 Photoproducts)
Photoprotection
Photoreaction
 (Photoreactions,
 Photoreactive,
 Photoresponse)
 X-Ref: Photoresponses
Photoreactivation
 (Photoreactivating)
Photoreceptors
 (Photoreception,
 Photoreceptive,
 Photoreceptor)
Photoreduction
 (Photoreductions,
 Photoreductive)
Photorefraction
 (Photorefractory)
Photoregeneration *see*
 Regeneration
Photoregulation
Photorespiration
 (Photorespiratory,
 Photorespire,
 Photorespiring)
Photoresponses *see*
 Photoreaction
Photoreversion
 (Photoreversal,
 Photoreversibility,
 Photoreversible)

Photosensitization
 (Photosensitisation,
 Photosensitive,
 Photosensitivity,
 Photosensitized,
 Photosensitizers,
 Photosensitizing)
 X-Ref: Photoinsensitivity
Photostability
Photostimulation
Photosynthesis
 (Photosynthate,
 Photosynthates,
 Photosynthesizing,
 Photosynthetic,
 Photosynthetical,
 Photosynthetically)
 X-Ref:
 Nonphotosynthetic
Photosystems *(Photosystem)*
Phototautomerism
Phototaxis *see*
 Phototropism
Photothermal
Phototoxicity
Phototransformation
 (Phototransformable)
Phototrophism
 (Phototrophic)
 X-Ref: Photoautotrophic,
 Photoorganotrophically
Phototropism *(Phototactic,*
 Phototropy)
 X-Ref: Photokinesis,
 Phototaxis
Photuris
Phoxim *(Valexone)*
 X-Ref: Valexon
Phragmanthera
Phragmidium
Phragmipedium
Phragmites
Phragmoporthe
Phragmosomes
Phreatic *(Phreato)*
Phreatophytes
 (Phreatophyte,
 Phreatophytic)
Phronia
Phrydiuchus
Phryganea
Phryganeidae
Phryganophilus
Phryma
Phrymarolin
Phryne
Phtalophos
Phthalates *(Phthalate)*
Phthaldehyde
Phthalic
Phthalide
Phthalocyanine
Phthalophos
Phthalysulfathiazole
Phthiracaridae
Phthiracarus

Phthiraptera
Phthirus *(Phthirius)*
Phthorimaea
Phumosia
Phyciodes
Phycita
Phycitidae *(Phycitid,*
 Phycitinae)
Phycobilins *(Phycobilin)*
Phycobiliproteids
Phycobiliproteins
Phycobilisomes
Phycobionts *(Phycobiont)*
Phycocolloids
Phycocyanin
Phycoerythrin
 (Phycoerythrine)
Phycologists *see* Algologists
Phycology *see* Algae
Phycomyces
Phycomycetes
 (Phycomycete)
Phycomycosis
Phygadeuon
Phyla *(Phylum)*
Phyllachora
Phyllactinia
Phyllanthus
Phyllantidine
Phyllarthron
Phyllitis
Phyllium
Phyllobates
Phyllobius
Phyllocacti *(Phyllocactus)*
Phyllocladus
Phyllocnistis
Phyllocoptes
Phyllocoptruta
Phyllodecta
Phyllodoce
Phyllodromia
Phyllody
Phylloerythrin
Phyllomorph
Phyllonoma
Phyllonorycter
Phyllophaga
Phyllophora
Phyllophoraceae
Phylloplane
Phyllospadix
Phyllosphere *(Phyllospheric)*
Phyllostachys
Phyllosticta
Phyllotaxy *(Phyllotaxis)*
Phylloteles
Phyllotopsis
Phyllotreta
Phylloxera
Phylloxeridae
Phylloxerina
Phylocentropus
Phylogeny *(Phylogenetic,*
 Phylogenetically,
 Phylogenic)

Phylolacca
Phylosamia
Phymaspermum
Phymata
Phymatolithon
Phymatotrichum
Physa
Physalis
Physaloptera
Physalopteridae
Physalospora
Physarella
Physarum
Physcia
Physcion
Physcomitrella
Physcomitrium
Physeter
Physicochemistry
　(Physicochemical)
Physics
Physiobiochemical *see*
　Biochemistry
Physiography
　(Physiographic)
Physiologists *(Physiologist)*
Physiology *(Physio,*
　Physiologic, Physiological,
　Physiologically)
Physiopathology
　(Physiopathologic,
　Physiopathological)
Physocarpus
Physocephala
Physochlaina
Physoderma
Physodes
Physogastry *(Physogastric)*
Physopella
Physopelta
Physoperuvine
Physostegia
Physostigmine *(Eserine)*
Phytase
Phytates *(Phytate)*
Phytenoyl
Phyteuma
Phythium
Phythophthora
Phytic
Phytin
Phytoactive *see*
　Phytochemistry
Phytoalexins *(Phytoalexin)*
Phytobaenus
Phytobia
Phytochemistry
　(Phytochemical)
　X-Ref: Phytoactive
Phytochromes
　(Phytochrome)
Phytocides *(Phytocide,*
　Phytoncide, Phytoncides)
Phytoclimatology
　(Phytoclimate)
Phytocoenoses

(Phytocenologic,
Phytocenological,
Phytocenology,
Phytocenoses,
Phytocenosis,
Phytocenotic,
Phytocoenological,
Phytocoenology,
Phytocoenosis,
Phytocoenotic)
Phytocoris
Phytodecta
Phytoecdysones
　(Phytoecdysone)
Phytoecology
　(Phytoecological)
Phytoene
Phytoestrogens
　(Phytoestrogen)
Phytoferritin
Phytoflagellates
　(Phytoflagellate)
Phytogenesis
Phytogenic
Phytogeography
　(Phytogeographic,
　Phytogeographical)
Phytohelminthology
Phytohemagglutinin
　(Phytohaemagglutin,
　Phytohaemagglutinin,
　Phytohaemagglutinins,
　Phytohemagglutinins)
Phytohormones
　(Phytohormone)
Phytokinins *see* Kinins
Phytol
Phytolacca
Phytolaccaceae
Phytolaccoside
Phytolectins
Phytoliths *(Phytolith)*
　X-Ref: Silicophytoliths
Phytolysosomes
　(Phytolysosome)
Phytomass
Phytomedicine
　(Phytomedical)
Phytomitogens
　(Phytomitogen)
Phytomonas
Phytomyza
Phytonematodes
Phytoparasitism
　(Phytoparasitic)
Phytopathogens
　(Phytopathogen,
　Phytopathogenic)
　X-Ref:
　　Antiphytopathogenic
Phytopathologists
　(Phytopathologist)
Phytopathology
　(Phytopathological)
Phytophaga
Phytophages *(Phytophage,*

Phytophagous)
Phytopharmacology
　(Phytopharmaceutical)
Phytophthora
　(Phytophthoras)
Phytophysiology
Phytoplanktons
　(Phytoplankton,
　Phytoplanktonic)
Phytoptidae
Phytoptus
Phytoquinoid
Phytosanitary
　(Phytosanitation)
Phytoscaphus
Phytoseiidae *(Phytoseiid)*
Phytoseiulus
Phytosociology
　(Phytosociological)
Phytosol *see* Trichloronate
Phytosphere
Phytosphingosine
Phytosterol *(Phytosterols)*
Phytotoxicity *(Phytotoxic)*
Phytotoxins *(Phytotoxin)*
Phytotron *(Phytotronic,*
　Phytotronics, Phytotrons)
Piaranthus
Piassava
Pica
Picea *(Spruces)*
　X-Ref: Spruce
Pichia
Picidae
Piciformes
Pickers *(Picker)*
Pickles *(Pickle, Pickled,*
　Pickling, Picklings)
Pickleworms *(Pickleworm)*
Pickup
Picloram
　X-Ref: Tordon
Picnometers *see*
　Pycnometer
Picolinamide
Picolinic
Picornaviruses
　(Picornavirus)
Picrasins
Picrasma
Picrorhiza
Picrotoxin
Pictipes
Pidan
Pidoplitchkoviella
Pied
Pieridae *(Pierid)*
Pieris
Pies *see* Pastry
Piesma
Piestopleura
Pietrain
Piezoelectricity
　(Piezoelectric)
Pigeonpeas *(Pigeonpea)*
　X-Ref: Arhar

Pigeonpox
Pigeons *(Pigeon)*
Pigfarming
Piggery *see* Pigsty
Piggotia
Piglets *(Piglet)*
Pigmeat *see* Pork
Pigmen
Pigments *(Nonpigmented,*
　Photopigment, Pigment,
　Pigmentary,
　Pigmentation, Pigmented,
　Pigmenter)
　X-Ref: Copigments,
　　Nonpigment,
　　Photopigments
Pigs *(Pig)*
Pigsty *(Piggeries, Pigpens,*
　Pigsties, Sties)
　X-Ref: Piggery, Sty
Pigweed
Pike
Pikonema
Pila
Pile *(Piles, Piling, Pilings)*
Pilea
Pileus *(Pileate)*
Pili
Pilobolus
Pilocarpine
Pilocarpus
Pilocereine
Pilogalumna
Pilophoron
Pilophorus
Pilose
Pilot
Pilularia
Pimaradiene
Pimaric
Pimaricin
Pimelea
Pimelia
Pimeliaphilus
Pimenta
Pimephales
Pimiento *(Pimento,*
　Pimientos)
Pimpinella
Pimpla
Pimplopterus
Pinaceae
Pinching
Pineal
Pinealectomy
　(Pinealectomized)
Pineapples *(Pineapple)*
Pinellia
Pinene
Pines *see* Pinus
Pinewood *(Pinewoods)*
Pinguicula
Pinguinain
Pinidine
Pinitol
Pinkeye

Pinnae
Pinocytosis
Pinoresinol
Pinosylvin
Pinus *(Pine)*
    X-Ref: Pines
Pinworm *(Pinworms)*
Pionea
Piophila
Piophilidae
Pipecolic
Piper
Piperaceae *(Piperaceous)*
Piperazine
Piperenone
Piperidine *(Piperidin,*
    Piperidino)
Piperidinium
Piperonyl
Piperonylbutoxide
Piperovatine
Pipes *(Pipe, Pipeline,*
    Pipelines, Piping)
Pips *(Pip)*
Piptadeniastrum
Piptanthus
Piptocalyx
Piptocephalis
Piptochaetium
Piptoporus
Pipturus
Pipunculidae
Piricularia *see* Pyricularia
Pirimidine
Pirimidinyldimethyl
Pirimiphos
Pirimor
Pirobasidium
Piroplasm *(Piroplasmas,*
    Piroplasms)
Piroplasmoses
    (Piroplasmosis)
Piroplasmosis *(Babesicidal,*
    Babesiosis)
    X-Ref: Babesiasis
Pirus *see* Pyrus
Pisatin
Pisaura
Pisauridae *(Pisaurinae)*
Pisces
Piscicide *(Piscicidal)*
Pisciculture
Piscidia
Pisolithus
Pissodes
Pistachio *(Pistachios)*
Pistacia
Pistia
Pistil *(Pistils)*
Pistillate
Piston
Pisum
Pitangueiras
Pitch *(Pitching)*
Pitchers *(Pitcher)*
Pith

Pithecellobium
Pithitis
Pithomyces
Pithophora
Pithyotettix
Pits *(Pit)*
Pitsaws *see* Sawing
Pittosporaceae
Pittosporum
Pituitary *(Pituitaries)*
Pityogenes
Pityokteines
Pityophthorus
Pityriasis
Pityrogramma
Pityrosporum
Pivot *(Pivotal)*
Pizza
Placenta *(Placentae,*
    Placental, Placentas,
    Placentation)
    X-Ref: Afterbirth,
    Transplacental
Placentophora
Placopecten
Plagianthus
Plagiochila
Plagiognathus
Plagiohammus
Plagiolepis
Plagiomnium
Plagiotrochus
Plagiotropism *(Plagiotropic,*
    Plagiotropy)
Plagiprospherysa
Plagithmysus
Plague *(Plagues)*
Plains *(Plain)*
Planaria
Planarian *(Planarians)*
Planavin
Plane *(Planar, Planer,*
    Planes, Planing)
Planetree *see* Platanus
Planimeter *(Planimetric)*
Planipennia
Plankton *(Microplankton,*
    Planktonic, Planktons)
    X-Ref: Microplanktons
Planktoniella
Planococcus
Planofix *see* NAA
Planorbarius
Planorbis
Planosols *(Planosol)*
Plantaginaceae
Plantago
Plantains *(Plantain)*
Plantations *(Plantation)*
Planters *(Planter)*
Planthoppers *(Planthopper)*
Planting *(Replant,*
    Replants)
    X-Ref: Preplanting,
    Replanting
Plantlets *(Plantlet)*

Plantules *(Plantule)*
Plantvax
Plaque *(Plaques, Plaquing)*
    X-Ref: Microplaque,
    Nonplaquing
Plasm *(Plasms)*
    X-Ref: Multiplasms
Plasma *(Plasmas,*
    Plasmatype, Plasmic)
Plasmacytoma
Plasmacytosis *(Plasmacytic)*
Plasmalemma
Plasmalogen
Plasmapheresis
Plasmatic *see* Protoplasm
Plasmids *(Plasmid)*
Plasmin
Plasminogen
Plasmocyte *(Plasmocytic)*
Plasmodesmata
Plasmodiophora
Plasmodium *(Plasmodia,*
    Plasmodial)
    X-Ref: Pseudoplasmodia
Plasmogamy
Plasmolysis *(Plasmolysed,*
    Plasmolytic)
    X-Ref: Deplasmolysis
Plasmon
Plasmopara
Plasmoptysis
Plastein
Plastics *(Plastic, Plasticity,*
    Plasticization, Plasticizer,
    Plasticizers, Plasticizing)
Plastids *(Plast, Plastid,*
    Plastidal, Plastidial,
    Plasts, Proplastid)
    X-Ref: Intraplastidial,
    Proplastids
Plastochron
Plastocyanin *(Plastocyanine,*
    Plastocyanins)
    X-Ref: Apoplastocyanin
Plastome *(Plastom)*
Plastoquinol
Plastoquinones
    (Plastoquinone)
Plastosciara
Platambin
Platambus
Platanaceae
Platanthera
Platanus *(Planetrees)*
    X-Ref: Planetree
Plate *(Plated, Plates,*
    Plating)
Plateau *(Plateaus)*
Platelets *(Platelet)*
    X-Ref: Antiplatelet
Platform *(Platforms)*
Plathemis
Plathypena
Platinum
Platitenia
Platycephala

Platycerium
Platycerus
Platycleis
Platycodin
Platycodon
Platycotis
Platydema
Platygaster
Platygasteridae
Platyhelminthes
Platyhypnidium
Platymeris
Platymonas
Platynosomum
Platynota
Platynothrus
Platynotini
Platypalpus
Platyparea
Platypezidae
Platyphytoptus
Platypodidae
Platyptilia
Platyrhacus
Platysma
Platystomatidae
Platystomidae
Platytelenomus
Platytylus
Playaspalangia
Plebeia
Plebeiogryllus
Plecia
Plecoptera
Plectascales
Plectenchyma
Plectomycetes
Plectonema
Plectranthus
Plectrocnemia
Plectroctena
Plegaderus
Pleganophorus
Pleioblastus
Pleiochaeta
Pleiotropy *(Pleiotrope,*
    Pleiotropic)
Pleistocene
Pleistophora
Plenodomus
Pleolophus
Pleomorphism
Pleopeltis
Pleospora
Pleosporaceae
Plerocercoid *(Plerocercoids)*
Plesiopsothrips
Pleurastrum
Pleuritis
Pleurochrysis
Pleuroclada
Pleurococcus
Pleurolithobius
Pleuronectes
Pleuropneumonia
    X-Ref:

Antipleuropneumonia
Pleurosorus
Pleurothallis
Pleurotus
Pleurozium
Pleuston
Plexus
Plictran
Plinthite
Pliocene
Plistophora
Plocamium
Ploceus
Plodia
Ploidy *(Ploid, Ploidic,*
    Ploidies)
    X-Ref: Mixoploidy
Plots *(Plot, Plotting)*
    X-Ref: Microplots
Plowing *(Ploughed,*
    Ploughing, Plowed,
    Underplowed, Unplowed)
    X-Ref: Nonplowing,
    Replowing,
    Underplowing,
    Unploughed
Plows *(Plough, Ploughs,*
    Plow)
Pluchea
Plucking
Plugs *(Plug, Plugging)*
Plumage *see* Feathers
Plumbagella
Plumbagin
Plumbaginaceae
Plumbago
Plume
Plumeria
Plums *(Plum)*
Plumule
Plurennial *see* Perennial
Plusia
Plusiidae
Plusiinae *(Plusiine)*
Plutella
Plutellidae
Pluteus
Plutonium
Plylloquinone
Plywood *(Plymill, Plywoods)*
Pneumatic
Pneumatopteris
Pneumococcus *see*
    Diplococcus
Pneumoconiosis
Pneumocystis
Pneumoencephalography
Pneumoenteritis
Pneumomycosis
Pneumonia *(Pneumonic,*
    Pneumonitis)
Pneumonyssus
Pneumostrongylus
Pneumothorax
Pnigalio
Pnyxia

Poa
Poaceae
Poanes
Poculum
Podagrica
Podagrion
Podapolipidae
Podaxis
Podisma
Podisus
Podocarpaceae
    (Podocarpaceous)
Podocarpus *(Podocarp)*
Podolampas
Podolide
Podophyllaceae
Podophyllum
Podosordaria
Podosphaera
Podospora
Podostemonaceae
    (Podostemaceae)
Podothrips
Podotrochlitis
Podotrochlosis
Pods *(Pod)*
Podura
Poduromorpha
Podzols *(Podsol,*
    Podsolization, Podsols,
    Podzol, Podzolic,
    Podzolitic, Podzolization,
    Podzolized,
    Pseudopodzolic,
    Pseudopodzolization)
    X-Ref: Pseudopodzol,
    Sodpodzolic
Poeas
Poecilia
Poecilimon
Poecilmitis
Poecilocera
Poecilocerus
Poecilopsis
Poekilocerus
Pogonatum
Pogonia
Pogonocherini
Pogonomyrmex
Pogonortalis
Pogonus
Pogostemon
Poikilohydrous
Poikilothermic
Poikiloxerophyte
    (Poikiloxerophytic)
Poinsettia *(Poinsettias)*
    X-Ref: Minipoinsettias
Poison *(Poisonous, Poisons)*
Poisoning *(Poisoned,*
    Poisonings)
    X-Ref: Unpoisoned
Pokeweed
Polanisia
Polarimetry *(Polarimeter)*
Polarography

(Polarographic)
Polarotaxis
Polders *(Polder)*
Polemoniaceae
Polemonium
Poles *(Pole)*
Polfa
Polfos
Polianthes
Polifungin
Polioencephalomalacia
Poliomyelitis *(Polio)*
Poliopastea
Poliovirus
Polistes
Polistidae *(Polistinae)*
Polled *(Polledness)*
Pollen *(Pollens)*
Pollenia
Pollenosis
Pollination *(Pollinate,*
    Pollinated, Pollinating,
    Pollinations)
    X-Ref: Transpollination,
    Unpollinated
Pollinators *(Pollinator)*
Pollutants *(Micropollution,*
    Pollutant)
    X-Ref: Antipollutants,
    Micropollutants
Pollution *(Polluted,*
    Pollutional)
Polonium
Poloxalene
Polwarth
Polyacetylene
    (Polyacetylenes,
    Polyacetylenic)
Polyachyrus
Polyacrylamide
    (Polyacrylamid)
Polyacrylamine
Polyacrylate
Polyacrylic
Polyacrylonitrile
Polyadenosine
Polyadenylate
    (Polyadenylated,
    Polyadenylation)
    X-Ref:
    Nonpolyadenylated
Polyadenylic
Polyaldehydexylan
Polyalthia
Polyamides *(Polyamide)*
Polyamines *(Polyamine)*
Polyamino
Polyanethol
Polyanthes
Polyanthin
Polyanthinin
Polyarginine
Polyarteritis *see* Arteritis
Polyarthritis *see* Arthritis
Polyaspis
Polybags

Polybia
Polybrominated
Polybromobiphenyls
Polycarbacin
    (Polycarbacine)
Polycarbonate
Polycarpicae
Polycentropus
Polychlorcamphene
Polychlorination
    (Polychlorinated)
Polychloronaphthalenes
Polychlorpinene
Polychrosis
Polyclada
Polycross *see* Crossbreeding
Polyctenidae
Polycyclic
Polycyclodiene
Polycystic
Polycythemia
    (Polycythaemia)
Polydactylism *(Polydactyly)*
Polydesmida
Polydesmidae
Polydesmoidea
Polydesmus
Polydimethylsiloxanes
Polydipsia *(Polydipsic)*
Polydrusus *(Polydrosus)*
Polyelectrolytes
    (Polyelectrolyte)
Polyembryony
    (Polyembryonic)
Polyenes *(Polyene)*
Polyenoic
Polyergus
Polyester *(Polyesters)*
Polyestrus *see* Estrus
Polyethism
Polyethylene *(Polyethylenes)*
    X-Ref: Polythene
Polyethyleneglycol
Polyethyleneimine
    (Polyethylenimine)
Polyfluoroethylene
Polygala
Polygalaceae
Polygalacturonases
    (Polygalacturonase)
Polygalacturonate
Polygalacturonic
Polygenic *see* Genes
Polygenis
Polyglucose *see* Glucoses
Polyglucosides
    (Polyglucoside)
Polyglutamyl
Polygonaceae
Polygonatum
Polygonella
Polygonia
Polygonum
Polygramma
Polygraphy *(Polygraphic)*
Polyguanylate

Polygyny *(Polygynous)*
Polyhalite
Polyhaploids *see* Haploids
Polyhedrosis
Polyhybrids *see*
  Hybridization
Polyhydroxybenzenes
Polyhydroxyphenylether
Polyisocyanate
Polyisoprenes
Polyisoprenoids
Polyisoprenol
Polyketides *(Polyketide)*
Polylepsis
Polylysine
Polymarcine
Polymerases *(Polymerase)*
Polymerization
  (Polymerisation,
  Polymerizations,
  Polymerized)
  X-Ref:
  Photopolymerization
Polymers *(Biopolymer,*
  Polymer, Polymeric,
  Polymerics)
  X-Ref: Biopolymers
Polymetaphosphate
Polymethyl
Polymixis
Polymorphism *(Dimorphic,*
  Dimorphism, Dimorphous,
  Polymorphic,
  Polymorphisms,
  Polymorphous)
  X-Ref: Dimorphs
Polymyxa
Polynuclear *see* Nucleus
Polynucleotidase
Polynucleotide *see*
  Nucleotides
Polyoldehydrogenases
Polyolefins *(Polyolefines)*
Polyols *(Polyol)*
Polyoxin
Polyoxyethylene
Polypedilum
Polypeptides *(Polypeptide)*
Polyphaenis
Polyphaga
Polyphagy *(Polyphagous)*
Polyphenoloxidases
  (Polyphenoloxidase)
Polyphenols *(Polyphenol,*
  Polyphenolic,
  Polyphenolics)
Polyphenylalanine
Polyphloretin
Polyphosphatase
Polyphosphates
  (Polyphosphate)
Polyphosphoric *see*
  Phosphoris
Polyplax
Polyploids *(Polyploid,*
  Polyploidization,

Polyploidized,
  Polyploidogenic,
  Polyploidy)
Polypodiaceae
  (Polypodiaceous)
Polypodium
Polyporaceae
Polyporellus
Polyporus *(Polypore,*
  Polypores)
Polyprenoids
Polyprenol
Polyprenyl
Polyprenylphenols
Polyprenyltoluquinols
Polypropyl
Polypropylene
Polyps
Polypteris
Polyram
Polyriboadenylic
Polyribocytidylic
Polyriboinosinic
Polyribosomes
  (Polyribosome)
Polysaccharases
Polysaccharides
  (Polysaccharide)
Polyscias
Polyserositis
Polysiblings *see* Siblings
Polysiphonia
Polysomes *(Polysomal,*
  Polysome)
Polyspermy
Polysphondylium
Polysponin
Polystichum
Polystictus
Polystigma
Polystyrenes *(Polystyrene)*
Polysulfides *(Polysulfide,*
  Polysulphide)
Polysynaptic *see* Synapsis
Polytene *(Polytenic,*
  Polyteny)
Polythene *see* Polyethylene
Polythiols
Polytoma
Polytomella
Polytrichales
Polytrichum
Polytrophic
Polytropic
Polytypism
Polyunsaturates
  (Polyunsaturated)
Polyurethanes
  (Polyurethane)
Polyuria
Polyuridylic
Polyuronides
Polyvinyl
Polyvinylidene
Polyvinylpyrrolidone
Polyxenus

Pomace
Pomaceae *see* Malaceae
Pomatostoma
Pome *(Pomaceous,*
  Pomiferous)
Pomegranate
  (Pomegranates)
Pomerantzia
Pomfret
Pomiculture
Pomologists
Pomology *(Pomological)*
Pompilidae *(Pompilid)*
Pompilus
Poncimarin
Poncirus
Pond *(Ponds)*
  X-Ref: Microponds
Ponded *(Ponding)*
Pondweed
Ponera
Ponerinae *(Ponerine)*
Pongamia
Ponies *(Pony)*
Ponkan
Pontamalota
Pontania
Pontederia
Pontederiaceae
Pontevedrine
Pontogeneia
Poodle *(Poodles)*
Popcorn
Popillia
Poplars *see* Populus
Popliteus *(Popliteal)*
Popped *(Popping)*
Poppy *(Poppies)*
Population *(Overpopulated,*
  Populational, Populations)
  X-Ref: Micropopulations,
  Overpopulation,
  Repopulation,
  Subpopulations
Populus *(Poplar)*
  X-Ref: Poplars
Porcellio
Porcellionidae
Porcellium
Porcine
Porella
Pores *(Pore)*
Poria
Porifera
Porina
Pork
  X-Ref: Pigmeat,
  Superpork
Porkers *(Porker)*
Porometer
Poronia
Poroporo
Porosin
Porosis
Porosity *(Porous)*
Porotermes

Porpax
Porphobilinogen
Porphyra
Porphyria
Porphyridiales
Porphyridium
Porphyrins *(Porphyrin)*
Porroglossum
Portal
Porthetria
Portulaca
Portulacaceae
Posidonia
Possum
Postacrosomal *see*
  Acrosomes
Postcoital *see* Copulation
Postcopulation *see*
  Copulation
Postdiapause *see* Diapause
Postecdysis *see* Ecdysis
Posteclosion *see* Eclosion
Postembyonic *see* Embryo
Postemergence
Postfumigation *see*
  Fumigation
Postgenital *see* Genitals
Postglacial *see* Glacial
Postharvest
Posthatch *see* Hatching
Posthitis
Postimaginal *see* Imaginal
Postimplantation *see*
  Implants
Postinfectional *see* Infection
Postingestional *see*
  Ingestion
Postirradiation *see*
  Irradiation
Postmating *see* Mating
Postmilking *see* Milking
Postmitotic *see* Mitosis
Postmortem *see* Autopsies
Postmyocardial *see*
  Myocardia
Postnatal *see* Birth
Postpartum *see* Birth
Postprandial
Postpubertal *see* Puberty
Postradiation *see* Radiation
Postrigor *see* Rigor
Postsegregation *see*
  Segregation
Postsowing *see* Sowing
Postsurgical *see* Surgery
Postsynaptic *see* Synapsis
Posture *(Postural, Postures)*
Postvaccination *see*
  Vaccination
Postweaning *see* Weaning
Potable
Potamogeton
Potamogetonaceae
Potamogetonin
Potamon
Potamophylax

Potash
Potassium *(Potassic)*
  X-Ref: Dipotassium,
    Radiopotassium,
    Tripotassium
Potatoes *(Potato)*
Potentilla
Potentiometry
  (Potentiometric,
  Potentionmeter)
Poterium
Potéxvirus *(Potexviruses)*
Pothos
Pots *(Flowerpot, Pot)*
  X-Ref: Flowerpots
Potting *(Potted)*
  X-Ref: Repotting
Potyvirus
Poultry
Poultryfarms
Poultrymen
Poults *(Poult)*
Poverty
Powders *(Powder,*
  Powdered)
Power *(Powered)*
Pox *(Poxes)*
Poxviruses *(Poxviral,*
  Poxvirus)
Practolol
Praecocaspis
Prairies *(Prairie)*
Pralidoxime
Prandiol
Prangos
Praseodymium
Prasinocladus
Prasinophyceae
Prasiola
Pratylenchidae
Pratylenchoides
Pratylenchus
Prawns *(Prawn)*
Prays
Preadolescent *see*
  Adolescence
Prealbumins *see* Albumins
Prealpine *see* Alpine
Precancerous *see* Cancer
Prechill *see* Cool
Precipitation *(Precipitable,*
  Precipitant, Precipitate,
  Precipitated, Precipitates,
  Precipitating,
  Precipitations)
  X-Ref: Coprecipitation,
    Nonprecipitating
Precipitins *(Precipitin)*
Precleavage *see* Cleavage
Precolostral *see* Colostrum
Preconditioning *see*
  Conditioning
Precooked *see* Cookery
Precooling *see* Cool
Precopulatory *see*
  Copulation

Precutting *see* Cuttings
Predation *(Predaceous,*
  Predacious)
Predator *(Predators,*
  Predatory)
Prediapause *see* Diapause
Prednisolone
Prednisone
Preemergence
Preemption
Preen
Prefabricated
  (Prefabrication)
Prefermentation *see*
  Fermentation
Prefiltration *see* Filtration
Preforan
Pregelatinized *see*
  Gelatinization
Pregermination *see*
  Germination
Pregnancy *(Gravid,*
  Pregnancies, Pregnant)
  X-Ref: Gravidity,
    Primiparous
Pregnandiol
Pregnane *(Pregnan,*
  Pregnanediol,
  Pregnanediols, Pregnanes)
Pregnene
Pregnenolone
Preharvesting *see* Harvest
Prehatch *see* Hatching
Preheated *see* Heat
Preillumination *see*
  Illumination
Preimaginal *see* Imaginal
Preimmunized *see*
  Immunization
Preimplantation *see*
  Implants
Preincubation *see*
  Incubation
Preinoculation *see*
  Inoculation
Preirradiation *see*
  Irradiation
Prekallikrein *see*
  Kallikreins
Premating *see* Mating
Prematurity *see* Maturity
Premedicated *see* Medicine
Premeiotic *see* Meiosis
Premelanoidines
Premerge *see* Dinoseb
Premitotic *see* Mitosis
Premixes *(Premix)*
Premonsoon *see* Monsoon
Premutation *see* Mutation
Prenatal *see* Birth
Prenectar *see* Nectar
Prenyl
Prenylation
Preoptic *see* Optics
Preoviposition *see*
  Oviposition

Preovulatory *see* Ovulation
Prepackaging *see* Packaging
Prepacking *see* Packing
Preparturient *see* Birth
Preplanting *see* Planting
Prepona
Prepuberty *see* Puberty
Prepubescent *see*
  Pubescence
Prepuce *(Preputial,*
  Preputiotomy)
Prepupae *see* Pupae
Prerefrigeration *see*
  Refrigeration
Prerigor *see* Rigor
Preruminant
Preschoolers *see* Children
Prescottiella
Presecamine
Preseeding *see* Seeding
Preservation *(Preservability,*
  Preservations, Preserve,
  Preserved, Preserving)
Preservatives *(Preservative)*
Preserves
Presettlement *see*
  Settlements
Preshrinkage *see* Shrinkage
Preslaughter *see* Slaughter
Presoaking *see* Soaking
Presowing *see* Sowing
Presprouted *see* Sprout
Pressing *(Press, Pressed,*
  Presses)
Pressure *(Pressures,*
  Pressurized)
Presterilized *see* Sterile
Prestressed *see* Stress
Presynaptic *see* Synapsis
Pretarsus *see* Tarsus
Pretraining *see* Training
Pretreatment *(Pretreat,*
  Pretreated, Pretreating)
Prevention *(Prevent,*
  Preventative, Prevented,
  Preventing, Preventive)
Preventol *see* Dichlorophen
Previllage *see* Villages
Previtelline *see* Vitelline
Preweaning *see* Weaning
Prey *(Preying)*
Prices *(Price, Priced,*
  Pricing)
Primaquine
Primates *(Primate)*
Primer
Priming
Primiparous *see* Pregnancy
Primitive
Primordia *(Primordial,*
  Primordium)
Primrose *(Primroses)*
Primula *(Primulas)*
Primulaceae
Primulagenin
Princisaria

Princisola
Pringsheimia
Prinsepia
Priocnemis
Prioninae
Prionomatis
Prionopoda
Prionotes
Prionoxystus
Prionus
Priopoda
Prisms *(Prism, Prismatic)*
Pristiphora
Pristocera
Pritchardia
Privet
Proacerella
Proacrosin
Proactinomyces
Proactinomycetes
  (Proactinomycete)
Proagopertha
Proanthocyanidins *see*
  Anthocyanidins
Proaporphines
  (Proaporphine)
Proatypus
Probability *(Probabilistic,*
  Probabilities)
Probing *(Probe, Probes)*
Proboscidea
Proboscis
Procaine
  X-Ref: Novocaine
Procambarua
Procambium *see* Cambium
Procanace
Procarboxypeptidase *see*
  Carboxypeptidases
Procarcinogenic *see*
  Carcinogenesis
Procecidochares
Proceras
Processors *(Processor)*
Prociphilus
Procladius
Procloeon
Procollagen *see* Collagen
Procollagenase
Procollina
Procryptotermes
Proctolaelaps
Proctolin
Proctophantasta
Proctophyllodes
Proctophyllodidae
Proctotrupoidea
Procyanidins *see* Cyanidins
Procyazine
Procyon
Prodenia
Prodoratus
Proembryo *see* Embryos
Proembryogenesis *see*
  Embryogeny
Proenzyme *see* Zymogen

Professions see Occupations
Profits (Profit, Profitability,
   Profitable, Profitableness,
   Profitably, Profitmaker)
Profluralin
Proformica
Progenitors (Progenitor)
Progeny (Ancestors,
   Progenies)
   X-Ref: Ancestral,
   Offspring
Progestagens (Progestagen)
Progesterone (Progesteron,
   Progestin, Progestins)
Progestogens (Progestogen)
Progoitrin
Progonadotropic see
   Gonadotropins
Programming (Programmer,
   Programmers)
Programs (Program)
Progurt
Progymnosperm
Progymnospermopsida
Prohemocytes see
   Hemocytes
Proinsulin see Insulin
Proisotoma
Projection (Projected,
   Projecting, Projections)
Prokaryotes (Prokaryote,
   Prokaryotic)
Prolactin
Prolamellar see Lamella
Prolamine see Pyrrolidine
Prolamins (Prolamin)
Prolans
Prolapse
Proleptonchus
Proliferation (Proliferating,
   Proliferative)
Proline
Promachus
Promacyl
Promazine
Promelittin
Prometaphase see
   Metaphase
Promethazine
Prometryne (Prometrin,
   Prometryn)
   X-Ref: Gesagard
Promintic see Methyridine
Pronamide
   X-Ref: Kerb,
   Propyzamide
Pronematus
Pronuclear see Nucleus
Pronucleus see Nucleus
Propaganda
Propagation (Propagated,
   Propagating,
   Propagations, Propagator)
Propane
Propanediol
Propanidid

Propanil (Propanide)
Propanol
Propanone
Proparathyroid see
   Parathyroid
Propazine
Propellants see Propulsion
Propenylether
Propeptides see Peptides
Properdin (Properdine)
Propham
Prophase
Prophylactic (Prophylaxis)
Prophyrin
Propiolactone
Propionamide
Propionates (Propionate)
Propionibacterium
   (Propionibacteria)
Propionic
Propionitrile
Propionyl (Propionylated)
Proplastids see Plastids
Propolis
Propoxur
   X-Ref: Baygon
Propranolol
Proprioception
   (Proprioceptors)
Proprioseiopsis
Propulsion (Propellant,
   Propelled)
   X-Ref: Propellants
Propygmephorus
Propylene
Propylphosphonate
Propylthiouracil
Propylur
Propyzamide see Pronamide
Prorocentrum
Prosapia
Prosapogenin
Prosimulium
Proso
Prosodes
Prosopanche
Prosopigastra
Prosopis
Prosopocera
Prospaltella
Prospects see Forecast
Prostaglandins
   (Prostaglandin)
Prostate (Prostatic)
Prosternon
Prostheses (Prosthesis,
   Prosthetic)
Prosthogonimus
Prostigmata
Prostin
Prostratin
Protacaval
Protaetia
Protalkaloid
Protamines (Protamine)
Protandrena

Protea (Proteas)
Proteaceae
Proteacidites
Proteases (Protease)
Protectants (Protectant)
Protection (Protect,
   Protecting, Protective,
   Protector, Protectors,
   Protects)
Proteinases (Proteinase)
Proteinemia
   X-Ref: Paraproteinaemia
Proteinogram
Proteinoplasts
   X-Ref: Proteoplasts
Proteins (Proteic, Proteid,
   Protein, Proteinaceous,
   Proteinate, Proteinic)
   X-Ref: Holoprotein,
   Isoprotein, Nonprotein
Proteinuria
Protelean
Proteoglycans
   (Proteoglycan)
Proteolipids (Proteolipid)
Proteolysis (Proteolytic)
   X-Ref: Antiproteolysis
Proteoplasts see
   Proteinoplasts
Proteosynthesis
Protermes
Proteus
Protezim
Prothallium (Prothallia,
   Prothallial, Prothallic,
   Prothallus)
Prothiocarb
Prothoate
Prothoracic
Prothrombin
Prothyma
Protists
Protium
Protoberberine
Protoblastenia
Protocerebrum
   (Protocerebral)
Protochlorophyll
Protochlorophyllides
   (Protochlorophyllide)
Protococcales
Protococcus
Protocorms see Corms
Protodioscine
Protolignin
Protomutilla
Protomyces
Protomycopsis
Protomyobia
Protonemata (Protonema,
   Protonemic)
Protonemura
Protonymph (Protonymphs)
Protoparce
Protopectin
Protopectinases

(Protopectinase)
Protoperithecial see
   Perithecium
Protophloem see Phloem
Protophormia
Protopine
Protoplasm (Protoplasmic)
   X-Ref: Plasmatic
Protoplasts (Protoplast)
Protopolybia
Protoporphyrin
Protopulvinaria
Protoryzin
Protosalvinia
Protosiphon
Protostrongylidae
Prototaxites
Prototheca
Prototoxin see Toxins
Prototrophy
Prototypes (Prototype)
Protoxaea
Protoxylem
Protoyuccoside see
   Yuccoside
Protozoa (Protozoal,
   Protozoan, Protozoans,
   Protozoon)
Protozoology
   (Protozoological)
Protracheoniscus
Protura
Proturentomon
Protyrosinase
Provenance
   X-Ref: Interprovenance,
   Intraprovenance
Proventriculus
   (Proventricular,
   Proventriculi)
Provirus see Viruses
Provitamin see Vitamin
Provolone
Proxenus
Proximpham
Proxvirus
Proxys
Prunasin
Prunes (Prune)
Pruning (Pruned, Pruners,
   Prunings)
Prunus
Pryauvsta
Przewalskia
Przhevalskiana
Psaenythia
Psalliota
Psalliotis
Psathyrella
Psathyrotes
Pselaphidae (Pselaphid)
Pselliopus
Psenulus
Psephenidae
Psephenus
Pseudacanthotermes

Pseudaletia
Pseudanabaena
Pseudanomalon
Pseudatomoscelis
Pseudaulacaspis
Pseudechinolaena
Pseudencyrtoides
Pseuderanthemum
Pseudeucoila
Pseudeurotium
Pseudevernia
Pseudobacterium
Pseudobetckea
Pseudocalymma
Pseudocastalia
Pseudocercospora
Pseudocercosporella
Pseudocholinesterases *see*
  Cholinesterases
Pseudococcidae
  (Pseudococcids)
Pseudococcus
Pseudocopsinine
Pseudocowpox
Pseudocryptorchidism *see*
  Cryptorchidism
Pseudocyclobuxin
Pseudocyesis
Pseudocyphellaria
Pseudodelitschia
Pseudodipsas
Pseudogagrella
Pseudogley *see* Gley
Pseudogonia
Pseudographiella
Pseudohemitaxonus
Pseudohermaphroditism *see*
  Hermaphroditism
Pseudoinfection *see*
  Infection
Pseudolarix
Pseudolithos
Pseudomethoca
Pseudomonas
  *(Pseudomonad,
  Pseudomonads)*
Pseudopalaucoris
Pseudopanax
Pseudoparmelia
Pseudoperonospora
Pseudopeziza
Pseudophaeolus
Pseudophilotes
Pseudophycus
Pseudopilocereus
Pseudopityophthorus
Pseudoplasmodia *see*
  Plasmodium
Pseudoplectania
Pseudopleuronectes
Pseudoplusia
Pseudopodzol *see* Podzols
Pseudopraon
Pseudopregnancy
  (Pseudopregnant)
Pseudopsichacra

Pseudorabies
Pseudorasbora
Pseudosabicea
Pseudosarcomatous *see*
  Sarcoma
Pseudosarcophaga
Pseudoscorpiones
Pseudoscorpionida
  *(Pseudoscorpion,
  Pseudoscorpionidea,
  Pseudoscorpions)*
Pseudosinella
Pseudotarphius
Pseudoterritorial *see*
  Territory
Pseudotrebouxia
Pseudotsuga
Pseudotuberculosis
  (Pseudotuberculous)
Pseudowintera
Pseudoxenasma
Pseudoxenos
Psidium
Psila
Psilidae
Psilocurus
Psilocybe
Psilocybin
Psilogramma
Psilopa
Psilostrophe
Psilotaceae
Psilotum
Psilus
Psithyrus
Psittacosis
Psittacus
Psoberan
Psocidae *(Psocid)*
Psocoptera
  X-Ref: Corrodentia
Psoculus
Psoralea
Psoralen *(Psoralens)*
Psorella
Psorergates
Psoroma
Psoromic
Psorophora
Psoroptes
Psoroptidae
Psorosis
Psychidae *(Bagworm)*
  X-Ref: Bagworms
Psychoda
Psychodelic
Psychodidae *(Psychodinae)*
Psychology *(Psychologic,
  Psychological)*
Psychopsidae
Psychosis *(Psychoses)*
Psychotria
Psychrometry
  *(Psychrometer,
  Psychrometers,
  Psychrometric,

Psychrometrical)
Psychrophilic *(Psychrophile,
  Psychrophiles,
  Psychrophilous)*
Psychrophobic
Psychrotrophic
  (Psychrotrophs)
Psylla
Psyllidae *(Psyllid, Psyllids)*
  X-Ref: Chermidae
Psylliodes
Psyllipsocus
Psylloidea
Ptarmigan
Ptelea
Pteleobius
Ptenidiophyes
Pterandrus
Pterella
Pteridaceae
Pteridines *(Pteridine)*
Pteridium
Pteridophyta *(Pteridophyte,
  Pteridophytes)*
Pterigynandrum
Pterin *(Pterins)*
Pteris
Pterobosca
Pterocarpans
  (Pterocarpanoid)
Pterocarpin *(Pterocarpinoid)*
Pterocarpus
Pterocarya
Pterocephalus
Pterochloroides
Pterocladia
Pterodon
Pterolichidae
Pterolophia
Pteromalidae *(Pteromalids,
  Pteromalinae)*
Pteromalus
Pterombrus
Pteronarcys
Pteronia
Pteronidea
Pterophoridae
Pterophorus
Pterophylla
Pterosides *(Pteroside)*
Pterosiphonia
Pterospermum
Pterostichinae
Pterostichus
Pterostigma
Pterothorax
Pteroylglutamates
  (Pteroylglutamate)
Pteroylpolyglutamyl
Pterygoneurum
Pterygosoma
Pterygosomidae
Pterygostegia
Pterygota *(Pterygote)*
Pteryla
Ptiliidae

Ptilimnium
Ptilinus
Ptilonyssus
Ptilotus
Ptinidae
Ptinus
Ptomaphaginus
Ptycholoma
Ptychomitrium
Ptychopteridae
Ptychosperma
Puberty *(Postpuberal,
  Prepuberal, Prepubertal,
  Pubertal)*
  X-Ref: Postpubertal,
  Prepuberty
Puberulin
Pubescence *(Pubescent)*
  X-Ref: Prepubescent
Puccinellia
Puccinia
Pucciniastrum
Pueraria
Puerariae
Puerperium *see* Birth
Puff *(Puffs)*
Puffins
Pulchellon
Pulex
Pulicidae
Pullets *(Pullet)*
Pullorosis
Pullulan
Pullularia
Pullus
Pulmonale
Pulmonaria
Pulmonary *(Pulmonic)*
  X-Ref: Extrapulmonary,
  Intrapulmonary
Pulmonata *(Pulmonate)*
Pulp *(Pulpability, Pulped,
  Pulping, Pulps)*
  X-Ref: Repulping
Pulpmill
Pulptimber *see* Pulpwood
Pulpwood *(Pulpwoods)*
  X-Ref: Pulptimber
Pulsatilla
Pulsation *(Pulsate,
  Pulsatile, Pulsating,
  Pulsator, Pulsators,
  Pulsatory)*
Pulse *(Pulsed, Pulses)*
Pulverization *(Pulverized,
  Pulverizers, Pulverizing,
  Pulverulent)*
Pulvinaria
Pulvinic
Pulvinones *(Pulvinone)*
Pulvinula
Pumice
Pumpkins *(Pumpkin)*
Pumps *(Pump, Pumping)*
Punctures *(Puncture)*
Pungent *see* Odor

Punica
Pupae *(Prepupa, Prepupal,*
  Pupa, Pupal, Pupation)
  X-Ref: Prepupae
Puparia *(Puparial,*
  Puparium)
Pupils *(Pupil, Pupillary)*
Puppies *see* Dogs
Purchasing *(Purchase,*
  Purchased, Purchases)
Purebreds ´*(Purebred,*
  Purebredness,
  Purebreeding)
Puree *(Purees)*
Purification *(Purified,*
  Purifying)
  X-Ref: Depuration,
  Epuration,
  Semipurified
Purines *(Purin, Purine)*
Purinyl
Purkinje
Puromycin
Purothionin
Purshia
Purslane
Puschkinia
Pustules *(Pustular, Pustule)*
Puto
Putoria
Putraflavone
Putranjiva
Putrescine
Pycnidia *(Pycnia, Pycnial,*
  Pycnidial, Pycnidiospore,
  Pycnidiospores,
  Pycnidium, Pycniospore,
  Pycniospores, Pycnium,
  Pycnospore)
Pycnometer
  X-Ref: Picnometers,
  Pyknometers
Pycnoporus
Pycnopsyche
Pycnoscelus
Pydanon
Pyelitis
Pyelonephritis
Pyemotes
Pyemotidae *(Pyemotid)*
Pygicornides
Pygiopsylla
Pygiopsyllidae
Pygmephorus
Pygostenini
Pygostolus
Pyknometers *see*
  Pycnometer
Pyla
Pylaiella
Pylorus *(Pyloric,*
  Pyloroplasty)
Pyobacillosis
Pyogenicity *(Pyogenic)*
Pyometra *(Pyometrial)*
Pyracantha

Pyracarbolid
Pyralidae *(Pyralid,*
  Pyralidid, Pyralididae)
Pyramin *see* Pyrazon
Pyran
Pyranocoumarin
Pyranometer
Pyranose
Pyrantel
Pyrausta
Pyraustidae
Pyraustinae
Pyrazinediium
Pyrazines *(Pyrazine)*
Pyrazoles *(Pyrazole)*
Pyrazoline
Pyrazolone
Pyrazolyl
Pyrazon *(Pyrazone)*
  X-Ref: Pyramin
Pyrene
Pyrenochaeta
Pyrenoids *(Pyrenoid)*
Pyrenomycetes
  (Pyrenomycete)
Pyrenophora
Pyrethrinoids
Pyrethrins *(Pyrethrin)*
Pyrethroids *(Pyrethroid,*
  Pyrethroidal)
Pyrethrum
Pyrexia
Pyrgomorphidae
Pyrgothrips
Pyrgus
Pyricularia
  X-Ref: Piricularia
Pyriculariosis
Pyridaphenthion
Pyridazin
Pyridazinones
  (Pyridazinone)
Pyridazinyl
Pyridine
Pyridinethiol
Pyridinium
Pyridinol
Pyridones *(Pyridone)*
Pyridoxal
Pyridoxalacetic
Pyridoxalphosphate
Pyridoxamine
Pyridoxine
Pyridoxyl
Pyridyl
Pyrilamine
  X-Ref: Mepyramine
Pyrilla
Pyrimethamine
Pyrimicarb
Pyrimidine *(Pyrimidines,*
  Pyrimidinyl)
Pyrimidinones
Pyrites *(Pyrite)*
Pyrobotrys
Pyrocatechol

Pyrochlorophyll *see*
  Chlorophyll
Pyrocin
Pyroclastic
Pyrocystis
Pyrogallol
Pyrogen *(Pyrogenic,*
  Pyrogens)
Pyroglutamate
Pyrola
Pyrolaceae
Pyroligneous
Pyrolysis *(Pyrolyses,*
  Pyrolytic, Pyrolyzates)
Pyrometry *(Pyrometric)*
Pyronema
Pyrones *(Pyrone)*
Pyronia
Pyronin
Pyrophosphatase *see*
  Diphosphatase
Pyrophosphate *see*
  Diphosphate
Pyrophosphohydrolase
Pyrophosphokinase
Pyrophosphorylase
  (Pyrophosphorylases)
Pyrostegia
Pyrosulfate *see* Disulfates
Pyroxychlor
Pyrrhalta
Pyrrhocactus
Pyrrhochlacia
Pyrrhocoridae
Pyrrhocoris
Pyrrhopappus
Pyrrolase *(Pyrrolases)*
Pyrrole *(Pyrroles)*
Pyrrolidine
  X-Ref: Prolamine
Pyrrolidino
Pyrrolidone
Pyrrolidonecarboxylic
Pyrroline
Pyrrolizidine
Pyrrolo
Pyrrophyta
Pyrrylmethane
Pyrus
  X-Ref: Pirus
Pyruvates *(Pyruvate)*
Pyruvic
Pythiaceae *(Pythiaceous)*
Pythium
Pyxidanthera
Pyxine
Pyxinia
Quackgrass
Quadraspidiotus
Quails *(Quail)*
Quarantine *(Quarantined)*
Quartz *(Quartzes)*
Quartzite
Quassia
Quassinoids *(Quassinoid)*
Quaternaria

Quaternary
Quebrachitol
Quebracho
Quedius
Quella
Quenching
Quercetagetin
Quercetin
Quercetinase
Quercimeritrin
Quercus *(Oaks)*
  X-Ref: Oak
Questopogon
Quiinaceae
Quinacrine
Quinaldofur
Quinazoline
Quinazolinones
Quince *(Quinces)*
Quindoxin
Quinic
Quinidine
Quinine
Quinoid
Quinol *see* Hydroquinones
Quinolinate
Quinoline *(Chinoline,*
  Quinolin, Quinolinic)
Quinolinol
Quinolinone
Quinolizidine
Quinomethionate
Quinone *(Quinones)*
Quinonoid
Quinovose
Quinoxaline *(Quinoxalines)*
Quinoxalinedithiol
Quinoxalinylmethylene
Quintozene *see*
  Pentachloronitrobenzene
Quisqualate
Quisqualis
Quisquiliae
Quotas *(Quota)*
Rabbitpox
Rabbits *(Hare, Rabbit)*
  X-Ref: Cottontail, Hares
Rabdochloa
Rabies *(Rabic, Rabid)*
  X-Ref: Antirabies
Raccoon
Race *(Races, Racial,*
  Racially)
  X-Ref: Interracial
Racehorse *see* Horses
Racemes *(Raceme,*
  Racemic)
Racemobambos
Rachelia
Rachilla
Rachiplusia
Rachis
Rachitis *see* Rickets
Raciborskiella
Racomitrium
Radamopsis

Radappertization
Radar
Radfordia
Radial
Radiant
Radiation *(Radiations,*
Radiative)
  X-Ref: Postradiation
Radicals *(Radical)*
Radicles *(Radicle,*
Radicular, Radicum)
Radio
Radioactivity *(Radioactive,*
Radioactively,
Radioactivities)
Radioassay *(Radioassays)*
Radioautography *see*
  Autoradiography
Radiobiology
(Radiobiological)
Radiocarbon *see* Carbons
Radiochemistry
(Radiochemical,
Radiochemicals)
  X-Ref: Radioelements
Radiochromatography *see*
  Chromatography
Radioecology
(Radioecological)
Radioelements *see*
  Radiochemistry
Radiofrequency
Radiography
(Microradiography,
Radiographic,
Radiographical,
Radiographs,
Roentgendiagnosis,
Roentgenogram,
Roentgenograms,
Roentgenographic)
  X-Ref:
    Microradiographic,
    Roentgenography
Radioimmunoassays
(Radioimmunoassay,
Radioimmunoassayable)
Radioimmunological
(Radioimmuno)
Radioimmunosorbent
Radioiodine *see* Iodine
Radioisotopes *(Radioisotope,*
Radioisotopic)
Radiolabeling
(Radiolabeled,
Radiolabelled,
Radiolabelling,
Radiotracer)
Radiology *(Radiologic,*
Radiological, Radiologist)
Radiolysis *(Radiolytic)*
Radiometry *(Radiometer,*
Radiometers,
Radiometric)
Radiomutagenesis *see*
  Mutagenesis

Radiomutants *see* Mutants
Radionuclides *see* Nuclides
Radiophosphorus *see*
  Phosphorus
Radiopotassium *see*
  Potassium
Radioprotection
(Radioprotective)
Radioreceptor
Radioresistance
(Radioresistant)
Radioscopy
Radioselenium *see* Selenium
Radiosensitivity
(Radiosensitive,
Radiosensitivities,
Radiosensitization,
Radiosensitizer,
Radiosensitizing)
Radiosodium *see* Sodium
Radiostimulation
Radiostrontium *see*
  Strontium
Radiosulphate *see* Sulfate
Radiotelemetry
(Radiotelemetric)
Radiotellurium *see*
  Tellurium
Radiotherapy
Radiotoxins
Radiowaves
Radishes *(Radish)*
Radium
Radius
Radon
Radopholus
Radula
Raffinose
Rafflesiaceae
Rafoxanide
Rafters
Ragimillet *(Ragi)*
Ragweed *(Ragweeds)*
Ragwort
Rahmani *(Rahmany)*
Rai *see* Raya
Raillietina
Railroads *(Rail, Railroad,*
Rails, Railway, Railways)
Rain *(Rainfall, Rainfalls,*
Rainfed, Rainlands,
Rains, Rainstorms,
Rainwater, Rainy)
  X-Ref: Downpours
Rainforests *(Rainforest)*
Raingauges *(Raingage,*
Raingages, Rainguage)
Raingrown
Raisin *(Raisins)*
Raking *(Rake)*
Ralgro *see* Zeranol
Ramalina
Raman
Ramaria
Rambouillet
Rambutan

Ramie
Ramonda
Rams *(Ram)*
Ramularia
Ramulispora
Rana
Ranales
Ranatra
Ranches *(Ranch, Rancher,*
Ranchers, Ranching)
Rancid *(Rancidity)*
Rangeland *(Rangelands)*
Ranges *(Range)*
Rangifer
Ranikhet
Ranizole
Ranunculaceae
Ranunculus
Rape *(Rapes)*
  X-Ref: Toria
Raperia
Rapeseed *(Rapeseeds)*
Raphanus
Raphia
Raphides *(Raphide)*
Raphidia
Raphidiidae
Raphidioptera
Raphidophyceae
Raphiolepis
Raspberries *(Raspberry)*
Raticides *(Raticide)*
Rations *(Ration, Rationing)*
Ratoon *(Ratooning,*
Ratoons)
Rats *(Rat)*
  X-Ref: Rattus
Rattan
Rattlesnake *see* Snakes
Rattus *see* Rats
Rauvolfia *(Rauwolfia)*
Ravenelia
Ravenelin
Ravensara
Ravines *(Ravine)*
Raya
  X-Ref: Rai
Rayon
Rays *see* Irradiation
rDNA*see* DNA
Reactors *(Reactor)*
Reaeration *see* Aeration
Realimentation *see*
  Alimentation
Reap *(Reaping)*
Reapers *(Reaper)*
Rearing *(Rear, Reared)*
Reboulia
Rebreeding *see* Breeding
Rebutia
Recapture *see* Captivity
Receptivity
Receptors *(Receptor)*
Recession
Recessive
Recharge

Recilia
Recipes *(Recipe)*
Recirculation *see*
  Circulation
Reclamation
(Agroreclamative,
Reclaim, Reclaimed,
Reclamations)
  X-Ref: Agroreclamation
Recolonization *see*
  Colonization
Recombinant
(Recombinants)
Recombination
(Recombinations,
Recombinogenic)
Reconnaissance
Reconstruction
(Reconstructing,
Reconstructive)
Recreation *(Recreational,*
Recreationists, Recreative)
Recrystallization *see*
  Crystallization
Rectum *(Rectal)*
  X-Ref: Intrarectally
Recurrent *(Recurrence,*
Recurring)
Recurvaria
Recycle *(Recycles,*
Recycling, Recyclization,
Recyclizations)
Redbud
Redcedar
Redox
Redpeppers *(Redpepper)*
Redrying *see* Drying
Reductases *(Reductase)*
Reductometric
(Reductometrical)
Reductones
Reduviidae *(Reduviid)*
Reduviolus
Redwood *(Redwoods)*
Reeds *(Reed)*
Reefs *(Reef)*
Reels *(Reel)*
Reesa
Reesimermis
Refeeding *(Refed, Refeed)*
Refermentation *see*
  Fermentation
Refiners *(Refiner)*
Refining *(Refined,*
Refinement, Refineries,
Refinery)
Reflectance *(Reflectant,*
Reflectants, Reflected,
Reflection, Reflective)
Reflectometer
Reflex *(Reflexes)*
Reflux
Reforestation
(Reafforestation,
Reforestations,
Reforested)

Refraction (Refractile,
   Refractive)
   X-Ref: Birefringence
Refractometers
   (Refractometer,
   Refractometric)
Refreezing see Freezing
Refrigeration (Refrigerated,
   Refrigerating,
   Refrigerator)
   X-Ref: Nonrefrigerated,
   Prerefrigeration
Refrozen see Frozen
Regeneration
   (Bioregenerative,
   Regenerants, Regenerate,
   Regenerated, Regenerates,
   Regenerating,
   Regenerations,
   Regenerative,
   Regenerator,
   Regenerators)
   X-Ref: Bioregeneration,
   Photoregeneration
Reglone (Reglon)
Regnellidium
Regrafting see Grafting
Regrassing see Grasses
Regreening see Greening
Regression (Regressing,
   Regressions, Regressive)
Regulation (Regulate,
   Regulated, Regulates,
   Regulating, Regulations,
   Regulative, Regulatory)
Regulators (Regulant,
   Regulants, Regulator)
   X-Ref: Morphoregulators
Regurgitation
Rehabilitation
   (Rehabilitating)
Reheat see Heat
Rehydration see Hydration
Reindeer (Reindeers)
Reinfection see Infection
Reinfestation see
   Infestation
Reinforcement (Reinforced,
   Reinforcing)
Reinoculated see
   Inoculation
Rejuvenation
Relapse (Relapses,
   Relapsing)
Relaxants (Relaxant)
   X-Ref: Myorelaxants
Relaxation (Relaxed)
Relaxin
Relbunium
Releaser
Relicina
Relishes
Remating see Mating
Remijia
Remineralization see
   Minerals

Remirea
Remnant
Remuneration
Renal see Kidneys
Renaturation
Rendering
Rendzina (Rendzinas)
Renilla
Renin
Rennet (Rennets)
Rennilase
Rennin
Renocera
Rents (Rent, Rentability,
   Rental, Rentals, Rented,
   Renting)
Reoselin
Reoviruses (Reoviridae,
   Reovirus)
Reoxidation see Oxidation
Repayment see Payments
Repellents (Antifeed,
   Antifeedant, Antifeeding,
   Antifeeds, Repellance,
   Repellant, Repellants,
   Repellence, Repellency,
   Repellent)
   X-Ref: Antifeedants
Repetobasidium
Replacers (Replacer)
Replanting see Planting
Replicase
Replication (Replicate,
   Replicated, Replicating,
   Replications, Replicative)
Replowing see Plowing
Repopulation see Population
Repositol
Repotting see Potting
Repression (Repressible)
Reprocessing
Reproduction
   (Reproducibility,
   Reproducible,
   Reproducing,
   Reproductions,
   Reproductive,
   Reproductivity)
Reptiles (Reptile, Reptilian)
Repulping see Pulp
Resazurin
Resection
Reseda
Resedaceae
Reseeding see Seeding
Reserpine (Reserpinized)
Reservation
Reserve (Reserves)
Reservoirs (Reservoir)
Reset see Set
Resettlement see
   Settlements
Residues (Residuality,
   Residuals, Residue)
Resinate
Resinoid

Resins (Resin, Resinous)
Resmethrin
Resol
Resonance (Resonant,
   Resonating)
Resorantel
Resorcin
Resorcinol (Resorcinols)
Respiration (Breath,
   Respiratory, Respiring)
   X-Ref: Breathing
Respirometry
   (Respirometer,
   Respirometric)
   X-Ref: Microrespirometry
Resseliella
Restacking see Stacking
Restaurants (Restaurant,
   Restaurateur)
Restionaceae
Restoration (Restitution,
   Restorer, Restores,
   Restoring)
Restraint (Restrained,
   Restrainer, Restraining)
   X-Ref: Unrestrained
Restriction (Restricted,
   Restricting, Restrictions,
   Restrictive)
   X-Ref: Unrestricted
Resupinate
Resuscitation
Resveratrol
Retail (Retailer, Retailers,
   Retailing)
Retardants (Retardancy,
   Retardant, Retarders)
Retardation (Retard,
   Retardance, Retarded,
   Retarding)
Reticularia
Reticuline
Reticulitermes
Reticulocytes (Reticulocyte)
Reticuloendothelial
Reticuloendotheliosis
Reticuloses (Reticulosis)
Reticulum (Reticular,
   Reticulate, Reticulated,
   Reticulation, Reticulo)
Retina (Retinae, Retinal,
   Retinas, Retino)
   X-Ref: Extraretinal
Retinitis
Retinoic
Retinol
Retinopathy
Retinula (Retinular)
Retinyl
Retirement
Retraining see Training
Retransplantation see
   Transplants
Retrieval
Retronecine
Retroperitoneal

Rettary
Retting (Retted)
   X-Ref: Unretted
Retzia
Reuteria
Revaccination see
   Vaccination
Revegetation (Revegetating)
Revenues
Revolution (Revolutionary,
   Revolutions)
Rewetted see Wet
Rhabdiopteryx
Rhabditidae
Rhabditis
Rhabdocline
Rhabdodendron
Rhabdomonas
Rhabdoms (Rhabdom)
Rhabdomyoma
   (Rhabdomyosarcoma)
Rhabdophaga
Rhabdoscelus
Rhabdoviruses
   (Rhabdovirus)
Rhachomyces
Rhacodium
Rhacomitrium
Rhadinaphelenchus
Rhadinopsylla
Rhagio
Rhagionidae
Rhagium
Rhagoletis
Rhamnaceae
Rhamnose
Rhamnoside (Rhamnosides)
Rhamnosylglucoside
Rhamnosylxyloside
Rhamnus
Rhamphomyia
Rhantus
Rhaphanus
Rhaphidozygaena
Rhaponticum
Rhea
Rheograms
Rheology (Rheologic,
   Rheological)
Rheometers (Rheometer,
   Rheometric, Rheometry)
Rhesus
Rheumatism (Rheuma,
   Rheumatic, Rheumatoid)
Rhexinia
Rhexothecium
Rhicnopeltella
Rhinacloa
Rhinanthus
Rhinaphena
Rhingia
Rhinitis
Rhinochenus
Rhinocladiella
Rhinocyllus
Rhinonyssidae

(Rhinonyssids,
   Rhinonyssinae)
Rhinopetalum
Rhinopneumonitis
Rhinosporidiosis
Rhinosporidium
Rhinotermitidae
Rhinotracheitis
Rhinoviruses (Rhinovirus)
Rhipicentor
Rhipicephalus
Rhipidolestes
Rhipiphoridae
Rhipsalis
Rhithropanopeus
Rhizina
Rhizobia (Rhizobial,
   Rhizobium)
Rhizobitoxine
Rhizobius
Rhizocarpon
Rhizocarpous
Rhizoclonium
Rhizoclosmatium
Rhizoctonia
Rhizogenesis
Rhizoglyphus
Rhizoid (Rhizoidal,
   Rhizoids)
Rhizomes (Rhizoma,
   Rhizomatic, Rhizomatous,
   Rhizome)
Rhizomorphs (Rhizomorph)
Rhizopertha see
   Rhyzopertha
Rhizophagus
Rhizophlyctis
Rhizophora
Rhizophoraceae
Rhizophydium
Rhizopin see IAA
Rhizopoda
Rhizopogon
Rhizopus
Rhizosphaera
Rhizosphere (Rhizoplane,
   Rhizospheres,
   Rhizospheric)
   X-Ref: Nonrhizosphere,
      Rootzone
Rhizotrogus
Rhizotron
Rhodacaridae
Rhodamine
Rhodanese
Rhodanine
Rhodesgrass
Rhodiola
Rhodnius
Rhodocybe
Rhododendron
   (Rhododendrons)
Rhodomelaceae
Rhodophysema
Rhodophysemopsis
Rhodophyta (Rhodophyceae,

Rhodophyte)
Rhodopseudomonas
Rhodopsin
   X-Ref: Metarhodopsin
Rhodospirillum
Rhodosporidium
Rhodotorula
Rhodymenia
Rhoeo
Rhogogaster
Rhombognathinae
Rhondnius
Rhopalidae
Rhopalidia
Rhopalocera
Rhopalophion
Rhopalosiphum
Rhopalum
Rhubarb
Rhus
Rhusflavanone
Rhyacia
Rhyacionia
Rhyacophila
Rhyacophilidae
Rhychosciara
Rhynchaenus
Rhynchopacha
Rhynchophora
Rhynchophorus
Rhynchophylline
   X-Ref:
      Isorhynchophylline
Rhynchosciara
Rhynchosia
Rhynchospora
Rhynchosporium
Rhynchota
Rhyncomya
Rhynia
Rhyparochrominae
Rhysodes
Rhyssa
Rhythm (Autorhythmic,
   Biorhythm,
   Biorhythmicity, Rhythmic,
   Rhythmically,
   Rhythmicity, Rhythmics,
   Rhythms)
   X-Ref: Autorhythm,
      Biorhythms
Rhytidiadelphus
Rhytisma
Rhyzopertha
   X-Ref: Rhizopertha
Ribes
Ribodeoxyviruses
Riboflavin (Riboflavine)
Riboflavinuria
Ribofuranosyl
Ribonuclease
   (Ribonucleases)
   X-Ref: RNase
Ribonucleates
Ribonucleic
Ribonucleoproteins

(Ribonucleoprotein)
Ribonucleosides
   (Ribonucleoside)
Ribonucleotides
   (Ribonucleotide)
   X-Ref:
      Ribooligonucleotides
Ribonucleotidyl
Ribooligonucleotides see
   Ribonucleotides
Ribose
Ribosides (Riboside)
Ribosomes (Ribosomal,
   Ribosome)
   X-Ref: Mitoribosomes,
      Nonribosomal
Ribosyl
Ribosylzeatin
Ribotide
Ribs (Rib)
Ribulose
Ribulosebiphosphate
Riccia
Rice (Paddies, Rices)
   X-Ref: Paddy
Ricefield see Ricelands
Ricelands (Riceland)
   X-Ref: Ricefield
Richardia
Ricin
Ricinocarpos
Ricinoleic
Ricinus
Rickets (Rachitic)
   X-Ref: Antirachitic,
      Rachitis
Rickettsia (Rickettsiae)
Rickettsiaceae
Rickettsiales (Rickettsial,
   Rickettsian)
Rickettsiella
Rickettsiosis (Rickettsioses)
Ricotta
Ridges (Ridge, Ridged,
   Ridging)
Rifampicin (Rifampicine,
   Rifampin, Rifamycin,
   Rifamycins)
Rigid (Rigidity)
Rigiopappus
Rigor
   X-Ref: Postrigor,
      Prerigor
Rinderpest
   X-Ref: Antirinderpest
Rinds (Rind)
Ring (Ringed, Ringing,
   Rings)
Ringspot
Ringworm (Ringworms)
Rinodina
Riodinidae
Riparian
Ripening (Ripe, Ripened,
   Ripener, Ripeners,
   Ripeness)

X-Ref: Nonripening,
   Unripe
Risama
Rishitin
Rivanol
Rivea
Rivers (River)
Rivinia
RNA (mRNAs, RNAs,
   rRNAs, tRNAs)
   X-Ref: mRNA, rRNA,
      tRNA
RNase see Ribonuclease
Roaches (Roach)
Roads (Road)
Roasting (Roast)
Robinia
Robinin
Robustic
Robustone
Roccella
Rock (Rockery, Rocks,
   Rocky)
Rodenticides (Rodenticidal,
   Rodenticide)
Rodents (Rodent, Rodentia)
Rodolia
Rodriguezia
Roegneria
Roentgen (Roentgenologic,
   Roentgenological,
   Roentgenology, Roentgens,
   Xradiation, Xray,
   Xraying)
   X-Ref: Xrays
Roentgenography see
   Radiography
Roesleria
Rogadinae
Rogor see Dimethoate
Romalea
Romano
Rompun
Roncus
Ronidazole
Ronit see Cycloate
Ronnel
   X-Ref: Fenchlorphos,
      Trolene
Ronstar
Roofs (Roof, Roofing)
Rook
Rooms (Room)
Roosters (Rooster)
Rooting (Rootability,
   Rooted, Rootings)
Roots (Root, Rootage,
   Rootcap, Rootlet,
   Rootlets, Taproots)
   X-Ref: Taproot
Rootstocks (Rootstock)
Rootworms (Rootworm)
Rootzone see Rhizosphere
Ropes (Rope)
Roquefort
Roquefortine

X-Ref: Unsaponifiable
Saponins *(Saponin)*
Saponite
Saponosides
Sapotaceae *(Sapotaceous)*
Sapotas *see* Sapodilla
Sappaphis
Saprinus
Saproglyphidae
Saprolegnia
Saprolegniaceae
Sapropel
Saprophagous
Saprophytes *(Saprophytic,*
    Saprophytism)
Sapwood
Sapyga
Saracha
Sarcina
Sarcocarp
Sarcococca
Sarcocystis *(Sarcocysts)*
Sarcocystosis
Sarcogyne
Sarcoid
Sarcolaena
Sarcolaenaceae
Sarcoma *(Sarcomas,*
    Sarcomata, Sarcomatous)
    X-Ref:
        Pseudosarcomatous
Sarcomatosis
Sarcomere
Sarcophaga
Sarcophagidae
    (Sarcophagid,
    Sarcophaginae)
Sarcophilus
Sarcoplasmic
Sarcoptes
Sarcoptidae *(Sarcoptic)*
Sarcoptiformes
    (Sarcoptiforme)
Sarcoscyphaceae
    (Sarcoscyphineae)
Sarcosporidia
Sarcosporidiosis
    (Sarcosporidiasis)
Sarcostemma
Sardine *(Sardines)*
Sardinella
Sargassaceae
Sargassan
Sargassum
Sarginae
Sarin
Sarka
Sarothamnus
Sarracenia
Sarraceniaceae
Sarracenin
Sarsazan
Sarson
Sasa
Sasaella
Sassafras

Satanicoptes
Satellites
    X-Ref: Skylab
Sathrobrota
Satiation *(Satiety)*
Saturation *(Saturated,*
    Unsaturated,
    Unsaturation)
    X-Ref: Subsaturated,
        Unsaturate
Satureja *(Satureia)*
Saturnia
Saturniidae *(Saturniid)*
Satyridae *(Satyrid)*
Satyrinae
Satyrium
Sauces *(Sauce)*
Sauerkraut
Saurauia
Saururaceae
Saururus
Sausage *(Sausages)*
    X-Ref: Wurst
Saussurea
Sauvagesia
Savannah *(Savanna,*
    Savannahs, Savannas,
    Semisavanna)
    X-Ref: Semisavannahs
Savings
Sawdust
Sawflies *(Sawfly)*
Sawgrass
Sawing *(Bandsaw, Sawn,*
    Saws)
    X-Ref: Bandsaws,
        Boxsaws, Pitsaws
Sawlogs *(Sawlog)*
Sawmill *(Sawmilling,*
    Sawmills)
Sawtimber
Saxicolous
Saxifraga
Saxifragaceae
Saxifrage
Saxinis
Sayomyia
Sayphos *see* Menazon
Scab
Scabies
Scabiosa
Scaevola
Scald
Scale *(Scales, Scaling)*
Scallion *(Scallions)*
Scallop
Scandium
Scanning *(Scan, Scanner,*
    Scans)
Scapania
Scape
Scapheremaeus
Scaphoideus
Scaphoidophytes
Scaphytopius
Scapteriscus

Scaptomyza
Scaptotrigona
Scapula
    X-Ref: Interscapular
Scar *(Scars)*
Scarabaeidae *(Scarab)*
Scarabaeoidea
Scarabaeus
Scarecrows
Scarification *(Scarified,*
    Scarifier, Scarifying)
Scaritini
Scatophaga *(Scathophaga)*
Scatophagidae
Scattering *(Scatter,*
    Scattered, Scatterometry)
Scavengers *(Scavenger)*
Sceletium
Scelionidae *(Scelionid)*
Sceliphron
Scelochilus
Scenedesmus
Scenopinidae
Scenopinus
Scent *see* Odor
Scepticus
Schedorhinotermes
Schedule *(Scheduled,*
    Schedules, Scheduling)
Schefflera
Schelhammera
Schelhammeridine
Schellackia
Scheuchzeria
Schima
Schinopsis
Schinus
Schisandra
    X-Ref: Schizandra
Schistidium
Schistocerca
Schistochila
Schistosoma *(Schistosome)*
    X-Ref: Bilharzia
Schistosomatidae
Schistosomiasis
    X-Ref: Bilharziasis
Schistosomicide
    (Schistosomicidal)
Schistosomules
Schistostega
Schizachyrium
Schizaea
Schizandra *see* Schisandra
Schizandrin
Schizandrol
Schizaphis
Schizocosa
Schizodiplogynium
Schizoflavins
Schizogony
Schizokinen
Schizolaena
Schizolobium
Schizomidae *(Schizomida)*
Schizomus

Schizomycetes
Schizonts
Schizophora
Schizophthirus
Schizophyllum
Schizoplasmodiopsis
Schizopteridae
Schizopyga
Schizosaccharomyces
Schizotetranychus
Schizura
Schlumbergera
Schneideria
Schoenoplectus
Schomburgkia
Schoolchildren *see* Children
Schools *(School, Schooling)*
Schouwia
Schraderia
Schwanniomyces
Sciadopitys
Sciapteryx
Sciara
Sciaridae *(Sciarid)*
Sciatic
Scilla
Scindapsus
Scintigraphy
Scintillation
Sciomyzidae
Scions *(Scion, Scionwood)*
Scirpophaga
Scirpus
Scirrhia
Scirtetellus
Scirtothrips
Scission
Scleranthus
Sclerenchyma *(Sclereids)*
Scleria
Sclerite *(Sclerites)*
Sclerocactus
Scleroderma
Scleroderris
Sclerolobium
Sclerophoma
Sclerophthora
Sclerophylls *(Sclerophyll)*
Scleroracus
Sclerosis *(Sclerosing)*
    X-Ref: Antisclerotic
Sclerospora
Sclerotia *(Sclerotial)*
Sclerotinia *(Sclerotinias)*
Sclerotiniaceae
Sclerotium
Sclerotization *(Sclerotic)*
Scolaban *see* Bunamidine
Scolex
Scoliidae
Scoliosis
Scolopendra
Scolopendrium
Scolopia
Scolopostethus
Scolothrips

Scolypopa
Scolytidae *(Scolytid)*
Scolytoidea
Scolytus
Scopolamine *(Scopolamin)*
Scopoletin
Scopolia
Scopolin
Scopula
Scopulariopsis
Scorbutic *see* Scurvy
Scorching *(Scorch)*
Score *(Scores, Scoring)*
Scorodocarpus
Scorpionflies *see* Mecoptera
Scorpionida *(Scorpionidea)*
Scorpions *(Scorpion)*
Scorzonera *(Scorsonera)*
Scotia
Scotogramma
Scotophase
Scotoplectus
Scototactic
Scours *(Scour, Scouring)*
Scrapie
Scratch
Screenhouses
Screening *(Screen,*
  Screened, Screenings,
  Screens)
Screwworm *(Screwworms)*
Scrippsiella
Scrobipalpa
Scrophularia
Scrophulariaceae
Scrotum *(Scrotal)*
  X-Ref: Intrascrotal
Scrub *(Scrubland,*
  Scrublands, Scrubs)
Scurf
Scurvy
  X-Ref: Antiscorbutic,
  Scorbutic
Scutacaridae
Scutacarus
Scutch *(Scutched,*
  Scutching)
Scutellarein
Scutellaria
Scutellarioside
Scutelleridae
Scutelleroidea
Scutellista
Scutellonema
Scutellum *(Scutella,*
  Scutellar)
Scutia
Scutigera
Scutigerella
Scutopilio
Scydmaenidae
Scymnus
Scyphophorus
Scytalidium
Scytalone
Scythes *(Scything)*

Scythris
Scytodidae
Scytonema
Sea *(Seas)*
Seafood *(Seafoods)*
Seagrasses *(Seagrass)*
Seagulls
Seal *(Sealant, Sealed,*
  Sealing, Seals)
Seam
Seasoning *(Seasoned,*
  Seasonings)
Seasons *(Season, Seasonal,*
  Seasonally)
Seawater
Seaweed *(Seaweeds)*
Sebastiania *(Sebastiana)*
Sebertia
Seborrhea *(Seborrheic)*
Sebum *(Sebaceous)*
Secale
Secamine
Sechium
Sechulin
Secohopane
Secoiridoids *(Secoiridoid)*
Secologanic
Secopyrrolizidine
Secretin
Secretion *(Secreted,*
  Secreting, Secretional,
  Secretions, Secretory)
  X-Ref: Antisecretory
Sectioning
Securinega
Sedaperone
Sedation *(Sedate, Sedated)*
Sedge *(Sedges)*
Sediments *(Sediment,*
  Sedimentary,
  Sedimentation,
  Sedimented, Sedimenting,
  Sedimentological)
Sedum
Seedbed *(Seedbeds)*
Seedborne
Seedcane
Seeders *(Seeder)*
Seedfall
Seeding *(Overseeded,*
  Reseed, Seeded, Seedings)
  X-Ref: Overseeding,
  Preseeding, Reseeding,
  Unseeded
Seedless
Seedlings *(Seedling)*
Seedpods *(Seedpod)*
Seeds *(Seed, Seedcoat,*
  Seedstone, Seedy)
  X-Ref: Monoseed,
  Nonseed
Seedstalk *see* Stalks
Seepage *(Seep, Seeps)*
  X-Ref: Antiseepage
Segregation *(Segregants,*
  Segregate, Segregated,

Segregating,
  Segregational,
  Segregations)
  X-Ref: Postsegregation
Sehima
Sehirus
Seira
Seismic
Seizure *(Seizures)*
Seladonia
Selaginella *(Selaginellae)*
Selatosomus
Selector
Selenaspidus
Selenastrum
Selenate
Selenicereus
Selenite
Selenium *(Seleniferous)*
  X-Ref: Radioselenium
Selenocystine
Selenodes
Selenomethionine
Selenomonas
Selenophoma
Selenosis
Selenothrips
Selfpollination *(Selfed,*
  Selfing, Selfpollinated)
Seligeria
Selliera
Selling *(Sell, Seller)*
Semanophorinae
Semanotus
Sematoneura
Semecarpus
Semen *(Seminal)*
Semeron
Semialdehyde
Semiaquatic *see* Aquatic
Semiarid *see* Arid
Semiautomatic *see*
  Automatic
Semicarbazide
Semidalis
Semidesert *see* Deserts
Semigametic *see* Gametes
Semihumid *see* Humidity
Semilethal *see* Lethal
Semilooper
Seminars *see* Conferences
Seminiferous
Semiothisa
Semipermeable *see*
  Permeability
Semipurified *see*
  Purification
Semisavannahs *see*
  Savannah
Semishrub *see* Shrubs
Semislotted *see* Slots
Semistarved *see* Starvation
Semisterility *see* Sterile
Semisynthetic *see*
  Synthetics
Semiuronides

Semiwild *see* Wild
Semliki
Semolinas *(Semolina)*
Sempervivum
Sencor
Sendanin
Senecio *(Senecios)*
Senecioneae
Senecionine
Senegin
Senescence *(Senescent,*
  Senescing, Senility)
  X-Ref: Senile
Senescene
Senile *see* Senescence
Senji *see* Clover
Senkirkine
Senna
Sennius
Senotainia
Senses *(Sensation, Sense,*
  Sensed, Sensing)
Sensilla
Sensitivity *(Hypersensitive,*
  Hypersensitively,
  Hypersensitivity, Sensitive,
  Sensitivities)
  X-Ref: Hypersensibility
Sensitization *(Sensitized,*
  Sensitizing)
Sensors *(Sensor)*
Sensory *(Sensorial,*
  Sensoric)
Separation *(Separating)*
Separators *(Separator)*
Separotheca
Sepedon
Sephadex
Sepharose
Sepiapterin
Sepiolite
Sepsis *see* Septicemia
Septage
Septic
Septicemia *(Septicaemia,*
  Septicaemic)
  X-Ref: Colisepticaemia,
  Sepsis
Septobasidium
Septonema
Septoria
Septum *(Septa, Septal,*
  Septate)
  X-Ref: Multiseptate
Sequenator
Sequences *(Sequence,*
  Sequential, Sequentially)
Sequestration
Sequestrene
Sequirin
Sequoia *(Sequoias)*
Sequoiadendron
Seral *see* Serum
Serala
Seraya
Serenoa

Serfs *(Serf)*
Sergentomyia
Serial *(Serially)*
Sericesthis
Sericin *(Sericins)*
Sericinae
Sericomyrmex
Sericostoma
Sericulture *(Sericultural)*
Serine
Seroallergic
Serodiagnosis
  (Serodiagnostic)
Seroepidemiologic *see*
  Epidemiology
Serogroups *(Serogroup)*
Serology *(Serologic,*
  Serological, Serologically)
Serosa *(Serosal)*
Serosurvey
Serotherapy
Serotonin *(Serotoninergic)*
Serotypes *(Serotype,*
  Serotypic, Serotyping)
Serous *see* Serum
Serozems *see* Sierozem
Serpentine *(Serpentinite)*
Serphoidea
Serpula
Serradella
Serratia
Serratocallis
Serratula
Serrodes
Serum *(Sera, Serums)*
  X-Ref: Seral, Serous
Serviceberry *see*
  Juneberries
Serviformica
Seryl
Sesame
Sesamia
Sesamoid
Sesamol
Sesamolin
Sesamum
Sesbania
Seseli
Sesiidae
Sesleria
Sesquioxide *(Sesquioxides,*
  Sesquioxidic)
Sesquiterpenes
  (Sesquiterpene,
  Sesquiterpenic,
  Sesquiterpenoid,
  Sesquiterpenoids)
Sessions *see* Conferences
Set *(Sets)*
  X-Ref: Reset
Setae *(Seta, Setal)*
Setaria *(Setarial)*
Setcreasea
Setoparmena
Setosphaeria
Settlements *(Settlement,*

Settlers)
  X-Ref: Presettlement,
    Resettlement
Seuratia
Seutelleridae
Sevidol
Sevin *see* Carbaryl
Sewage *(Sewages, Sewer,*
  Sewerage)
Sewing
Sex *(Asexually, Sexed,*
  Sexes, Sexing, Sexual,
  Sexuality, Sexually)
  X-Ref: Agamic, Asexual,
    Bisexual,
    Hypersexuality
Seychellene
Seymeria
Shade *(Shaded, Shading,*
  Shady)
Shadow *(Shadows)*
Shakers *(Shaker)*
Shale *(Shales)*
Shallot *(Shallots)*
Shank
Sharefarmer
Sharks *(Shark)*
Shashlik
Shattering
Shearing *(Shear, Sheared,*
  Shears)
  X-Ref: Shorn
Sheath *(Sheathed,*
  Sheathing, Sheaths)
Sheds *(Cowshed, Shed)*
  X-Ref: Cowsheds
Sheep
Sheepfarming
Sheepmen *(Sheepman)*
Sheepskins
Shelf *(Shelving)*
Shelflife
Shelfordella
Shell *(Shells)*
Shellfish
Shelling *(Shelled, Sheller,*
  Shellers)
Shelter *(Sheltered, Shelters)*
Shelterbelt *(Shelterbelts)*
Shelterwood
Shepherd *(Shepherds)*
Shepherdia
Shepherdspurse
Sherbets *(Sherbet)*
Sherries *(Sherry)*
Shielding *(Shield, Shielded,*
  Shields)
Shigella
Shigellosis
Shiitake
Shikimate
Shikimic
Shikonin
Shiner *(Shiners)*
Shipping *(Ship, Shipboard,*
  Shipment, Shipments,

Shipped, Shipper,
  Shippers, Ships,
  Transshipment)
  X-Ref: Transshipments
Shipworms *(Shipworm)*
Shives
Shock *(Shocked, Shocking,*
  Shocks)
Shoeing *see* Horseshoeing
Shoes *(Shoe)*
Shogaol
Shoot *(Shoots)*
Shootfly
Shopping
Shorea
Shorelines *(Shore,*
  Shoreline, Shores)
Shorn *see* Shearing
Shortages *(Shortage)*
  X-Ref: Shortfalls
Shortening *(Shortenings)*
Shortfalls *see* Shortages
Shortgrass
Shorthorn
Shoulder *(Shoulders)*
Shovel *(Shovels)*
Shredder *(Shredded,*
  Shredding)
Shrews *(Shrew)*
Shrimps *(Shrimp)*
Shrinkage *(Shrink,*
  Shrinking)
  X-Ref: Preshrinkage
Shrinkproofing
Shrubs *(Shrub, Shrubbery,*
  Shrubby, Shrubland,
  Shrublands)
  X-Ref: Semishrub,
    Undershrubs
Shucks *(Shuck)*
Shuckworms *(Shuckworm)*
Shunt *(Shunting, Shunts)*
Sialic
Sialidae
Sialis
Sialoglycoproteins
Sialoproteins
Siamin
Siaresinolic
Sibaria
Sibine
Sibinia
Siblings *(Sib, Sibling,*
  Sibships)
  X-Ref: Polysiblings
Sibthorpia
Sicklepod
Sickles *(Sickle)*
Sicyos
Sida
Sideramines
Sideritis
Siderochromes
Siduron
Sierozem *(Sierozems)*
  X-Ref: Serozems

Sieve *(Sieved, Sieves,*
  Sieving)
Sigara
Sigatoka
Sigmoid
Signals *(Signal, Signaling)*
Signiphora
Silage *(Ensilaged, Ensilages,*
  Ensilaging, Ensiled,
  Ensiling, Silaged, Silages,
  Silaging)
  X-Ref: Desilaging,
    Ensilage
Silanes *(Silane)*
Silene
Silica *(Siliceous,*
  Silicification)
  X-Ref: Desilicified
Silicate *(Silicated, Silicates)*
Silicic
Silicofluoride
Silicon *(Siliconized)*
Silicone
Silicophytoliths *see*
  Phytoliths
Silk *(Silks)*
Silkglands *(Silkgland)*
Silkmoths *(Silkmoth)*
Silkworm *(Silkworms)*
Silos *(Silo)*
Silpha
Silphidae
Silphium
Silt *(Siltation, Silted,*
  Silting, Silty)
Silurian
Silver
Silverberry
Silverfish
Silverside
Silvex
Silvichemicals
Silvicides *(Arboricide)*
  X-Ref: Arboricides
Silviculture *(Silvicultural,*
  Sylvicultural)
  X-Ref: Sylviculture
Silviculturist
  (Silviculturalist)
Silybine
Silybum
Silychristine
Silydianine
Silyl
Silymarins
Simarouba
Simaroubaceae
Simaroubolides
Simazine *(Simazin)*
  X-Ref: Gesatop
Simian
Simmental *(Simmenthal)*
  X-Ref: Fleckvieh
Simmondsia
Simocybe
Simplocaria

Simulation *(Simulants,*
Simulate, Simulated,
Simulating, Simulations,
Simulator)
Simuliidae *(Simuliid,*
Simuliids)
Simulium
Sinapine
Sinapis
Sinapoyl
Sindbis
Sindhi *(Sindhis)*
Sinella
Singapora
Sinigrin
Sinningia
Sinopimelodendron
Sinopteridaceae
Sinthusa
Sintonius
Sinus *(Sinuses)*
Sinusitis
Sinusoidal *(Sinusoidally)*
Siparuna
Sipha
Siphona
Siphonaptera
    X-Ref: Aphaniptera
Siratro
Sire *(Sired, Sires)*
Sirex
Siricidae *(Siricid)*
Sirogonium
Siroheme
Sisal
Sister *(Sisters)*
Sisymbrium
Sisyphus
Sisyra
Sisyridae
Sisyrinchium
Sitanion
Site
    X-Ref: Microsites
Siteroptes
Sitobion
Sitodiplosis
Sitona
Sitophaga
Sitophagus
Sitophilus
Sitosterol
Sitotroga
Skeletomuscular
Skeleton *(Skeletal,*
Skeletons)
    X-Ref: Extraskeletal
Skeletonema
Skidding *(Skid, Skidder,*
Skidders, Skids)
Skiing
Skim *(Skimmed,*
Skimming)
Skimmia
Skimmilk
Skin *(Dermoids, Skinned,*

Skinning, Skins,
Subcutaneously)
X-Ref: Corium,
    Cutaneous, Dermal,
    Dermis, Dermoid,
    Intracutaneous,
    Intradermal,
    Percutaneous,
    Subcutaneous,
    Subcutis, Subdermal,
    Subepidermal
Skototropism
Skull *(Skulls)*
Skunks
Skylab *see* Satellites
Skyrin
Slab *(Slabbing, Slabs)*
Slag *(Slags)*
Slash *(Slashing)*
Slats *(Slat)*
Slaughter *(Abattoir,*
Abbattoir, Abbattoirs,
Butchered, Butchery,
Slaughtered,
Slaughterhouse,
Slaughterhouses,
Slaughtering, Slaughters)
    X-Ref: Abattoirs,
    Butchering,
    Preslaughter
Sleep *(Sleeping)*
Sleet
Sleeves *(Sleeve)*
Slide *(Slides, Sliding)*
Slime
Slip *(Slippery, Slipping,*
Slips)
Sloanea
Slope *(Sloped, Slopes,*
Sloping)
Slots *(Slot, Slotted, Slotting)*
    X-Ref: Semislotted
Sloughs
Sludge *(Sludges)*
Slugs *(Slug)*
Sluices
Slums
Slurry *(Slurries)*
Smallwood
Smear *(Smearing, Smears)*
Smectites *(Smectite)*
Smelling *see* Olfactory
Smelowskia
Smicraulax
Smicromyrme
Smicromyrmilla
Smiela
Smilacaceae
Smilax
Sminthuridae
Sminthurides
Sminthurididae
Sminthurinus
Sminthurus
Smithiogaster
Smithora

Smittia
Smittium
Smog
Smoke *(Smoked, Smoker,*
Smokes, Smoking,
Smoky)
Smolnitsa *see* Grumusols
Smuts *(Smut)*
Snacks *(Snack, Snacking)*
Snails *(Snail)*
Snakebite
Snakegourds *(Snakegourd)*
Snakeroot
Snakes *(Snake)*
    X-Ref: Rattlesnake
Snakeweed
Snapbean *(Snapbeans)*
Snapdragon *(Snapdragons)*
Sneezeweed
Snowdrops
Snowmobiling *(Snowmobile,*
Snowmobiles)
Snowmold *(Snowmolds)*
Snows *(Snow, Snowcovered,*
Snowdrift, Snowfall,
Snowing, Snowloads,
Snowmelt, Snowy)
    X-Ref: Oversnow
Snyderol
Soaking *(Presoaked)*
    X-Ref: Presoaking
Soap *(Soaps, Soapy)*
Soapwort
Social
Socialism *(Socialist)*
Socialization
Socioeconomic
    (Socioeconomical)
Sociology *(Sociologic,*
Sociological)
Sod *(Sods)*
Soda
Sodium
    X-Ref: Radiosodium
Sodpodzolic *see* Podzols
Sodseeding
Softwood *(Softwoods)*
Sogata
Sogatella
Sogatodes
Soilborne
Soils *(Edaphologic,*
Edaphological, Soil)
    X-Ref: Edaphic
Soladulcidin
Solakrol
Solamargine
Solanaceae *(Solanaceous)*
Solanidine
Solanine
Solanocapsin
Solanum
Solar *see* Sun
Solarimeter
Solarium
Solasodine *(Solasodin)*

Soldanella
Solenobia
Solenopotes
Solenopsis
Solenostoma
Solid *(Solidification,*
Solidified, Solids)
Solidago
Solierella
Solipeds
Solonetz *(Solonetzes,*
Solonetzic)
Solutes *(Solute)*
Solva
Solvent *(Solvents)*
Solvolysis
Somatic
Somatomedin
Somatometric
Somatostatin
Somatotropin
Somera
Sonchus
Sonerila
Sonic
Sonication *(Sonicated)*
    X-Ref: Unsonicated
Sophophora
Sophora
Sophoramine
Sophoreae
Sophoroside
Sophronitis
Sophrorhinus
Sopubia
Sorbaria
Sorbate
Sorbic
Sorbitan
Sorbitol *(Sorbit)*
    X-Ref: Glucitol
Sorbose
Sorbus
Sordaria
Sordariaceae
Sordariella
Sorex
Sorghastrum
Sorgho *(Sorgo)*
Sorghum *(Sorghums)*
Sorosporium
Sorption *(Sorbed, Sorbent,*
Sorbents)
Sorrel
Sort *(Sorter, Sorting)*
Sorus *(Sori)*
Souffles *(Souffle)*
Soulamea
Sound *(Sounds)*
Soursop
Sourveld *see* Veld
Sourwood
Souse
Sowbane
Sowerbaea
Sowing *(Oversown, Sowings,*

Sown, Undersown)
X-Ref: Oversowing,
Postsowing, Presowing,
Undersowing
Sows *(Sow)*
Sowthistle
Soyasaponin
Soybeans *(Soy, Soya,*
Soyabean, Soyabeans,
Soybean)
Soymeal
Soymilk
Spacecraft
Spacing *(Spaced, Spacings,*
Spatial, Spatially)
Spades *(Spade)*
Spadix *(Spadixes)*
Spaghetti
Spalangia
Spalax
Span *(Spanning, Spans)*
Spaniel *(Spaniels)*
Spanworms *(Spanworm)*
Sparassis
Sparaxis
Sparganiaceae
Sparganium
Sparganothinae
Sparganothini
Sparganothis
Sparmannia
Sparrows *(Sparrow)*
Spartina
Spasms *(Spasm,*
Spasmolytic, Spastic)
X-Ref: Antispasmodic
Spathosternum
Spathularia
Spavin
Spawning *(Spawn)*
Spearmint
Specialization *(Specialisation)*
Specializations *(Specialised)*
Speciation
Specimen *(Specimens)*
Speciosine
Speck
Spectin
Spectinomycin
Spectrochemical
Spectrofluorometry
(Spectrofluorimetric,
Spectrofluorometric,
Spectrophotofluorometric)
X-Ref:
Microspectrofluorimetric,
Spectrophotofluorometry
Spectrofotometric *see*
Spectrophotometry
Spectrography
(Spectrographic)
Spectrometry *(Spectrometer,*
Spectrometric,
Spectrometrical)
Spectrophotofluorometry
see Spectrofluorometry

Spectrophotometry
(Microspectrophotometric,
Microspectrophotometrical,
Spectrophotometer,
Spectrophotometers,
Spectrophotometric,
Spectrophotometrical)
X-Ref:
Microspectrophotometry,
Spectrofotometric
Spectroradiometers
(Spectroradiometer)
Spectroscopy *(Spectroscopic)*
Spectrum *(Spectra, Spectral,*
Spectrally)
Specularia
Speedwells
Speleognathinae
Speleognathopsis
Spenthop
Speophyes
Speovelia
Spergula
Spergularia
Spergulatriol
Spermateliosis *see*
Spermatogenesis
Spermatheca *(Spermathecal)*
Spermatids *(Spermatid)*
Spermatocyte
(Spermatocytes)
Spermatocytogenesis *see*
Spermatogenesis
Spermatogenesis
(Spermateleosis,
Spermatogenic,
Spermatogenous,
Spermiogenesis)
X-Ref: Spermateliosis,
Spermatocytogenesis
Spermatogonia
(Spermatogonial,
Spermatogonium)
Spermatophores
(Spermatophore)
Spermatophyta
(Spermatophyte,
Spermatophytes)
Spermatozoa
(Heterospermic, Sperm,
Spermatic, Spermatologic,
Spermatological,
Spermatozoal,
Spermatozoan,
Spermatozoon, Sperms)
X-Ref: Heterosperm
Spermatozoid
(Spermatozoids)
Spermine *(Spermidine)*
Spermoderm
Spermolepis
Spermophilus
Spermosin
Spermosphere
Spermotoxic
Speyeria

Sphinx
Spicaria
Sphaceloma
Sphacelotheca
Sphaenolobium
Sphaerella
Sphaeriaceae
Sphaeriales
Sphaeridiinae
Sphaerobolus
Sphaerocarpos
Sphaerocaulus
Sphaeroceridae
Sphaerococcenol
Sphaerococcopsis
Sphaerococcum
Sphaerococcus
Sphaerodema
Sphaerogastra
Sphaerolichus
Sphaeroma
Sphaeromeria
Sphaeronostoc
Sphaerophoria
Sphaerophorus
Sphaerophysa
Sphaeroplast *(Sphaeroplasts)*
Sphaeroplea
Sphaeropsidaceae
Sphaeropsidales
(Spheropsidal)
Sphaeropsis
Sphaeropteris
Sphaerosperma
Sphaerostilbe
Sphaerotheca
Sphaerozosma
Sphaerularia
Sphaerulariidae
(Sphaerulariid)
Sphaerulina
Sphagnaceae
Sphagnum
Sphalerite
Sphallomorpha
Sphanostemon
Sphecidae *(Sphecids)*
Sphecoidea
Sphegigaster
Sphenocleaceae
Sphenomeris
Sphenophorus
Sphenophyllales
Sphenophyllum
Spherophorus
Spheroplasts *(Spheroplast)*
Spherosomes
Spherule *(Spherules)*
Sphex
Sphingidae *(Sphingid)*
Sphingolipids
Sphingomorpha
Sphingomyelin
Sphingonotus
Sphingophilous
Sphingosine *(Sphingosines)*

Sphacelaria
Sphacelia
Spices *(Spice, Spicing)*
Spicules *(Spicule)*
Spiders *(Spider)*
Spigelia
Spike *(Spikelet, Spikelets,*
Spikes)
Spikenard
Spilasma
Spillways *(Spillway)*
Spilocaea
Spilonota
Spilopsyllus
Spin *(Spinnability, Spinners,*
Spinning, Spins)
X-Ref: Spun
Spinach
Spinacia *(Spinacea)*
Spindle *(Spindles)*
Spine *(Spinal, Spined,*
Spines)
Spinifex
Spintherophyta
Spinturnix
Spiracles *(Spiracle,*
Spiracular)
Spiraea
X-Ref: Spirea
Spiral *(Spirals)*
Spiralum
Spiramycin *(Spiramycine)*
Spiranthes
Spirea *see* Spiraea
Spirillum
Spirits
Spirobenzylisoquinoline
Spirocerca
Spirocercosis
Spirochaeta
Spirochaetales *(Spirochete,*
Spirochetes)
Spirochetosis
(Spirochaetosis)
Spirodela
Spirogyra
Spirometra
Spirooxindole
Spiroplasma *(Spiroplasmas)*
Spiroplasmosis
Spirostan
Spirostanol
Spirostreptidae
Spirostreptus
Spirulina
Spiruridae
Spissistilus
Spittlebug *(Spittlebugs)*
Spleen *(Spleens, Splenic)*
Splenectomy *(Splenectomized)*
Splenomegaly
Splints *(Splint)*
Splitting *(Split, Splits,*
Splitter, Splitters, Splittings)
Spodoptera
Spodosol *(Spodosols)*

Spofa
Spoilage *(Spoilages)*
Spoils
Spondylitis
Spondylosis
Sponge *(Sponges)*
Spongilla
Spongiococcum
Spongiophyton
Spongospora
Spongovostox
Sporale
Sporangia *(Sporangial,*
   Sporangials, Sporangium)
Sporangiophores
   (Sporangiophore)
Sporangiospores
   (Sporangiospore)
Sporelings *(Sporeling)*
Sporendonema
Spores *(Diaspore, Spora,*
   Sporal, Spore, Spored,
   Sporeforming, Sporidial,
   Sporo)
   X-Ref: Diaspores,
      Mitospores,
      Multispored
Sporicides *(Sporicidal)*
Sporidesmin *(Sporidesmins)*
Sporidesmium
Sporidiobolus
Sporidium *(Sporidia)*
Sporobolomyces
Sporobolomycetaceae
Sporobolus
Sporocarp *(Sporocarpic)*
Sporocysts *(Sporocyst)*
Sporocytes *(Sporocyte)*
Sporocytophaga
Sporoderm
Sporogenesis *(Asporogenous,*
   Sporogenic)
   X-Ref: Asporogenetic
Sporogony *(Sporogonic)*
Sporophores *(Sporophore)*
Sporophytes *(Sporophyte,*
   Sporophytic)
Sporopollenin
Sporormia
Sporormiella
Sporothrix
Sporotrichosis
Sporotrichum
Sporozoa
Sporozoites
Sporulation *(Sporulated,*
   Sporulating, Sporulations)
   X-Ref: Antisporulant
Spot *(Spots, Spotted,*
   Spotting)
Spraing
Sprayers *(Sprayer)*
Spraying *(Spray, Sprayable,*
   Sprayed, Sprayings,
   Sprays)
   X-Ref: Unsprayed

Spreaders *(Spreader)*
Spring
Springtails
Springwood
   X-Ref: Earlywood
Sprinklers *(Sprinkle,*
   Sprinkler, Sprinkling)
Sprout *(Chitted, Sprouted,*
   Sprouting, Sprouts)
   X-Ref: Chitting,
      Nonsprouting,
      Presprouted
Spruce *see* Picea
Sprue
Spun *see* Spin
Spurge
Spyridia
Squalenes *(Squalene)*
Squamatic
Square *(Squares)*
Squash *(Squashes)*
Squid
Squill
Squirrels *(Squirrel)*
Stabilization *(Stabilisation,*
   Stabiliser, Stabilisers,
   Stabilising, Stabilities,
   Stabilize, Stabilized,
   Stabilizer, Stabilizers,
   Stabilizing)
Stables *(Stable, Stabled,*
   Stabling)
Stachybotryotoxicosis
Stachybotrys
Stachydrine
Stachyose
Stachys
Stachysic
Stacking *(Stacked, Stacks)*
   X-Ref: Restacking
Staehelina
Staggers
Stagmatoptera
Stagnation *(Stagnant)*
Stagonospora
Staining *(Stain, Stainability,*
   Stainable, Stained, Stains)
Stakes *(Staking)*
Staleness *(Stale)*
Stalks *(Stalk, Stalked,*
   Stalking)
   X-Ref: Seedstalk
Stallions *(Stallion)*
Stalls *(Stall)*
Stamen *(Stamenless,*
   Stamens, Staminal,
   Staminate)
   X-Ref: Multistaminate
Stamp *(Stamping, Stamps)*
Stanchions *(Stanchion)*
Standards *(Standardisation,*
   Standardised,
   Standardization,
   Standardizations,
   Standardized,
   Standardizing)

Stands
Stanhopea *(Stanhopeas)*
Stannic
Stapelia
Staphylea
Staphyleaceae
Staphylinidae *(Staphylinid,*
   Staphylinids)
Staphylococcosis
Staphylococcus
   (Staphylococcal,
   Staphylococci,
   Staphylococcic)
Staple *(Stapled, Staples)*
Starches *(Starch, Starchy)*
Stargrass
Starlings *(Starling)*
Starthistle
Starvation *(Starve, Starved,*
   Starving)
   X-Ref: Semistarved
Stathmopoda
Static
Statice
Stations *(Station)*
Statisticians *(Statistician)*
Statistics *(Statistic,*
   Statistical, Statistically)
Statocytes
Statoliths
Stator
Statutes *see* Laws
Staurothele
Steaks *(Steak)*
   X-Ref: Beefsteak
Steam *(Steaming, Steams)*
Stearic
Stearoyl
Stearoyllactylate
Stearyl
Steatitis
Steatoda
Steatonyssus
Steatorrhoea
Steatosis
Steckling *(Stecklings)*
Steel *(Steels)*
Steepland
Steepness
Steers *(Steer)*
Steganacarus
Stegasta
Stegobium
Stegodyphus
Stegomyia
Steironema
Stele *(Steles)*
Stelis
Stella
Stellaria
Stelopolybia
Stem *(Culmicolous, Culms,*
   Interstems, Stems)
   X-Ref: Culm, Interstem
Stemarin
Stemborer

Stemodia
Stemona
Stemonitis
Stempellia
Stemphylium
Stemwood
Stenacris
Stenamma
Stenella
Steneotarsonemus
Steninae
Stenobothrus
Stenocarpa
Stenochlaena
Stenocranus
Stenodema
Stenogaster
Stenolemus
Stenolophini
Stenomorpha
Stenonema
Stenopa
Stenoperla
Stenophagous
Stenophara
Stenopoda
Stenoponia
Stenopsocus
Stenopsyche
Stenoptilia
Stenosis
Stenosmylus
Stenotabanus
Stenotaphrum
Stenothemus
Stenus
Stephania
Stephanitis
Stephanoderes
Stephanofilaria
Stephanopyxis
Stephanosphaera
Stephanotis
Stephanurus
Steppes *(Steppe)*
Sterculia
Sterculiaceae
Stereocaulon
Stereochemistry
Stereoscopy *(Stereoscopic)*
Stereoselective
Stereospecific
   (Stereospecificity)
Stereostructure
Stereotaxic
Stereotydeus
Stereum
Sterigmatocystin
Sterigmatomyces
Sterilants *(Sterilant)*
Sterile *(Sterilisation,*
   Sterilised, Sterility,
   Sterilization, Sterilized,
   Sterilizing, Unsterilized)
   X-Ref: Androsterility,
      Cytosterile, Nonsterile,

Presterilized,
Semisterility,
Substerilizing,
Thermosterilization,
Unsterile
Sterility *(Androsterile)*
Sterilizers *(Sterilizer)*
Sterines
Sterna
Sternochetus
Sternostoma
Sternum *(Sternal)*
Sterocaulon
Steroids *(Steroid, Steroidal,*
Steroides, Steroidogenesis,
Steroidogenic)
Sterol *(Sterolic, Sterols)*
X-Ref: Nonsterol
Steropleurus
Steryl
Stethorus
Stevia
Steviol
Stevioside
Stewartia *(Stewartias)*
Stews *(Stew)*
Stibadocera
Stibadocerella
Stibophen
Stichopogon
Sticta
Stictaceae
Stictane
Stictis
Stictyosiphon
Stigeoclonium
Stigma *(Stigmas, Stigmata,*
Stigmatic)
Stigmaeidae
Stigmasteryl
Stigmella
Stigmellidae
Stigmina
Stilbenequinone
Stilbenes *(Stilbene)*
Stilbestrol *see*
Diethylstilbestrol
Stilbus
Stilesia
Stillage
Stillbirths *see* **Birth**
Stilpnotia
Stimulation *(Stimulates)*
Stings *(Sting, Stinging,*
Stingless)
Stinkbug
Stipa
Stipe *(Stiped, Stipes)*
Stipule *(Stipules)*
Stiretrus
Stizolobic
Stizolobinic
Stizolobium
Stizolophus
Stizus
Stochastic *(Stochastics)*

Stock *(Stocked, Stocking,*
Stocks)
X-Ref: Understocked
Stockbreeding
(Stockbreeders)
Stockmen
Stockpiles *(Stockpile,*
Stockpiling)
Stockyards
Stoichiometry
(Stoichiometric)
X-Ref: Substoichiometric
Stolbur
Stolonophora
Stolons *(Stolon,*
Stoloniferous)
Stolotermes
Stolotermitinae
Stomach *(Forestomach,*
Gastro, Intragastrically,
Stomached, Stomachs)
X-Ref: Forestomachs,
Gastric, Intragastric
Stomata *(Stoma, Stomal,*
Stomatal, Stomate,
Stomates, Stomatic)
Stomatanthes
Stomatitis
Stomatogastric
Stomopteryx
Stomoxys
Stoneflies *(Stonefly)*
Stones *(Stone, Stony)*
Stools *see* **Feces**
Storage *(Storability,*
Storable, Storages,
Storaging, Store, Stored,
Storehouse, Storehouses,
Storerooms, Stores,
Storing, Stowage)
X-Ref: Stow
Storms *(Storm,*
Thunderstorm)
X-Ref: Thunderstorms
Stover
Stoves *(Stove)*
Stow *see* **Storage**
Strabismus
Straelensia
Stragania
Straightbred *see* **Breeding**
Straighthead
Strain *(Strained, Straining,*
Strains)
X-Ref: Substrains
Strainer
Strand *(Stranded,*
Strandedness, Strands)
Strangalia
Strangles
Strangulation *(Strangulate)*
Strategus
Stratification
(Interstratified, Strata,
Stratifications, Stratified,
Stratify, Stratigraphic,

Stratigraphy, Stratum)
X-Ref: Biostratigraphy,
Interstratification
Stratiomyidae
Stratosphere *(Stratospheric)*
Straw *(Strawed, Straws)*
Strawberries *(Strawberry)*
Strawboard
Streak *(Streaks)*
Stream *(Streambed,*
Streamflow, Streamflows,
Streaming, Streams,
Streamside)
X-Ref: Upstream
Streambanks *(Streambank)*
Streblidae *(Streblid)*
Strelitzia
Strelitziaceae
Strepsicrates
Strepsiptera
Streptanthera
Streptanthus
Streptobacillus
Streptocarpus
Streptococcosis
Streptococcus
(Streptococcal,
Streptococci)
Streptocycline
Streptolysin
Streptomyces
Streptomycetaceae
(Streptomycete,
Streptomycetes)
Streptomycin
Streptonigrin
Streptopus
Streptosorangium
Streptothricosis
(Streptotrichosis)
Streptoverticillium
Stresnil
Stress *(Stressed, Stresses,*
Stressing, Stressors)
X-Ref: Antistress,
Nonstressed,
Prestressed
Stridulation *(Stridulating,*
Stridulatory)
Striga
Strigiphilus
Strigol
Strigops
Strigula
Strike
Stringhalt
Strip *(Stripped, Stripper,*
Strippers, Stripping,
Strippings, Strips)
Strobilar
Strobili *see* **Cones**
Strobliomyia
Stroboscopy *(Stroboscopic)*
Stroke
Stroma *(Stromal, Stromata)*
Stromateus

Stromatinia
Strongylidae *(Strongyle,*
Strongyles, Strongylid,
Strongylids)
Strongylognathus
Strongyloidea *(Strongyloid)*
Strongyloides
Strongyloidiasis
(Strongyloidosis)
Strongylosis
Strongylus
Strontium
X-Ref: Radiostrontium
Strophanthidin
Strophanthin
Strophanthus
Stropharia
Strumeta
Struvite
Strychnine
(Strychninization)
Strychnos
Strymon
Stryphnodendron
Stryracaeae
Stubble *(Stubbles)*
Students *(Student)*
Studs *(Stud)*
Stump *(Stumps)*
Stumpage
Stumpwood
Stunt
Sturmiopsis
Sturnidoecus
Sty *see* **Pigsty**
Styles *(Stylar)*
Stylets *(Stylet)*
Stylidiaceae
Stylidium
Styloconops
Styloctetor
Stylogaster
Stylogomphus
Stylopidae
Stylosanthes
Stylotermes
Styrax
Styrene
Styringomyia
Styrofoam
Styrol
Suaeda
Subalpine
Subaortic *see* **Aorta**
Subaquatic *see* **Aquatic**
Subarid *see* **Arid**
Subbaromyces
Subcapsular *see* **Capsule**
Subcellular *see* **Cells**
Subchloroplast *see*
Chloroplasts
Subchromatid *see*
Chromatids
Subclavian *see* **Clavicle**
Subclones *see* **Clone**
Subcoccinella

Subcooling *see* Cool
Subcoxa *see* Coxa
Subcutaneous *see* Skin
Subcutis *see* Skin
Subdermal *see* Skin
Subepidermal *see* Skin
Suberenol
Suberin *(Suberins)*
Suberization *(Suberized)*
Subesophageal *see*
　Esophagus
Subfertility *see* Fertility
Subfossil *see* Fossils
Subfreezing *see* Freezing
Subgeneric *see* Generic
Subhumid *see* Humidity
Subirrigation *see* Irrigation
Sublethal *see* Lethal
Sublimatioh
Sublittoral *see* Littoral
Subluxations *see* Luxations
Submandibular *see*
　Mandibles
Submaxillary *see* Maxillary
Submembranous *see*
　Membranes
Submergence *(Submerged,*
　Submersion)
Submicroscopic *see*
　Microscopy
Submicrosomal *see*
　Microsome
Submitochondrial *see*
　Mitochondria
Submontane
　(Submontanous)
Subnecrotic *see* Necrosis
Suboccipital *see* Occipital
Suboesophageal *see*
　Esophagus
Suboperculate *see*
　Operculum
Subpasteurization *see*
　Pasteurization
Subpermafrost *see*
　Permafrost
Subpopulations *see*
　Population
Subsampling *see* Sampling
Subsaturated *see* Saturation
Subsidence
Subsidy *(Subsidies,*
　Subsidization, Subsidized,
　Subsidizing)
Subsistence *(Subsisting)*
Subsoils *(Subsoil, Subsoiler,*
　Subsoiling)
Substerilizing *see* Sterile
Substitution *(Substituents,*
　Substitute, Substituted,
　Substitutes, Substituting,
　Substitutions)
Substoichiometric *see*
　Stoichiometry
Substrains *see* Strain
Substrate *(Substrates,*

　Substratum)
Subterranean
Subtidal *see* Tide
Subtilisins *(Subtilisin)*
Subtilled *see* Tillage
Subtoxic *see* Toxicity
Subtropical *(Subtropic,*
　Subtropics)
Subulura
Suburbs *(Suburban,*
　Suburbanization)
Subviral *see* Viruses
Subzero *see* Cold
Succession *(Successional,*
　Successions)
Succinamic
Succinate
Succinic
Succinoxidase
Succinyl *(Succinylated,*
　Succinylation)
Succinylcholine
Succulents *(Succulent)*
Suckers *(Sucker, Suckering)*
Sucking *(Suctorial)*
Suckling *(Suckled, Suckler,*
　Sucklings)
　X-Ref: Multisuckling
Sucrase *see* Invertase
Sucrose
　X-Ref: Nonsucrose,
　　Saccharose
Suction
Suctobelba
Suctobelbidae
Suctobelbila
Sudangrass
Sudax
Suessenguthia
Sufetula
Suffocating *see* Asphyxia
Sugar *(Nonsugars, Sugars)*
　X-Ref: Anhydrosugars,
　　Gur, Monosugars,
　　Nonsugar
Sugarbeets *(Sugarbeet)*
Sugarbelt
Sugarcane *(Sugarcanes)*
Sugarmills *(Sugarmill)*
Sugi
Sugiol
Suillus
Suisynchron *(Suisinchron)*
Sulcatol
Sulcicnephia
Sulfadiazine *(Sulphadiazine)*
Sulfadimethoxine
　(Sulphadimethoxine)
Sulfadimidine
　(Sulfadimidin,
　Sulphadimidine)
Sulfadoxine *(Sulphadoxine)*
Sulfallate
Sulfamethazine
　(Sulfamezathine)
Sulfamide *(Sulfamides)*

Sulfamonomethoxine
Sulfamoyl
Sulfanilamide
　(Sulphanilamide)
Sulfanitran
Sulfapyridine
Sulfapyrimidine
　(Sulphapyrimidine)
Sulfaquinoxaline
　(Sulfaquinoxalin)
Sulfatases *(Sulfatase)*
　X-Ref: Sulphatase
Sulfate *(Monosulfate,*
　Radiosulfate, Sulfated,
　Sulfates, Sulfation,
　Sulphate, Sulphated,
　Sulphates, Sulphation)
　X-Ref: Monosulfates,
　　Radiosulphate
Sulfathiazole
　(Sulphathiazole)
Sulfatophosphate
　(Sulphatophosphate)
Sulfene
Sulfhydrase
Sulfhydryl *(Sulfhydril,*
　Sulfhydryls, Sulphhydryl,
　Sulphydryl)
Sulfide *(Sulfides, Sulfidic,*
　Sulfidity, Sulphide,
　Sulphides)
　X-Ref: Monosulphide
Sulfinate
Sulfisoxazole
Sulfitation *(Sulphitation)*
Sulfite *(Sulfited, Sulfites,*
　Sulphite, Sulphiting)
Sulfobromophthalein
　(Sulphobromophthalein)
Sulfohydrolase
Sulfolane
Sulfolipids *(Sulfolipid)*
Sulfonamide *(Sulfonamides,*
　Sulphonamide,
　Sulphonamides)
Sulfonate *(Sulfonates,*
　Sulphonate, Sulphonates)
Sulfonation *(Sulfonated,*
　Sulphonated)
Sulfone *(Sulfon, Sulfones,*
　Sulphone, Sulphones)
Sulfonic
Sulfonium
　X-Ref: Sulphonium
Sulfonolipid
Sulfonyl *(Sulphonyl)*
Sulfonylureas
　(Sulphonylureas)
Sulfosuccinate
Sulfotransferase
Sulfoxide *(Sulfoxides,*
　Sulphoxide, Sulphoxides)
Sulfoximine
Sulfur *(Sulfo, Sulfurated,*
　Sulfurcoated,
　Sulfurization, Sulphur,

　Sulphured, Sulphuring)
Sulfuric *(Sulphuric)*
Sulfurous *(Sulphurous)*
Sulfurtransferases
Sulfurylase
Sulmtaler
Sulochrin
Sulphatase *see* Sulfatases
Sulphonium *see* Sulfonium
Sultone
Suluguni
Sumithion *see* Fenitrothion
Summer *(Summering,*
　Summers, Summertime)
　X-Ref: Midsummer
Summerwood
Sump
Sun *(Sundried, Sunlight,*
　Sunlit, Sunny, Sunpower,
　Sunrise, Sunset, Sunshine)
　X-Ref: Insolation, Solar
Sunburn
Sundew
Sunfish
Sunflowers *(Sunflower)*
Sunnhemp
Sunscald
Sunscreen
Sunspots *(Sunspot)*
Supella
Superamine *see* Amines
Superboraphidia
Superchill *see* Cool
Supercooling *see* Cool
Superfetation *see* Fetus
Supergenes *see* Genes
Superheated *see* Heat
Superinfection *see* Infection
Superinfestation *see*
　Infestation
Superlutin *(Superlutine)*
Supermarkets
　(Supermarket)
Superobesity *see* Obesity
Superovulation *see*
　Ovulation
Superoxide
Superparasitism *see*
　Parasites
Superphosphates *see*
　Phosphates
Superphosphoric *see*
　Phosphoris
Superpork *see* Pork
Supersuppressors *see*
　Suppression
Supervision *(Supervisor,*
　Supervisors, Supervisory)
Supinidine
Supinine
Supplements *(Supplement,*
　Supplementation,
　Supplementations,
　Supplementing)
Suppression
　(Supersuppressor,

Suppressants, Suppressed.
Suppresses, Suppressing.
Suppressive, Suppressor.
Suppressors, Supressing)
  X-Ref: Antisuppressor,
    Supersuppressors
Suppurative
Supracide
Supralethal *see* **Lethal**
Supralittoral *see* **Littoral**
Suprarenal *see* **Kidneys**
Suramin
Suregada
Surfactant *(Surfactants)*
Surgeons *(Surgeon)*
Surgery *(Surgeries,*
  Surgical, Surgically)
  X-Ref: Nonsurgical,
    Postsurgical
Surplus *(Surpluses)*
Surra
Surti
Surveying *(Surveyor,*
  Surveyors)
Survival *(Survivability,*
  Survivals, Survive,
  Survived, Surviving,
  Survivors)
Sus *see* **Swine**
Susabinori
Susceptibility *(Nonsuscept,*
  Susceptibilities,
  Susceptible)
  X-Ref: Nonsusceptible
Suspension *(Suspended,*
  Suspending, Suspensions,
  Suspensor)
Sutera
Suturaspis
Suture *(Suturable, Sutured,*
  Sutures, Suturing)
Suxamethonium
Svastrides
Swainsona
Swallows *(Swallow)*
Swallowtail *see* **Papilionidae**
Swamp *(Swamping,*
  Swamplands, Swamps,
  Swampy)
Swan *(Swans)*
Sward *(Swards)*
Swarms *(Swarm, Swarmers,*
  Swarming, Swarmings)
Swartzia
Swazine
Sweat *(Sweated, Sweating)*
Swedes *see* **Rutabaga**
Sweep *(Sweeping, Sweeps)*
Sweet *(Sweetened,*
  Sweetener, Sweeteners,
  Sweetening, Sweetness)
  X-Ref: Unsweetened
Sweetclover *see* **Clover**
Sweetcorn
Sweetfern *see* **Comptonia**
Sweetgum

Sweetia
Sweetolethaeus
Sweetpea *(Sweetpeas)*
Sweetpotatoes *(Sweetpotato)*
Sweets *see* **Desserts**
Sweetsop
Swelling *(Swellings,*
  Swollen)
  X-Ref: Tumefacient
Sweroside
Swertia
Swertisin
Swietenia
Swill
Swine *(Hog)*
  X-Ref: Hogs, Sus
Swinepox
Switchgrass
Syagrus
Sycamore *(Sycamores)*
Sycanus
Sycophaga
Sylvestrene
Sylviculture *see* **Silviculture**
Symbiocladius
Symbiosis *(Symbiont,*
  Symbiontic, Symbionts,
  Symbioses, Symbiote,
  Symbiotes, Symbiotic,
  Symbiotically)
  X-Ref: Aposymbiotic,
    Asymbiotic,
    Mutualism,
    Nonsymbiotic
Symmerista
Symmocidae *(Symmocid)*
Sympathectomy
Sympathomimetic
  (Sympathomimetics)
Sympatry *(Sympatric)*
Sympetrum
Symphanes
Sympherobius
Symphoricarpos
  (Symphoricarpus)
Symphorionts
Symphoromyia
Symphyla *(Symphylan,*
  Symphylid, Symphylids)
Symphypleona
Symphyta
Symphytum
Sympiesis
Sympiezorhincus
Sympistis
Symplasm *(Symplasmic)*
Symplast *(Symplastic)*
Symplectromyces
Symplocos
Sympodial
Symptoms *(Symptomless)*
  X-Ref: Asymptomatic
Symtriazine
Synadenium
Synanthedon
Synanthropic

Synapsis *(Asynaptic.*
  Synapse, Synaptic,
  Synaptically)
  X-Ref: Asynapsis,
    Polysynaptic,
    Postsynaptic,
    Presynaptic,
    Transsynaptic
Synaptomys
Synaptonemal
Syncephalis
Synchronization
  (Asynchronously,
  Asynchrony,
  Synchronisation,
  Synchronize,
  Synchronized,
  Synchronizing,
  Synchronous,
  Synchronously,
  Synchrony)
  X-Ref: Asynchronous
Synchytrium
Syncytium *(Syncytia,*
  Syncytial)
Syndactylism *(Syndactyly)*
Syndicalism *(Syndical,*
  Syndication)
Syndrome *(Syndromes)*
Syndyas
Synechococcus
Synechocystis
Synecology *(Synecological)*
Synergism *(Synergetic,*
  Synergic, Synergisms,
  Synergist, Synergistic,
  Synergists, Synergized)
Synergus
Syneta
Syngameon
Syngamiasis *(Syngamosis)*
Syngamidae
Syngamus
Syngonium
Syngramma
Synharmonia
Synnema
Synonyma
Synopeas
Synostosis
Synovex
Synovia *(Synovial)*
Synovitis
Synsepalum
Syntanol
Syntaxonomy *see*
  Taxonomy
Syntelia
Synteliidae
Syntexis
Synthase *(Synthases)*
Synthesiomyia
Synthetases *(Synthetase)*
  X-Ref: Cosynthetase
Synthetics *(Synthetic,*
  Synthetical, Synthetically)

  X-Ref: Semisynthetic
Syntomidae
Syntomis
Syntomus
Syntormon
Syntretus
Syntrichalonia
Syntropis
Synura
Synusia *(Synusium)*
Syphacia
Syrenia
Syringa
Syringes *(Syringe)*
Syringodium
Syringopais
Syringophilidae
Syringophiloidus
Syringospora *see* **Candida**
Syringoxide
Syringyl
Syriogenin
Syrista
Syrphidae *(Syrphid,*
  Syrphids, Syrphyd)
Syrphini
Syrphus
Syrup *(Sirup, Syrups)*
Systelloderes
Systematics *see* **Taxonomy**
Systemic *(Systemically,*
  Systemics)
  X-Ref: Nonsystemic
Systena
Systenus
Systole *(Systolic)*
Syston
Systox *see* **Demeton**
Systropus
Syzeuctus
Syzygium
Tabanidae *(Tabanid,*
  Tabanids)
Tabanus
Tabebuia
Tabernaemontana
Tabernanthe
Tabersonine
Tablecloths
Tablelands *(Tableland)*
Tachina
Tachinaephagus
Tachinidae *(Tachinid,*
  Tachinids)
  X-Ref: Larvaevoridae
Tachinomorphus
Tachinus
Tachycardia
Tachycines
Tachydromia
Tachypleus
Tachyporinae
Tachys
Tachysphex
Tachyta
Tachytes

Tactile *see* Touch
Taenia
Taeniarhynchus
Taeniatherum
Taeniidae *(Taeniid)*
Taeniolella
Taenioma
Taeniopteris
Taeniopteryx
Taeniostigma
Taeniothrips
Taenitis
Taenoglyptes
Tafasan
Tageteae
Tagetes
Tagging *(Tag, Tagged, Tags)*
Tahyna
Taiga
Tail *(Tailed, Tails)*
    X-Ref: Cercal, Tailless
Tailings *(Tailing)*
Tailless *see* Tail
Taint
Taiwania
Taiwanin
Taiyutyla
Takadiastase
Takama
Takeall
Takyrs
Talaromyces
Talcum *(Talc)*
Talfan *see* Teschen
Talicmine
Talinella
Tallgrass *see* Grasses
Tallow *(Tallows)*
Talpa
Talpidae
Tamarack
Tamaricaceae
Tamarillos *(Tamarillo)*
Tamarind
Tamarindus
Tamarix *(Tamarisk)*
Tamarixetin
Tamoxifen
Tamponade
Tamus
Tanacetum
Tanagers
Tanarthrus
Tangelos *(Tangelo)*
Tangerines *(Tangerine)*
Tankers *(Tanker)*
Tanks *(Tank)*
Tannase
Tannic
Tannide *(Tannides)*
Tanning *(Tanned, Tanneries, Tannery)*
Tannins *(Tanniferous, Tannin)*
Tanoak

Tansonemoidea
Tansy
Tantalum
Tanydesmus
Tanymecosticta
Tanymecus
Tanypeza
Tanypezidae
Tanypodinae
Tanyproctus
Tanypus
Tanytarsini
Tanytarsus
Tape *(Tapes)*
Taper *(Tapered, Tapering)*
Tapetum *(Tapetal)*
Tapeworm *(Tapeworms)*
Taphrina
Taphrinales
Taphronota
Tapinoma
Tapinopterus
Tapioca
Tapping *(Tap, Tapped)*
Taproot *see* Roots
Tapura
Tar *(Tars)*
Tarantula
Tarantulidae
Taraxacum
Taraxasterol
Tardigrada *(Tardigrade, Tardigrades)*
Tare *(Tares)*
Tarenna
Target *(Targets)*
    X-Ref: Nontarget
Targhee
Targionia
Tariff *(Tariffing, Tariffs)*
Taro
Tarragon
Tarrietia
Tarsitis
Tarsocera
Tarsocheylidae
Tarsonemidae *(Tarsonemid, Tarsonemida)*
Tarsonemina
Tarsonemini
Tarsonemus
Tarsus *(Pretarsal, Tarsal, Tarsi, Tarso)*
    X-Ref: Ankle, Intertarsal, Pretarsus
Tartaric *(Tartar)*
Tartrate
Tassels *(Tasseling)*
Taste *(Palatable, Taster, Tasters, Tastes, Tasting)*
    X-Ref: Gustatory, Palatability
Tattoo
Taught *see* Education
Taurine
Taurocholate

Tautomer *(Tautomerism)*
Taxaceae
Taxation *(Tax, Taxable, Taxed, Taxes)*
Taxillus
Taxiphyllin
Taxir
Taxodiaceae
Taxodium
Taxomyia
Taxonomists *(Taxonomist)*
    X-Ref: Biosystematists
Taxonomy *(Biosystematic, Biosystematical, Biosystematy, Biotaxonomic, Syntaxonomic, Systematology, Systematy, Taxa, Taximetric, Taxogenetics, Taxon, Taxonomic, Taxonomical, Taxonomically, Taxons)*
    X-Ref: Biosystematics, Biotaxonomy, Organotaxic, Syntaxonomy, Systematics
Taxus *(Yew)*
    X-Ref: Yews
Tazettine
Tea *(Teas)*
Teachers *see* Educators
Teaching *see* Education
Teak
Tear *(Tearing)*
Teaser
Teats *(Teat, Teatcup, Teatcups)*
    X-Ref: Mammilla, Nipple
Tebenna
Tebuthiuron
Technetium
Technicians *(Technician)*
Teclea
Tecleanone
Tecoma
Tecomella
Tectaria
Tectocepheus
Tectocoris
Tectona
Tectonic *(Tectonical)*
Tectorigenin
Tectum *(Tectal)*
Tedion *see* Tetradifon
Teens *see* Adolescence
Teeth *(Tooth)*
    X-Ref: Odontogenesis
Teferin
Teff
Teflon
Tegenaria
Tegrodera
Tegument *see* Integuments
Teiidae
Telamonia

Telaranea
Telaxerophytoids
Telecommunications *see* Communication
Telemechanization *see* Mechanization
Telemetry *(Telemetered, Telemetering, Telemetric)*
Telencephalon
Telenominae
Telenomus
Teleogryllus
Teleonemia
Telephone *(Telephony)*
Telescope *(Telescopic, Telescoping)*
Teleutospores
Television
Teliapsocus
Teliospores *(Teliospore)*
Telipogons
Tellima
Tellurite
Tellurium
    X-Ref: Radiotellurium
Telmatoscopus
Telodrin *see* Isobenzan
Teloidine
Telomeric
Telone *see* Dichloropropenes
Telopea
Telophase
TEM *see* Tretamine
Temephos *(Temefos)*
Temik *see* Aldicarb
Temnochila
Temnorhinus *(Temnorrhinus)*
Temnothorax
Tempeh
Temperate
Temperature *(Temperatures)*
Tempering
Template
Tenagobia
Tenants *(Tenancy, Tenant)*
Tenderization *(Tenderizer, Tenderizers, Tenderizing)*
Tenderness
Tenderometer
Tendinitis
Tendipedidae *see* Chironomidae
Tendipes *see* Chironomus
Tendon *(Tendons)*
Tenebrio
Tenebrionidae *(Tenebrionid, Tenebrionids)*
Tenebroides
Tenoran *see* Chloroxuron
Tenosynovitis *(Tendosynovitis)*
Tenotomy
Tensile

Tensiometers *(Tensiometer,*
  Tensiometric)
Tension *(Tensions)*
Tent *(Tents)*
Tenthredinoidea
  (Tenthredinidae)
Tenthredo
Tenthredoides
Tenthredopsis
Tentoxin
Tenuazonic
Tenuipalpidae *(Tenuipalpid)*
Tenuipalpus
Tenuirostritermes
Tenure *(Tenures, Tenurial)*
Teosinte
TEPA
Tepary
Tephrinopsis
Tephritidae *(Tephritid,*
  Tephritids, Trypetids)
  X-Ref: Trypetidae
Tephrosia
TEPP
Terabol
Teratisms *see* Abnormality
Teratocarcinoma
Teratogenesis *(Teratogen,*
  Teratogenic,
  Teratogenicity,
  Teratogens)
  X-Ref: Nonteratogenic
Teratology *(Teratologic,*
  Teratological)
Teratolytta
Teratomas *(Teratoma,*
  Teratomatous)
Teratomyces
Teratoppia
Teratosphaeria
Terbacil
Terbufos
Terbutaline
Terbutryn *(Terbutrin)*
Terebenthifolic
Teredinidae
Teredo
Terenol
Terephthalic
Tergum *(Tergal)*
Termatomiris
Termatophylidae
Termatophylidea
Termatophylum
Terminalia
Terminology
  (Terminological,
  Terminologies)
Termitaria
Termite *(Termites)*
Termiticides *(Termiticidal,*
  Termiticide)
Termiticolous
Termitidae *(Termitinae)*
Termitomyces
Termitophiles

(Termitophilous)
Termitosagma
Terniflorin
Terns *(Tern)*
Terpenes *(Monoterpene,*
  Monoterpenic,
  Monoterpenoid, Terpene)
  X-Ref: Monoterpenes,
  Monoterpenoids
Terpenoid *(Terpenoids)*
  X-Ref: Nonterpenoid
Terphenyls
Terphis
Terpolymers
Terpsinoe
Terpyridine
Terrace *(Terraced, Terraces,*
  Terracing)
Terraclor *see*
  Pentachloronitrobenzene
Terrain *(Terrains)*
Terramycin
Terrariums *(Terrarium)*
Terrasytam
Terrazole *(Terrazol)*
Terrein
Terrestrial
Terrier *(Terriers)*
Territory *(Territorial,*
  Territoriality, Territories)
  X-Ref: Microterritorial,
  Pseudoterritorial
Tersilochus
Tertbutylbenzyl
Tertiary
Terylene
Teschen
  X-Ref: Talfan
Tesguino
Tesota
Tessaratoma
Tessaratominae
Tessaratomini
Testa *(Testae, Testal)*
Testacida
Testes *(Testicle, Testicles,*
  Testicular, Testis)
Testosterone
  X-Ref:
  Methyltestosterone
Testudo
Tetanocera
Tetanops
Tetanus *(Tetanic)*
Tetany *(Tetanies)*
Tethering *(Tether,*
  Tethered)
Tethina
Tetraacetate
Tetraacetic
Tetraacetylglucoside
Tetraalkylammonium
Tetrablemmidae
Tetraborate
Tetrabrachys
Tetrabromide

Tetracanthella
Tetracarpidium
Tetrachloride
Tetrachlorobiphenyl
  (Tetrachlorodiphenyl)
Tetrachlorodibenzofuran
Tetrachlorodiphenylethane
Tetrachloroethane
  X-Ref: Cellon
Tetrachloroethylene
Tetrachloronitrobenzene
  X-Ref: Fusarex
Tetrachlorophenol
Tetrachloroterephthalate
Tetracladium
Tetracoccosporium
Tetracoccus
Tetracyclic
Tetracycline *(Tetracyclin,*
  Tetracyclines)
Tetrad *(Tetrads)*
Tetradecadien
Tetradecane
Tetradecenal
Tetradecene
Tetradecenyl
Tetradepside *see* Depsides
Tetradifon
  X-Ref: Tedion
Tetradymia
Tetraenes *(Tetraene)*
Tetraenols *(Tetraenol)*
Tetragnathidae
Tetragoneuria
Tetragonia
Tetragonoderus
Tetrahomoterpene
Tetrahydrocannabinol
Tetrahydrofolate
Tetrahydrofuran
Tetrahydrofurfuryl
Tetrahydroisoquinolines
  (Tetrahydroisoquinoline)
Tetrahydropyranyl
Tetrahydropyrimidine
Tetrahydroquinoline
Tetrahydrosecamine
Tetrahydroxanthyletins
Tetrahydroxychalcone
Tetrahymanol
Tetrahymena
Tetraiodofluorescein
Tetrakis
Tetralogy
Tetralonia
Tetralopha
Tetram
Tetramer *(Tetrameric,*
  Tetramers)
Tetrameres
Tetramethoxyflavone
Tetramethylputrescine
Tetramethylthiuram
Tetramisole *(Tetramizol)*
  X-Ref: Levamisole,
  Nilverm

Tetramorium
Tetramyxa
Tetranactin
Tetraneura
Tetranitrobiphenyl
Tetranitromethane
Tetranortriterpenoids
Tetranychidae *(Tetranychid)*
Tetranychus
Tetranycopsis
Tetraopes
Tetrapanax
Tetrapeptide
Tetraphalerus
Tetraphenylborate
Tetraphenylboron
Tetraphis
Tetrapion
Tetraplasandra
Tetraploids *(Tetraploid,*
  Tetraploidal, Tetraploidy)
Tetrapodili
Tetraponera
Tetrasaccharides
  (Tetrasaccharide)
Tetrasomic *(Tetrasomics)*
Tetraspores
  (Tetrasporangial,
  Tetraspore, Tetrasporic,
  Tetrasporocyst)
Tetrastichus
Tetrasul
Tetrathyridia
Tetratrichomonas
Tetrazole
Tetrazolium
Tetricidae
Tetrigidae
Tetrix
Tetrodontophora
Tetrodotoxin
Tetrol
Tetronic
Tetropium
Tetrose
Tetroxide
Tettigonia
Tettigoniidae *(Tettigoniids)*
Tettigonioidea
Tetyra
Teucrium
Textiles *(Cloth, Textile)*
  X-Ref: Cloths
Texture *(Textural,*
  Texturation, Textured,
  Textures, Texturized,
  Texturizing,
  Texturometer)
Thalamus *(Thalamic)*
Thalassia
Thalassin
Thalassiosira
Thalassodendron
Thalassodes
Thalassotrechus
Thalia

Thalibrunimine
Thalictine
Thalictrum
Thalidomide
Thallium
Thallophyta *(Thallophyte)*
Thallus *(Thalli, Thallic,*
  Thalloid, Thalluses)
Thamnidiaceae
Thamnidium
Thamnium
Thamnosma
Thamnurgides
Thanasimus
Thanatephorus
Thanite
Tharoopama
Tharparkar
Thatch
Thaumalea
Thaumaleidae
Thaumetopeinae
Thaumetopoea
Thaumetopoeidae
Thaw *(Thawed, Thawing)*
Thea
Theaceae
Theaflavins
Thearubigins
Theasapogenol
Thebaine
Theca *(Thecal)*
Thecamoebae
Thecaphora
Theclinesthes
Thecodiplosis
Thecomyia
Thecophora
Theileria
Theileriidae
Theileriosis *(Theilerial,*
  Theileriasis)
Thelazia
Thelaziasis
Thelaziidae
Thelephora
Thelephoraceae
Thelidiaceae
Theligonum
Thelobania
Thelocactus
Thelocarpon
Thelohania
Thelydesmus
Thelyphonidae
Thelypteridaceae
Thelypteris
Themeda
Themos
Theobaldia
Theobroma
Theobromine
Theophylline *(Theophyllin)*
Theopsis
Theraphosidae
Therapy *(Therapeusis,*

Therapeutic,
Therapeutical,
Therapeutics, Therapies)
X-Ref: Biotherapy,
  Cotherapy
Theretra
Thereuonema
Thereva
Therevidae *(Therevine)*
Theridiidae *(Theridiid)*
Theridion
Therioaphis
Therion
Thermal *(Thermally,*
  Thermic, Thermo)
Thermarol
Therminol
Thermionic
Thermistor
Thermoactinomyces
Thermoanalysis
Thermoascus
Thermobia
Thermochemistry
  (Thermochemical)
Thermocouple
  X-Ref: Thermopile
Thermodormancy
Thermoduric
Thermodynamics
  (Thermodynamic)
Thermoelectric
Thermofiltration *see*
  Filtration
Thermogenesis
Thermogradient
Thermographic
Thermogravimetry
  (Thermogravimetric)
Thermoinactivation
Thermoluminescence
  (Thermoluminescent)
Thermolysis *(Thermolytic)*
Thermomechanical
Thermometer
  (Thermometers,
  Thermometric,
  Thermometry)
Thermomonospora
Thermomyces
Thermomycolase
Thermonectus
Thermonuclear
Thermonuclease
Thermooxidative
Thermoperiods
  (Thermoperiod,
  Thermoperiodic,
  Thermoperiodism)
Thermophilic
  (Thermophilous)
Thermophysical
Thermopile *see*
  Thermocouple
Thermoplastics
  (Thermoplastic,

Thermoplasticity)
Thermoprocessing
  (Thermoprocessed)
Thermopsis
Thermoreceptor
Thermoregulation
  (Thermoregulatory)
Thermoresistance
  (Thermoresistence)
Thermosensitivity
  (Thermosensitive)
Thermosetting
Thermostability
  (Thermostabilization)
Thermostat
Thermosterilization *see*
  Sterile
Thermotaxis
Thermotherapy
Thermotolerant
  (Thermotolerance)
Thesaurismosis
Thesaurus
Thesium
Thespesia
Thevetia
Thiabendazole
  (Thiabendazol)
  X-Ref: Thibenzole
Thiadiazinethione
Thiadiazole *(Thiadiazol)*
Thiadiazolyl
Thiaminase
Thiamine *(Thiamin)*
  X-Ref: Aneurin,
  Antithiamine
Thiamylal
Thianaphthen
Thiazafluron
Thiazine
Thiazoles *(Thiazole)*
Thiazolidinones
  (Thiazolidinone)
Thiazolines
Thiazolyl
Thiazone *see* Dazomet
Thibenzole *see*
  Thiabendazole
Thickening *(Thicken,*
  Thickened, Thickeners,
  Thickenings)
Thickets *(Thicket)*
Thickness *(Thick, Thicker,*
  Thicknesses)
Thielavia
Thielaviopsis
Thigmotropism
  (Thigmotropic)
Thimet *see* Phorate
Thinning *(Thinned,*
  Thinnings)
  X-Ref: Unthinned
Thinobius
Thinopinus
Thinox
Thinusa

Thioacetamide
Thiobacillus
Thiobarbiturate
Thiobarbituric
Thiocarbamates
  (Thiocarbamate)
Thiocarbamyl
Thiocarbonate
Thiocyanates *(Thiocyanate,*
  Thiocyanato)
Thiodan *see* Endosulfan
Thioesterases
Thioethers *(Thioether)*
Thiofanox
Thiogalactose
Thioglucose
Thioglucoside
Thioglycolic *(Thioglycollic)*
Thioguanine
Thiohempa
Thiohydantoin
Thioketal
Thiol *(Thiolic, Thiols)*
Thiolases *(Thiolase)*
Thiolignin
Thiometon
  X-Ref: Ekatin
Thiomolybdates
Thionazin
Thioneins *(Thionein)*
Thiones *(Thione)*
Thionine
Thionocarbonates
Thionophosphate
Thionyl
Thiopental
Thiopentone
Thiophanate *(Thiophanates)*
  X-Ref: Topsin
Thiophene *(Thiophenes,*
  Thiophens)
Thiophenols
Thiophos *see* Parathion
Thiophosphate
Thiopropanal
Thioptera
Thiopyrophosphate
Thioredoxin
Thiosemicarbazide
Thiosemicarbazone
  (Thiosemicarbazones)
Thiosulfate *(Thiosulphate)*
Thiosulphinates
Thiosulphonates
Thiotepa
Thiothreonine
Thiouracil
Thiourea *(Thioureas)*
Thioureido
Thiouridine
Thiram
  X-Ref: TMTD
Thirst
Thistles *(Thistle)*
Thiuram
Thixotropy

Thlaspi
Thomasiniana
Thomisidae
Thomisus
Thomomys
Thompsonella *see* **Echeveria**
Thora
Thoracolumbar
Thorax *(Thoracic. Thoraco)*
　X-Ref: Chest
Thorium
Thornea
Thoroughbred
　(Thoroughbreds)
Thozetella
Thracophilus
Thrashed *see* **Thresh**
Thrassis
Thraulodes
Thread *(Threads)*
Threitol
Threonine *(Threonin)*
Threonyl
Thresh *(Threshability.
　Threshed, Threshing)*
　X-Ref: Thrashed
Threshers *(Thresher)*
Threshold *(Subthreshold.
　Thresholds)*
Threskiornis
Thripidae *(Thrips)*
Thrixspermum
Thrombidiidae
　(Thrombidiinae)
Thrombophlebitis
Thrombus
　*(Antithrombogenic.
　Thrombin, Thrombo.
　Thrombocyte.
　Thrombocytes,
　Thrombocythemic.
　Thrombocytic.
　Thrombocytopenia,
　Thrombocytopenic.
　Thromboembolic.
　Thromboembolism.
　Thrombolytic.
　Thromboplastic.
　Thromboplastin.
　Thrombopoietic.
　Thrombosic, Thrombosis.
　Thrombosthenin)*
　X-Ref: Antithrombin
Throscidae
Throughfall
Thuidium
Thuja *(Thuya)*
Thujopsis
Thunderstorms *see* **Storms**
Thunia
Thuricide
Thuris
Thylakoids *(Thylakoid)*
Thylene
Thyme
Thymectomy

(Thymectomized)
Thymelaea
Thymelaeaceae
　(Thymeleaceae)
Thymelicus
Thymidine
Thymidylate
Thymine *(Thymineless)*
　X-Ref: Methyluracil
Thymocytes *(Thymocyte)*
　X-Ref: Antithymocyte
Thymol
Thymoma *(Thymomas)*
Thymus *(Thymic)*
Thynnascaris
Thyraeella
Thyreus
Thyridopteryx
Thyrocalcitone
　*(Thyrocalcitonin.
　Thyrocalcitonine)*
Thyroglobulin
Thyroid *(Antithyroidal.
　Thyro,. Thyroidal.
　Thyroids)*
　X-Ref: Antithyroid
Thyroidectomy
　(Thyroidectomized)
Thyroiditis
Thyroparathyroidectomy
Thyroproteins
　(Thyroprotein)
Thyrotoxicosis
Thyrotropin *(Thyrotrophic.
　Thyrotropic)*
Thyroxine *(Thyroxin)*
Thysaniezia
Thysanocercinae
Thysanoptera
　(Thysanopteran)
Thysanosoma
Thysanura *(Thysanuran)*
Tiaja
Tiarosporella
TIBA *see* **Triiodobenzoic**
Tibia *(Tibiae, Tibial.
　Tibias)*
Tibicen
Tibiozus
Tickbean
Tickclover
Ticks *(Tick, Tickborne)*
Tide *(Tidal, Tides)*
　X-Ref: Intertidal,
　Nontidal, Subtidal
Tie *(Ties)*
Tieghemella
Tieghemiomyces
Tigloyloxytropane
Tigogenin
Tigridia
Tigridieae
Tigriopus
Tikka
Tilapia
Tilehorned

Tiles *(Tile)*
Tilia *(Linden)*
　X-Ref: Basswood,
　Lindens
Tiliaceae
Tiliacora
Tilingia
Tillage *(Till, Tillability.
　Tillable, Tillages, Tilled.
　Tilling, Tills, Tilth)*
　X-Ref: Intertilled,
　Nontillage, Subtilled,
　Untilled
Tillam *see* **Pebulate**
Tillandsia
Tillers *(Tiller, Tillering)*
　X-Ref: Nontillering
Tilletia
Tilletiaceae
Tilsiter *(Tilsit)*
Tilt
Timarcha
Timber *(Timbers,
　Timberyard)*
Timberland *(Timberlands)*
Timberline
Timet *see* **Phorate**
Timia
Timolol
Timomenus
Timonius
Timothy
Tin *(Tinplate)*
Tinaminyssus
Tinea
Tineidae *(Tineid)*
Tineola
Tingidae
Tingis
Tinned *see* **Canning**
Tinobregmus
Tinocallis
Tinospora
Tins *see* **Canning**
Tinus
Tipburn
Tiphia
Tiphiidae
Tipula
Tipulidae *(Tipulid.
　Tipulinae)*
Tipuloidea
Tiquilia
Tirathaba
Tires *(Tire, Tyre, Tyred)*
　X-Ref: Tyres
Tirpate
Tisiphone
Tissues *(Tissue)*
Titanium
Tithonia
Tithonin
Tittmannia
Tityus
Tmesipteris
Tmetocera

TMTD *see* **Thiram**
Toadflax
Toads *(Toad)*
　X-Ref: Bufo
Toadstools *see* **Mushrooms**
Tobacco *(Tobaccos)*
Tobosa
Tobravirus
Tocopherols *(Tocopherol)*
Tocopheryl
Tocotrienol *(Tocotrienols)*
Toddler *see* **Children**
Todea
Toe *(Toed, Toenail, Toes)*
Tofu
Togaviruses
Tolbutamide
Tolerance *(Tolerances,
　Tolerant, Toleration)*
　X-Ref: Intolerances
Tolpis
Toluamide
Toluene
Toluidine
Toluquinones *(Toluquinone)*
Tolypella
Tolyposporium
Tolypothrix
Tomanol
Tomatidine
Tomatillidine
Tomatine *(Tomatin)*
Tomatoes *(Tomato)*
Tomentella
Tomentin
Tomentogenin
Tomicomerus
Tomicus
Tomocarabus
Tomocerus
Tomography
Tongue *(Tongues)*
Tonnage
Tonometer *(Tonometry)*
Tonoplasts *(Tonoplast)*
Tonsil *(Tonsillar, Tonsils)*
Tonsillectomy
Toolbar
Toona
Topcross *(Topcrossing)*
Topdressings *(Topdressed,
　Topdressing)*
Topiary
Topinambour *(Topinambur)*
Topochemistry
　(Topochemical)
Topogard
Topography *(Topographic,
　Topographical)*
Topology *(Topological)*
Topomyia
Topophysis
Toposequence
Topping *(Topped, Toppers)*
Toprina
Topsin *see* **Thiophanate**

Topsoils *(Topsoil)*
Topworking see **Grafting**
Torbidan
Tordon see **Picloram**
Torenia
Toria see **Rape**
Torilis
Tormona
Tornabenia
Torque
Torrendia
Torrents *(Torrent)*
Torreya
Torreyal
Torreyol
Torsiometer
Torsion *(Torsional)*
Tortella
Tortillas *(Tortilla)*
Tortricidae *(Tortricid.*
  Tortricids, Tortricinae)
Tortrix
Tortula
Torula
Torularia
Torulinium
Torulopsis
Torus
Torymidae *(Torymid)*
Torymus
Tosylate
Totipotency *(Totipotent)*
Totolapanus
Touch *(Touches, Touching)*
  X-Ref: Tactile
Toughness *(Tough.*
  Toughening)
Toumeyella
Tourism *(Tourist, Touristic.*
  Tourists, Traveled.
  Traveler, Travelers.
  Traveller, Travellers.
  Travelling, Travels)
  X-Ref: Travel
Tournefortia
Towers *(Tower)*
Toxaphene
Toxascaris
Toxemia *(Toxaemia,*
  Toxaemic)
Toxicity *(Toxic, Toxicities,*
  Toxicobiological,
  Toxicoses, Toxicosis)
  X-Ref: Antitoxic,
  Nontoxic, Pathotoxin,
  Subtoxic
Toxicodendron
Toxicology *(Toxicologic.*
  Toxicological)
Toxigenicity *(Toxigenic)*
Toxins *(Antitoxicant.*
  Antitoxin, Biotoxin.
  Toxicant, Toxicants.
  Toxin)
  X-Ref: Antitoxicants,
  Biotoxins, Prototoxin

Toxocara
Toxocariasis
Toxochernes
Toxodera
Toxoid *(Toxoids)*
Toxomerus
Toxomyia
Toxoplasma *(Toxoplasm.*
  Toxoplasmic, Toxoplasms)
Toxoplasmatidae
Toxoplasmosis
  (Antitoxoplasma)
  X-Ref: Antitoxoplasmic
Toxoptera
Toxorhynchites
Tozzia
Trabeculae *(Trabeculated)*
Trace *(Traced, Tracer.*
  Tracers, Traces, Tracing)
Tracheal *(Trachea.*
  Tracheae, Tracheary.
  Tracheas, Tracheation)
  X-Ref: Endotracheal,
  Intratracheal,
  Transtracheal
Tracheids *(Tracheid)*
Tracheitis
Trachelas
Trachelomonas
Trachelopachys
Tracheobronchial
Tracheobronchitis
Tracheogenesis
Tracheoles
Tracheomycosis
  (Tracheomycotic)
Tracheoniscus
Tracheotomy
Trachoma
Trachycarpidium
Trachycarpus
Trachyphloeus
Trachys
Trachysoma
Trachyspermum
Trachysphaera
Trachysphyrus
Tract *(Tracts)*
Traction *(Tractive)*
Tractors *(Tractor)*
  X-Ref: Microtractors
Tracyella
Tracyina
Trade *(Traders, Trades.*
  Trading, Tradings)
Trademarks *(Trademark)*
Tradescantia
Tragacanth
Tragia
Tragocephala
Tragopogon
Tragus
Trailer *(Trailers)*
Trails *(Trail)*
Training *(Train, Trained.*
  Trainee, Trainees.

Trainer, Trainers.
  Trainings, Trains)
  X-Ref: Pretraining,
  Retraining, Untrained
Trajectory *(Trajectories)*
Trakephon
Trama
Trametes
Tramisol
Trampling
Tranexamic
Tranquilizers
  (Tranquilization.
  Tranquilizer.
  Tranquilizing.
  Tranquillization.
  Tranquillizing)
Transacylase
Transacylation
Transamidinase
Transaminases
  (Transaminase)
Transamination
Transcapillary see **Capillary**
Transcarbamoylase
Transcarbamylase
Transcinnamic see **Cinnamic**
Transcortin
Transcriptase
Transcriptions *(Transcribed.*
  Transcript, Transcription.
  Transcriptional,
  Transcripts)
Transcuticular see **Cuticles**
Transdiol
Transducers *(Transducer.*
  Transducing.
  Transduction.
  Transductional,
  Transductions)
Transect
Transeliminase
Transfection *(Transfecting)*
Transfer *(Transferability.*
  Transferable.
  Transference.
  Transferring, Transfers)
Transferases *(Transferase)*
Transferrin *(Transferrins)*
Transformation
  (Biotransformations.
  Transform.
  Transformable.
  Transformations.
  Transformed.
  Transforming.
  Transforms)
  X-Ref:
  Biotransformation,
  Nontransformed
Transfusion *(Transfusions)*
Transgenosis
Transglucosylamylase
Transhepatic see **Hepatic**
Transhydrogenase
Transhydroxymethylase

Transketolase
Translation *(Translational)*
Translocation *(Translocate.*
  Translocated.
  Translocatee.
  Translocating.
  Translocations)
Translucence *(Translucency.*
  Translucent)
Transmammary see
  Mammary
Transmembrane see
  Membranes
Transmethylases see
  Methyltransferases
Transmethylation
Transmission
  (Transmissability,
  Transmissibility.
  Transmissible.
  Transmissions.
  Transmissivity, Transmit.
  Transmits, Transmittable.
  Transmittal,
  Transmittance.
  Transmitters,
  Transmitting)
  X-Ref: Nontransmissible
Transosomes
Transovarial see **Ovary**
Transparency *(Transparent)*
Transpeptidase
Transpeptidation
Transphosphorylases
Transphosphorylation see
  Phosphorylation
Transpirants
  (Antitranspirant.
  Transpirant)
  X-Ref: Antitranspirants
Transpiration
  (Transpirational.
  Transpiring)
  X-Ref: Antitranspiration
Transplacental see **Placenta**
Transplants
  (Allotransplants.
  Autotransplanted.
  Autotransplants,
  Transplant,
  Transplantability.
  Transplantable.
  Transplantation.
  Transplantations,
  Transplanted.
  Transplanter.
  Transplanters,
  Transplanting)
  X-Ref:
  Allotransplantation,
  Autotransplantation,
  Retransplantation
Transpollination see
  Pollination
Transponder
Transportation *(Transport,*

Transportability,
Transportable,
Transportations,
Transported, Transporter,
Transporters,
Transporting, Transports)
Transposition
(Transpositions)
Transshipments *see*
 Shipping
Transsynaptic *see* **Synapsis**
Transtracheal *see* **Tracheal**
Transudation
Transuterine *see* **Uterus**
Transvaginal *see* **Vagina**
Transvascular *see* **Vascular**
Transverse *(Transversal)*
Tranylcypromine
Tranzschelia
Trapa
Trapaceae
Trapella
Trapeolum
Trapezonotus
Trappers *(Trapper)*
Traps *(Ovitrap, Trap,*
 Trapped, Trapping)
 X-Ref: Ovitraps
Trash *see* **Waste**
Trauma *(Traumas,*
 Traumatic, Traumatism,
 Traumatisms,
 Traumatized,
 Traumatogenic,
 Traumatological,
 Traumatology)
Traumatomutilla
Traumatotactic
Travel *see* **Tourism**
Travertine
Trays *(Tray)*
Tread *(Treading)*
Treadmill
Treatia
Trebouxia
Trechiama
Trechinae
Trechnites
Trechus
Treehoppers *(Treehopper)*
Trees *(Tree, Treeless)*
Treflan *see* **Trifluralin**
Trefoil *(Trefoils)*
Trehalase *(Trehalases)*
Trehalose
Trellises *(Trellis, Trellised,*
 Trellising)
Trema
Trematoda *(Trematode,*
 Trematodes)
Trematodiasis
Trematosphaeria
Tremella
Tremellaceae
Tremellales
Tremor *(Tremorgenic,*

Tremors)
Trenbolone
Trench *(Trenched,*
 Trenches, Trenching,
 Trenchless)
Trenomyces
Trentepohlia *(Trentepohlias)*
Trephine *(Trephining)*
Treponema
Tretamine
 X-Ref: TEM,
 Triethylenemelamine
Triacanthoside
Triacetate
Triacetelus
Triacetinase
Triacetylcellulose
Triacryloyl
Triacylglycerols
 (Triacylglycerol)
Triaenodes
Triaenophorus
Trialeurodes
Triallate
Triamcinolone
Triamide
Triamine
Trianea
Trianthema
Trianthophora
Triassic
Triassoblatta
Triatoma *(Triatomid)*
Triatominae *(Triatomine)*
Triazine *(Triazin, Triazines)*
Triazinobenzimidazole
Triazole *(Triazol, Triazoles)*
Triazoline
Triazophos
 X-Ref: Hostathion
Tribal *(Tribalism, Tribe,*
 Tribes)
Tribolium
Tribonema
Tribrissen
Tribromsalan
Tribulus
Tribunil *see*
 Methbenzthiazuron
Tributyrin
Tributyrinase
Tricalysia
Tricarbanilate
Tricarbonyl
Tricarboxylate
Tricarboxylic
Tricentrus
Triceratium
Tricetin
Trichaster
Trichiinae
Trichilia
Trichinella *(Trichina,*
 Trichinae)
Trichinoscope
 (Trichinoscopy)

Trichinosis *(Trichinelliasis,*
 Trichinellosis,
 Trichiniasis, Trichinous)
Trichlorfon *(Trichlorophon,*
 Trichlorphon)
 X-Ref: Chlorofos,
 Chlorophos, Dipterex,
 Foschlor, Neguvon
Trichloris
Trichloroacetate
Trichloroacetic
 (Trichloracetic)
Trichloroacetyl
Trichloroanisole
Trichlorobenzoic
Trichloroethane
Trichloroethylene
Trichlorofluoromethane
Trichloromethyl
Trichloromethylbenzyl
Trichloromethylthio
Trichloronate *(Trichloronat)*
 X-Ref: Phytosol
Trichlorophen
Trichlorophenol
Trichlorophenoxy
Trichlorophenoxyacetic
Trichlorophenyl
Trichloropicolinic
 (Trichloropycolinic)
Trichobezoars
Trichobilharzia
Trichobius
Trichoblasts
Trichobothria
Trichocarpus
Trichocephalus *see*
 Trichuris
Trichocera
Trichocereus
Trichoceridae
Trichocladius
Trichocladus
Trichocline
Trichocolea
Trichoconiella
Trichoconis
Trichocorixa
Trichocoronis
Trichodectidae
Trichodelitschia
Trichoderma
Trichodes
Trichodesma *(Trichodesmic)*
Trichodesmium
Trichodiene
Trichodorus
Trichoglossum
Trichoglossus
Trichoglottis
Trichogorytes
Trichogramma
Trichogrammatidae
Trichogrammatoidea
Trichohelea
Trichoid

Tricholaena
Tricholioproctia
Tricholipeurus
Tricholith
Tricholoma
Tricholomopsis
Trichomanes
Trichomes *(Trichoma,*
 Trichome)
Trichomma
Trichomonas *(Trichomonad)*
Trichomoniasis
Trichomycetes
Trichomycin
Trichonematidae
Trichoneura
Trichoniscidae
Trichoniscus
Trichophaea
Trichophagia
Trichophyton
Trichophytosis
Trichoplusia
Trichopoda
Trichopria
Trichoptera *(Trichopteran)*
Trichopterygidae
Trichosanthes
Trichoseptoria
Trichosia
Trichosphaeria
Trichospilus
Trichosporon
Trichostema
Trichostomum
Trichostrongylidae
 (Trichostrongylid)
Trichostrongyloidea
 (Trichostrongyloid)
Trichostrongylosis
Trichostrongylus
 (Trichostrongyle)
Trichosurus
Trichotanypus
Trichotaphe
Trichothecene
Trichothecin
Trichothecium
Trichotichnus
Trichouropodini
Trichrysis
Trichuriasis
 (Trichocephaliasis,
 Trichocephaliosis,
 Trichocephalosis)
Trichuridae
Trichuris
 X-Ref: Trichocephalus
Trichuroidea
Trichurus
Tricimba
Trickle
Triclisia
Tricorynus
Tricuspid
Tricyclazole

Tricyrtis
Tridacnin
Tridactylus
Tridax
Tridentoquinone
Tridentosmia
Trienbolone
Trienoic
Trientalis
Triesters *(Triester)*
Triethanolamine
Triethylamine
Triethylene
Triethylenemelamine *see*
   Tretamine
Trifenmorph *see*
   Tritylmorpholine
Triflubazam
Trifluoride
Trifluoroacetylation
   (Trifluoroacetylated)
Trifluoromethyl
Trifluralin *(Trifluraline)*
   X-Ref: Treflan
Trifoliastrum *see* Trigonella
Trifolirhizin
Trifolium
Triforine
Trigalactosyl
Trigeminal
Trigilletimine
Triglochin
Triglochinin
Triglyceride *(Triglycerides)*
Trigona
Trigonalidae *(Trigonalid)*
Trigonella
   X-Ref: Trifoliastrum
Trigonia
Trigonidium
Trigonotyliscus
Trigonotylus
Trigonurus
Trihabda
Trihydrate
Trihydrochloride
Triiodobenzoic
   X-Ref: TIBA
Triiodothyronine
Triisopropanolamine
Trilete
Trillium *(Trilliums)*
Trilobomyza
Triloboxylon
Trimedlure
Trimekain
Trimepranol
Trimer
Trimerotropis
Trimethoprim
Trimethoxyflavone
Trimethoxystyrene
Trimethylamine
Trimethylaminoxide
Trimethylammonium
Trimethylene

Trimethyllysine
Trimethylsilyl
   (Trimethylsilylated)
Trimmatostroma
Trimming *(Trimmer)*
Trinectes
Trinervitene
Trinervitermes
Trinitrotoluene
Trinucleate *see* Nucleus
Triodanis
Triodia
Triol
Triorganotins
Triorthocresyl
Triorthotolyl
Triose
Triosephosphate
Trioxide
Trioxygenated *see*
   Oxygenation
Trioza
Triparanol
Tripe
Tripeptides *(Tripeptide)*
Tripetaleia
Triphaena
Triphasia
Triphenyl
Triphenylphosphine
Triphenyltetrazolium
Triphenyltin
Triphora
Triphosphatase
   (Triphosphatases)
Triphosphate
   (Triphosphates)
Triphosphopyridine
Triplets *(Triplet)*
Tripleurospermum
Triplochiton
Triploids *(Triploid,*
   Triploidy)
Triplosporium
Tripogandra
Tripolyphosphate
Triposporium
Tripotassium *see* Potassium
Tripsacum
Tripterocalyx
Tripteroides
Tripterospora
Trirhabda
Trisaccharides
   (Trisaccharide)
Trisetacus
Trisetaria
Trisetum
Trishormomya
Trisodium
Trisome *(Trisomic,*
   Trisomics, Trisomy)
Trisopsis
Trispora
Trissocladius
Trissolcus

Trissoleus
Tristania
Tristeza
Tristicha
Tristria
Tristyly *(Tristylous)*
Trisulfide
Triteleia
Triterpenes *(Triterpene,*
   Triterpenic)
Triterpenoids *(Triterpenoid)*
Trithyreus
Triticale *(Triticales)*
Triticinae
Triticosecale
Triticum *(Triticums)*
Tritirachium
Tritium *(Tritiated,*
   Tritiation)
Tritomegas
Triton
Tritonia
   X-Ref: Montbretia
Tritrichomonas
Tritrichosiphum
Trituration
Tritylmorpholine
   X-Ref: Trifenmorph
Tritylpyridinium
Triumfetta
Trixagus
Trizygia
tRNA*see* RNA
Trochodendron
Trochosa
Trodax
Trofosfamide
Troginae
Troglodromus
Troglohyphantes
Troglops
Troglorhynchus
Troglotrematidae
Trogoderma
Trogositidae
Trogulidae
Trogus
Troides
Trolene *see* Ronnel
Trollius
Trollixanthin
Trombicula
Trombiculidae *(Chigger,*
   Trombiculid)
   X-Ref: Chiggers
Trombidiformes
   (Trombidiform)
Trombidiidae
Trombidium
Tromethamine
Tropaeolaceae
Tropaeolum
Tropane
Trophallaxy *(Trophallactic)*
Trophic
Trophobiosis

Trophoblast *(Trophoblastic)*
Trophocytes *(Adipocytes,*
   Trophocyte)
   X-Ref: Adipocyte
Trophyallaxy
Tropics *(Neotropic,*
   Neotropical, Tropic,
   Tropical)
   X-Ref: Intertropical,
   Neotropics
Tropidishia
Tropiduchidae
Tropine
Tropism *(Tropisms)*
   X-Ref: Biotropism
Tropisternus
Tropoelastin
Tropolone *(Tropolones)*
Tropomyosin
Troponin
Troposphere
Trossulus
Trotter *(Trotters, Trotting)*
Trouessartiinae
Troughs *(Trough)*
Trout *(Trouts)*
Trubidimetry
Trucks *(Truck, Trucking,*
   Truckloading)
   X-Ref: Lorry
Truffles *(Truffle)*
Trunk *(Trunks)*
Trunkwood
Truss *(Trusses, Trussing)*
Trynocoris
Trypanocidal
Trypanosoma
Trypanosomatidae
   (Trypanosomatid,
   Trypanosomatids)
Trypanosomes
   (Trypanosomal,
   Trypanosome)
Trypanosomiasis
   (Trypanosomiases)
Trypanozoon
Trypargilum
Trypetidae *see* Tephritidae
Tryphonoidae
Trypocalliphora
Trypodendron
Tryporyza
Trypoxylon
Trypsin *(Antitryptic,*
   Trypsinization, Trypsins,
   Tryptic)
   X-Ref: Antitrypsin
Trypsinogen
Tryptamine *(Tryptamines)*
Tryptophan *(Tryptophane)*
Tryptophanyl
Tryptophol
Tryptophyl
Tryptoquivaline
Tryptoquivalone
Tryptose

Tsetse
Tsigai
Tsuga *(Hemlocks)*
    X-Ref: Hemlock
Tsushimycin
Tubaic
Tube *(Tubal, Tubes,*
    Tubing, Tubular)
    X-Ref: Intratubal
Tuber *(Tuberous, Tubers)*
    X-Ref: Nontuber
Tubercles *(Tubercle,*
    Tuberculate)
Tubercularia
Tuberculariaceae
Tuberculina
Tuberculins *(Tuberculin,*
    Tuberculinic,
    Tuberculinization)
Tuberculolytic
    (Tuberculocidal)
Tuberculosis *(Tuberculous)*
    X-Ref: Antituberculous
Tuberiferine
Tuberization *(Tuberisation)*
Tuberolachnus
Tuberose
Tubewells *see* Wells
Tubiflorae
Tubocurarine
Tubules *(Microtubular,*
    Microtubule, Tubule)
    X-Ref: Antimicrotubule,
    Microtubules
Tubulifera *(Tubuliferan)*
Tubulin
Tubulosine
Tuckerella
Tufted
Tukerella
Tularemia
Tulipa *(Tulip, Tulips)*
Tulipalins
Tulipeae
Tuliposides
Tullbergia
Tulostoma
Tumbleweed
Tumefacient *see* Swelling
Tumidiscapus
Tumor *(Antitumour,*
    Nontumoral, Oncogenesis,
    Oncogenicity, Oncolytic,
    Tumoral, Tumorigenesis,
    Tumorigenic,
    Tumorization, Tumorous,
    Tumors, Tumour,
    Tumours)
    X-Ref: Antitumor,
    Intratumor,
    Nononcogenic,
    Nontumorigenic,
    Oncogenic, Oncologic
Tuna *(Tunas)*
    X-Ref: Tunny
Tundra *(Tundras)*

Tung
Tungro
Tungstate
Tungsten
Tunica
Tunicamycin
Tunneling *(Tunnel,*
    Tunnelling, Tunnels)
Tunny *see* Tuna
Tupelo
TUR *see* Cycocel
Turbellaria *(Turbellarian)*
Turbidimetry
    (Turbidimetric)
Turbidity *(Turbid)*
Turbine *(Turbines)*
Turbinicarpus
Turbulence *(Turbulent)*
Turf *(Turfgrass,*
    Turfgrasses, Turfing,
    Turfs)
Turgidity *(Turgid, Turgor)*
Turion
Turisynchron
Turkey *(Turkeys)*
    X-Ref: Meleagris
Turmeric
Turneraceae
Turnips *(Turnip)*
Turpentine *(Turpentining)*
Tursiops
Turtles *(Turtle)*
Turtula
Tussah
Tussilago
Tussock *(Tussocks)*
Tussockmoth
Tutin
Tutorial *see* Education
Tutu
Tween *(Tweens)*
Twigs *(Twig)*
Twilight
Twins *(Twin, Twinning)*
Tychius
Tydeidae *(Tydeid, Tydeids)*
Tydeus
Tylan *see* Tylosin
Tylenchida
Tylenchida *(Tylenchid)*
Tylenchoidea
Tylencholaimellus
Tylenchorhynchus
Tylenchulus
Tylenchus
Tylocerus
Tylopelta
Tylophora
Tylophorine
Tylopilus
Tylose *see* Methylcellulose
Tylosema
Tyloses *(Tylosis)*
Tylosin
    X-Ref: Tylan
Tymoviruses *(Tymovirus)*

Tympanites *see* Bloat
Tympanitis
Tympanophora
Tympanotomy
Tympanum *(Tympanal,*
    Tympanic)
Tympany
Typha
Typhaceae
Typhlitis
Typhloceras
Typhlocyba
Typhlocybinae
    (Typhlocybine)
Typhlocybini
Typhlodromus
Typhlogastrura
Typhoctinae
Typhoid
Typhoons *(Typhoon)*
Typhula
Typhus
    X-Ref: Antityphus
Typhylodromus
Typology
Tyramine
Tyres *see* Tires
Tyria
Tyroglyphidae *see* Acaridae
Tyroglyphus *see* Acarus
Tyrolichus
Tyrophagus
Tyrosinases *(Tyrosinase)*
Tyrosine
Tyrosol
Tyrosone
Tyrosyl
Tytthus
Tyzzer
Ubiquinone *(Ubiquinones)*
Udders *(Udder)*
Udo
UDP
Ulcers *(Ulcer, Ulcerated,*
    Ulceration, Ulcerative)
    X-Ref: Antiulcerogenic
Ulex
Ulidia
Ulidiidae
Ulmaceae
Ulmus *(Elm)*
    X-Ref: Elms
Ulna *(Ulnar)*
Uloboridae
Uloborus
Ulota
Ulothrix
Ulotrichales
Ultisols *(Ultisol)*
Ultracentrifuge *see*
    Centrifuge
Ultrafiltration
    (Ultrafiltrable,
    Ultrafiltrate,
    Ultrafiltrated)
Ultrafine *see* Fineness

Ultrahistochemical *see*
    Histochemistry
Ultralow
Ultramicroanalysis
Ultramicrotomy *see*
    Microtomy
Ultrared *see* Infrared
Ultrasonics *(Ultrasonic,*
    Ultrasonically,
    Ultrasonicated,
    Ultrasound, Ultrasounds)
Ultrasonography
Ultrasterilization
    (Ultrasterilized)
Ultrastructure
    (Ultrastructural,
    Ultrastructures)
Ultraviolet
Uluganine
Ulugbekia
ULV
Ulva
Ulvaceae
Ulvales
Ulvaline
Ulvaria
Umbel
Umbellales *see*
    Umbelliflorae
Umbelliferae *(Umbellifer,*
    Umbellifera,
    Umbelliferous,
    Umbellifers)
Umbelliferone
Umbelliflorae
    X-Ref: Umbellales
Umbellularia
Umbilical
Umbilicaria
Umbonia
Unanesthetized *see*
    Anesthesia
Unaspis
Unbaited *see* Baits
Unbarked *see* Bark
Unbleached *see* Bleaching
Unburned *see* Burning
Uncaria
Uncastrated *see* Castration
Unchelated *see* Chelates
Unciger
Unciliated *see* Cilia
Uncinia
Uncinula
Uncoating *see* Coating
Unconfined *see*
    Confinement
Unconjugated *see* Conjugate
Uncropped *see* Crops
Uncultivated
Uncured *see* Curing
Undecane
Undecanoic
Undecanol
Undecaprenol
Undecylenic

Undenatured *see*
 Denaturation
Undercrops
Underdeveloped
 (Underdevelopment)
Underdrainage *see* Drainage
Underfeeding *see*
 Malnutrition
Underflooding *see* Floods
Underfloor *see* Floors
Underground
Undergrowth
Undernourishment *see*
 Malnutrition
Undernutrition *see*
 Malnutrition
Underplanting
 (Underplanted)
Underplowing *see* Plowing
Underprivileged
Underseeding *(Underseeds)*
Undershrubs *see* Shrubs
Undersowing *see* Sowing
Understocked *see* Stock
Understory *(Understorey)*
Underweight *see* Weight
Underwing *see* Wing
Undiluted *see* Dilution
Undisturbed *see*
 Disturbance
Undrained *see* Drainage
Undried *see* Drying
Undulation *(Undulant,*
 Undulating)
Unemployment *see*
 Employment
Unesterified *see*
 Esterification
Unextracted *see* Extraction
Unfermentable *see*
 Fermentation
Unfertilized *see*
 Fertilization
Unflooded *see* Floods
Unforested *see* Forest
Unfrozen *see* Frozen
Ungerminated *see*
 Germination
Ungernia
Ungraded *see* Grading
Ungrafted *see* Grafting
Ungrazed *see* Grazing
Ungulates *(Ungulate)*
Unheated *see* Heat
Unhedged *see* Hedges
Unhulled *see* Hulling
Unhusked *see* Husks
Unhydrogenated *see*
 Hydrogenation
Unialgal *see* Algae
Unicellular
Unifloral *see* Flowers
Unifoliate *see* Foliage
Uninfected *see* Infection
Uninjured *see* Injury
Uninspected *see* Inspection

Uninsulated *see* Insulation
Uninucleate *see* Nucleus
Uninucleolate *see* Nucleolus
Uniola
Uniparental *see*
 Parthenogenesis
Unirradiated *see* Irradiation
Unirrigated *see* Irrigation
Unisexuality *see* Dioecism
Unlimed *see* Lime
Unloader *(Unloaders)*
Unloading *see* Loading
Unmalted *see* Malt
Unpacking *see* Packing
Unparasitized *see* Parasites
Unpeeled *see* Peels
Unploughed *see* Plowing
Unpoisoned *see* Poisoning
Unpollinated *see* Pollination
Unrestrained *see* Restraint
Unrestricted *see* Restriction
Unretted *see* Retting
Unripe *see* Ripening
Unsanitary *see*
 Contamination
Unsaponifiable *see*
 Saponification
Unsaturate *see* Saturation
Unseeded *see* Seeding
Unsonicated *see* Sonication
Unsprayed *see* Spraying
Unsterile *see* Sterile
Unsweetened *see* Sweet
Unthinned *see* Thinning
Untilled *see* Tillage
Untrained *see* Training
Unvaccinated *see*
 Vaccination
Unvegetated *see* Vegetation
Unvented *see* Ventilation
Unweaned *see* Weaning
Unwilted *see* Wilting
Unxia
Upholstering *(Upholstered,*
 Upholstery)
Upis
Uplands *(Upland)*
Upmilling *see* Milling
Upstream *see* Stream
Upwelling
Urachus
Uracil
Uranin
Uranium
Uranotaenia
Uranyl
Uraria
Urate
Urban *(Urbanism,*
 Urbanistic, Urbanite,
 Urbanites, Urbanization,
 Urbanized, Urbanizing)
Urbanus
Urd *see* Mung
Urea *(Ureas)*
 X-Ref: Carbamide

Ureaform
Ureaplasmas *(Ureaplasma)*
Urease
 X-Ref: Antiurease
Urecholine
Uredial
Uredinales
Uredo
Uredospores *(Urediospore,*
 Urediospores, Uredospore)
Ureides *(Ureid, Ureide)*
Ureidobenzoic
Ureidohexahydropyrimidine
Uremia *(Uraemia, Uremic)*
Urena
Urentius
Ureolysis *(Ureolytic)*
Ureter *(Ureteral, Uretero,*
 Ureters)
Urethane *(Urethan,*
 Urethanes)
Urethra *(Urethral)*
Urethritis
Urethrogram
Urethroplasty
Urethrostomy
Urginea
Urial
Uric
Uricase
Uricemia
Uricolysis
Uridine
Uridylic
Urinalysis
Urine *(Urinary, Urination,*
 Urines)
Urobilinogen
Uroblaniulus
Urocanase
Urocerus
Urochloa
Urocystis
Urogenital
Urography
Uroleucon
Urolithiasis
Urology *(Urologic,*
 Urological)
Uromacquartia
Uromelan
Uromenus
Uromyces
Uronema
Uronic
Uropedium
Urophlyctis
Urophora
Uroplata
Uropodidae *(Uropodid)*
Uropodina
Uropodinae
Uroporphyrinogen
Uropygi
Uropygial
Uroscelio

Urothripini
Ursene
Ursia
Ursinia
Ursolic
Ursus
Urtica
Urticaceae
Urticaria *(Urticarial)*
Urubumiris
Urushiol *(Urushiols)*
Usnea
Usneaceae
Usnic
Ussuriella
Ustilaginales
Ustilago
Utaperla
Uteroglobin
Uteroplacental
Uterotropic
Uterotubal
Uterovaginal
Uterus *(Uteri, Uterine,*
 Utero)
 X-Ref: Extrauterine,
 Intrauterine,
 Transuterine
Utetheisa
Utricularia
Uukuniemi
Uvaol
Uvaretin
Uvaria
Uvariodendron
Uveitis
Uzarigenin
Vacations *(Vacation)*
Vaccaria
Vaccina *see* Cowpox
Vaccination *(Postvaccinal,*
 Vaccinal, Vaccinate,
 Vaccinated, Vaccinating,
 Vaccinations, Vaccinator)
 X-Ref: Nonvaccinated,
 Postvaccination,
 Revaccination,
 Unvaccinated
Vaccines *(Vaccine)*
Vaccinia
Vaccinium
Vachanic
Vacsegoside
Vacuole *(Vacuolar,*
 Vacuolated, Vacuolation,
 Vacuoles, Vacuolization,
 Vacuolized)
Vacuum
Vadonia
Vadose
Vagina *(Intravaginal,*
 Vaginal, Vaginally)
 X-Ref: Intravaginally,
 Perivaginal,
 Transvaginal
Vaginitis

Vagotomy *(Vagotomized)*
Vagus *(Vagal)*
Vahliaceae
Vaillantine
Valanga
Valepotriates *(Valepotriate)*
Valerate
Valeriana
Valerianaceae
Valerianella
Valeric
Valexon *see* Phoxim
Valgus
Validamycin *(Validacin)*
Valinate
Valine
Valinomycin
Valley *(Valleys)*
Vallisneria
Vallota
Valonia
Valsa
Valsaria
Valuation *(Valuations)*
Valve *(Valval, Valves, Valvular)*
Vamidothion
Vampyrostenus
Vanadate
Vanadium
Vanaspati
Vanda *(Vandas)*
Vandalism *(Vandal)*
Vandeae *(Vandinae)*
Vandex
Vanessa
Vanilla
Vanillin
Vanilmandelic
Vapam *see* Carbathione
Vapona *see* Dichlorvos
Vapor *(Vaporisation, Vaporization, Vaporized, Vapors, Vapour, Vapourers, Vapourizing, Vapours)*
Vararia
Variance *(Covariances, Variances)*
   X-Ref: Covariance
Variegated *(Variegation)*
Varieties *(Varietal, Varietals, Variety)*
   X-Ref: Intervarietal, Intravarietal, Multivarietal
Variolaric
Varnish *(Varnishes, Varnishing)*
Varroa
Varuna
Vas
Vasates
Vasca
Vascular *(Microvasculature, Vascularisation,*

Vascularised, Vascularity, Vascularization, Vasculature, Vdsculiferous)
   X-Ref: Extravascular, Intravascular, Microvascular, Perivascular, Transvascular
Vasculitis
Vasectomy *(Vasectomies, Vasectomised, Vasectomized)*
Vaseline
Vasoactive *(Vasoactivity)*
Vasoconstriction *(Vasoconstrictor, Vasopressor)*
   X-Ref: Vasopressors
Vasodilation *(Vasodilator)*
Vasomotor *(Vasomotion)*
Vasoneurosis *see* Angioneurosis
Vasopressin
   X-Ref: Antivasopressin
Vasopressors *see* Vasoconstriction
Vasotocins *(Vasotocin)*
Vateria
Vatesus
Vats *(Vat)*
Vaucheria
Vaucheriaceae
Veal
Vealers *(Vealer)*
Vectorcardiography *(Vectorcardiogram)*
Vectors *(Vector, Vectored, Vectorial)*
   X-Ref: Intervector
Vegetable *(Vegetables)*
Vegetarians *(Vegetarian, Vegetarianism)*
Vegetation *(Vegetal, Vegetated, Vegetational, Vegetations)*
   X-Ref: Macrovegetation, Unvegetated
Vegetative *(Vegetating, Vegetatively)*
Vehicle *(Vehicles, Vehicular)*
Veillonella
Vein *(Veinal, Veined, Veins, Vena, Venipuncture, Venous)*
Veitchia
Vejovidae
Vejovis
Velamen
Velarifictorus
Veld
   X-Ref: Highveld, Sourveld
Veliidae
Vellozia

Velloziaceae
Velpar
Velvetbeans *(Velvetbean)*
Velvetleaf
Venalstonine
Venatiolaspis
Venation *(Venations)*
Vendex
Vending
Vendor
Veneers *(Veneer, Veneered, Veneering)*
Venereal
Venison
Venography
Venom *(Venomous, Venoms)*
Ventilago
Ventilation *(Ventilate, Ventilated, Ventilating, Ventilator, Ventilators, Ventilatory)*
   X-Ref: Unvented
Ventricles *(Ventricle, Ventricular)*
   X-Ref: Interventricular, Intraventricular
Ventriculitis
Ventriculus
Venturia
Venzar *see* Lenacil
Vepris
Verachthonius
Veraguensin
Veralodisine
Verapamil
Veratreae
Veratrilla
Veratrine *(Veratrinic)*
Veratrum
Verbascose
Verbascum
Verbena
Verbenaceae *(Verbenaceous)*
Verbenol
Verbenone
Verbesina
Verindal
Vermicelli
Vermicular
Vermiculite *(Vermiculites, Vermiculitic)*
Vermifugation
Vermin
   X-Ref: Antivermal
Verminosis *(Verminoses, Verminous)*
Vernalization *(Vernalisation, Vernalized)*
Vernodesmine
Vernoflexuoside
Vernolate
Vernolepin
Vernolic
Vernonia

Veromessor
Veronaea
Veronica
Veronicastrum
Verrallia
Verruculotoxin
Versiol
Vertabopus
Vertaline
Vertebrae *(Vertebra, Vertebral)*
   X-Ref: Intervertebral
Vertebrates *(Vertebrate)*
Verticicladiella
Verticillium
Vertisols *(Vertisol)*
Vervet
Vesicle *(Vesical, Vesicles, Vesicula, Vesicular, Vesiculated, Vesiculation)*
Vesiculectomy
Vespa
Vespertilionidae
Vesperus
Vespidae *(Vespid)*
Vespoidea
Vespula
Vessel *(Vessels)*
Vestibular
Vetadex
Vetalar
Vetch *(Vetches)*
Veterinarians *(Veterinarian, Veterinaries)*
Veterinary
Vetiver
Vetiveria
Vezdaea
Viability *(Viabilities, Viable)*
   X-Ref: Inviability
Vibrations *(Vibrating, Vibration, Vibrational, Vibratory)*
Vibrators *(Vibrator)*
Vibrio *(Vibrion, Vibrionic, Vibrions, Vibrios)*
Vibriosis
Viburnum *(Viburnums)*
Vicia
Vicilin
Victorin
Vicuna
Videin
Video
Vigna
Villa
Villages *(Village, Villagers)*
   X-Ref: Previllage
Villarsia
Villi *(Villus)*
   X-Ref: Microvilli
Vinasse
Vinblastine
   X-Ref: Vincaleukoblastine
Vinca

Vincadifformine
Vincaleukoblastine *see*
  Vinblastine
Vincamine
Vincanidine
Vincanine
Vincathicine
Vincetoxicum
Vincovaline
Vincristine
Vindoline
Vinegar *(Vinegars)*
Vineridine
Vinerine
Vines *(Vine)*
Vineyards *(Vineyard,*
  Vineyardist)
Viniculture *see* Viticulture
Vinification *(Vinifications)*
  X-Ref: Autovinificators
Vinyl
Vinyldiethylphosphate
Vinylpyridines
Vinylpyrrolidone
Vinylsulfonic
Viola
Violaceae
Violaxanthin
Violets *(Violet)*
Viologen
Virachola
Viremia *(Viraemia)*
Virescent *(Virescence)*
Virgin *(Virgins)*
Virginiamycin
Viricide
Viridicatin
Virin
Viroids *(Viroid)*
Virola
Virologists *(Virologist)*
Virology *(Virologic,*
  Virological)
Virulence *(Avirulent,*
  Virulences, Virulent)
  X-Ref: Antivirulent,
    Avirulence
Viruliferous *see* Infection
Viruses *(Antivirus, Viral,*
  Virion, Virions, Virogenic,
  Virose, Viroses, Virosis,
  Virus, Viruslike,
  Virusological)
  X-Ref: Antiviral,
    Provirus, Subviral
Viscaceae
Viscaria
Visceral *(Viscera)*
Viscoelasticity
Viscometry *(Viscometer,*
  Viscometers, Viscometric)
Viscosity *(Viscose,*
  Viscosities, Viscous)
  X-Ref: Hyperviscosity,
    Microviscosity
Viscum

Visna
Visnadin
Visual *(Visibility, Visible,*
  Vision, Visions,
  Visualized, Visually)
Vitaceae
Vitalboside
Vitamin *(Vitamine,*
  Vitaminized,
  Vitaminology, Vitamins)
  X-Ref: Megavitamin,
    Multivitamin,
    Neovitamin,
    Provitamin
Vitavax *see* Carboxin
Vitelline *(Vitellin)*
  X-Ref: Perivitelline,
    Previtelline
Vitellogenesis *(Vitellogenic)*
Vitellogenin *(Vitellogenins)*
Vitellus
Vitex
Vitexin
Viticosterone
Viticulture *(Vinicultural,*
  Viticultural)
  X-Ref: Viniculture
Vitis
Vitrandepts
Vitro
Vittadinia
Vitula
Viviparous *(Viviparity,*
  Vivipary)
Vivo
Vizella
Voacanga
Voacangine
Voaketone
Voandzeia
Vobasine
Vocal *(Vocalisation,*
  Vocalization,
  Vocalizations, Voice)
Vocations *(Vocation,*
  Vocational)
Vochysia
Vogeloside
Vogtia
Volatile *(Volatiles,*
  Volatilisation, Volatility,
  Volatilization, Volatilized)
  X-Ref: Nonvolatile
Volcanoes *(Volcanic,*
  Volcano)
Voles *(Vole)*
Voltage *(Volt, Volts)*
Voltammetry
Voltinism *(Voltine)*
  X-Ref: Bivoltine
Volubilinin
Volumetry *(Volumetric)*
Volutaphis
Volutella
Volvaria
Volvariella

Volvocaceae
Volvocales
Volvox
Vomifoliol
Vomiting
Vorlex
Vortex
Vriesea
Vulcanization *(Vulcanisate,*
  Vulcanisates,
  Vulcanisation,
  Vulcanizates, Vulcanized,
  Vulcanizing)
Vulgarone
Vulpes
Vulpia
Vultures *(Vulture)*
Vulva *(Vulval, Vulvar,*
  Vulvo)
  X-Ref: Bivulvarity
Vulvovaginitis
Vydate
Wachtliella
Wafer *(Wafered, Wafering,*
  Wafers)
Waffles
Wages *(Wage)*
Wagons *(Wagon)*
Walking
Wall *(Walled, Walls)*
Wallaby
Wallflower
Wallichoside
Wallrothiella
Walnuts *(Walnut)*
Walshiidae
Waltheria
Wankeiella
Warble
Warblefly
Warblers *(Warbler)*
Wardenia
Wardomyces
Warehouses *(Warehouse,*
  Warehousing)
Warfarin
Warmth *(Warm, Warmed,*
  Warmer, Warming)
Warts *(Wart)*
Washers *(Washer)*
Washing *(Wash,*
  Washability, Washable,
  Washed, Washes,
  Washings)
Washout
Wasmannia
Wasps *(Wasp)*
Waste *(Wastage, Wasted,*
  Wasteflow, Wasteful,
  Wastes, Wasting)
  X-Ref: Trash
Wastelands *(Wasteland)*
Wastepaper
Wastewaters *(Wastewater)*
Watchdogs *see* Dogs
Water *(Waterborne, Waters,*

  Watery)
  X-Ref: Meltwater
Waterchestnuts
Watercourses *(Watercourse)*
Watercress
Waterculture *see*
  Hydroponics
Waterers *(Waterer)*
Waterfowl
Waterhyacinths
  (Waterhyacinth)
Watering *(Watered)*
Waterlily
Waterlogging *(Waterlogged)*
Watermelons *(Watermelon)*
Waterpower
Waterproofing *(Waterproof)*
Watershed *(Watersheds)*
Waterways *(Waterway)*
Waterweeds
Watophilus
Wattle *(Wattles)*
Wavelength *(Wavelengths)*
Waves *(Wave)*
Wax *(Waxed, Waxes,*
  Waxing, Waxless, Waxy)
  X-Ref: Nonwaxy
Waxmoth
Weaning *(Wean, Weaned,*
  Weaner, Weaners,
  Weanling)
  X-Ref: Ablactation,
    Postweaning,
    Preweaning, Unweaned
Wear *(Wearing)*
Weasels *(Weasel)*
Weather
Weathering *(Weathered)*
Weaving *(Weave, Woven)*
  X-Ref: Nonwoven,
    Wovens
Web *(Webs)*
Webworms *(Webworm)*
Wedelia
Wedge *(Wedges)*
Weed *(Weeded, Weediest,*
  Weediness, Weeding,
  Weeds, Weedy)
Weedazol *see* Amitrole
Weeders *(Weeder)*
Weedicides *see* Herbicides
Weedkillers *see* Herbicides
Weevils *(Bollweevil, Weevil)*
  X-Ref: Bollweevils
Weigela
Weight *(Weighable,*
  Weighed, Weighing,
  Weighted, Weighting,
  Weightings,
  Weightlessness, Weights)
  X-Ref: Underweight
Weinmannia
Weirs *(Weir)*
Weissia
Welchia
Welding *(Weld, Welder,*

Welders, Welds)
Welfare
Wellingtonia
Wells *(Well)*
  X-Ref: Boreholes,
    Tubewells
Welwitschia
Wesmaelia
Wesmaelius
Westiellopsis
Wet *(Rewetting, Wetness,
  Wettability, Wettable,
  Wetting)*
  X-Ref: Nonwettability,
    Rewetted
Wetas *(Weta)*
Wethers *(Wether)*
Wetlands *see* Marshes
Wettinia
Wetwood
Whales *(Whale)*
Wheat *(Wheats)*
Wheatbelt
Wheatgerm
Wheatgrass *(Wheatgrasses)*
Wheatings
Wheatmeal
Wheel *(Wheeled, Wheels)*
Whelk
Whelps *(Whelp, Whelping)*
Whetzelinia
Whey *(Wheys)*
Whippet
Whipscorpion
Whipworms
Whirligig
Whirling *(Whirl, Whirls)*
Whirlwinds
Whiskeys *(Whiskey,
  Whiskies, Whisky)*
Whitefish *(Whitefishes)*
Whiteflies *(Whitefly)*
Whiting *(Whitings)*
Whitlow
Wholesale *(Wholesalers,
  Wholesales, Wholesaling)*
Whortleberries
  (Whortleberry)
Wiborgia
Wicks *(Wick)*
Wieners *(Wiener)*
Wife *see* Women
Wikstroemia
Wilcoxia
Wild
  X-Ref: Semiwild
Wilderness *(Wildernesses)*
Wildfires *see* Fire
Wildflowers *(Wildflower)*
Wildfoods
Wildfowl
Wildlands *(Wildland)*
Wildlife
Wilhelmia
Willdenowia
Williamsonia

Willows *see* Salix
Wills
Wilt *(Wilts)*
Wilting *(Wilted)*
  X-Ref: Unwilted
Winch *(Winching)*
Wind *(Winds, Windward)*
Windbarriers *see*
  Windbreaks
Windborne
Windbreaks *(Windbreak,
  Windbreaking)*
  X-Ref: Windbarriers
Windfalls *(Windfall)*
Windmill *(Windmills)*
Window *(Windowed,
  Windowless, Windows)*
Windrow *(Windrowed,
  Windrowers, Windrowing)*
Windstorms
Windswept
Windthrow *(Windthrown)*
Wine *(Winemaking, Wines)*
Winegrape *see* Grapes
Wineries *(Winery)*
Wing *(Wingbeat, Winged,
  Wingless, Wings)*
  X-Ref: Apterous,
    Underwing
Winter *(Overwintered,
  Overwintering, Wintered,
  Wintering, Winters,
  Wintertime)*
  X-Ref: Overwinter
Winteraceae
Winterberry
Winterfat
Wintergrazing *see* Grazing
Wintergreen
Winterhardiness
  (Winterhardy)
Winteria
Winterizing *(Winterization,
  Winterized)*
Winthemia
Wire *(Wired, Wires,
  Wiring)*
Wiregrass
Wireworms *(Wireworm)*
Wisanine
Wiseana
Wisteria *(Wistaria)*
Witches
Witchgrass
Witchweed
Withaferin
Withania
Withanolides *(Withanolide,
  Withanolids)*
Withering *(Wither,
  Withered)*
Withers
Witloof *see* Endive
Wives *see* Women
Wohlfahrtia
Wolbachiae

Wolffia
Wollastonite
Wolverine
Wolves *(Wolf)*
Women *(Woman)*
  X-Ref: Wife, Wives
Wood *(Wooden, Woodiness,
  Woody)*
Woodchips *(Woodchip,
  Woodchipping)*
Woodcock
Woodland *(Wooded,
  Woodlands, Woods)*
Woodlice *(Woodlouse)*
Woodlot *(Woodlots)*
Woodpeckers *(Woodpecker)*
Woodruff
Woodsia
Woodsorrel
Woodwardia
Woodwasps
Woodwork *(Woodworking)*
Woodworm
Woodyard
Wool *(Wooled, Woolen,
  Woolens, Woolgrower,
  Woolly, Wools, Wooly)*
Work *(Worked, Worker,
  Workers, Workforce,
  Working, Workman,
  Workmen)*
Workshops *(Workshop)*
Wormer *(Worming)*
Worms *(Worm, Wormlike)*
Wormwood
Worsted
Worts *(Wort)*
Wound *(Wounded,
  Wounding, Wounds)*
Wovens *see* Weaving
Wrap *see* Packing
Wrens *(Wren)*
Wrightia
Wrinkle *(Wrinkled,
  Wrinkledness, Wrinkling)*
Wuchereria
Wurmbaeoideae
Wurst *see* Sausage
Wyeomyia
Wyerone
Xanthan
Xanthates *(Xanthate,
  Xanthation)*
Xanthene
Xanthine *(Xanthines)*
Xanthinin
Xanthins *(Xanthin)*
Xanthinuria *(Xanthuria)*
Xanthium
Xanthocanace
Xanthocercis
Xanthochymol
Xanthocillin
Xanthogenate
Xanthogramma
Xantholinus

Xanthomegnin
Xanthommatin
Xanthomonas
  X-Ref: Zanthomonas
Xanthones *(Xanthone)*
Xanthophyceae
Xanthophylls *(Xanthophyll)*
Xanthophyllum
Xanthophyta
Xanthopimpla
Xanthoria
Xanthorrhiza
Xanthorrhoea
Xanthosine
Xanthosoma
Xanthotoxin
  X-Ref: Methoxsalen
Xanthoxin
Xanthyletin
Xenia
Xenillidae *see* Carabodidae
Xenillus
Xenobesuchetia
Xenobiotics *(Xenobiotic)*
Xenocastor
Xenochila
Xenodromius
Xenografting *see*
  Heterograft
Xenoleleupia
Xenon
Xenopacarus
Xenopsylla
Xenopus
Xenorhiza
Xenoryctes
Xenos
Xenotypa
Xenylla
Xeric
Xerodraba
Xerography
Xerophilic *(Xerophilous)*
Xerophthalmia
Xerophytes *(Xerophyte,
  Xerophytic)*
Xerosis
Xerostictia
Xerotherm *(Xerothermic,
  Xerothermophylic)*
Xestoblatta
Ximenia
Xiphidium
Xiphidorus
Xiphinema
Xiphium
Xipholimnobia
Xiphosomella
Xiphozele
Xiphydriidae *(Xiphydriid)*
Xrays *see* Roentgen
Xubida
Xyela
Xyelidae
Xylanases *(Xylanase)*
Xylans *(Xylan)*

Xylaplothrips
Xylaria
Xylariaceae *(Xylariaceous)*
Xylastodoris
Xylazine *(Xylazin)*
Xyleborus
Xylem *(Metaxylem)*
Xylene *see* Xylol
Xyletinus
Xyleutes
Xyligen
Xylitol
Xylobolus
Xylobotryum
Xylocarpus
Xylocopa
Xylocopinae
Xylocoris
Xyloctonus
Xylol
    X-Ref: Xylene
Xylonic
Xylophaga
Xylophagidae
Xylophagous
Xylophilous
Xylopia
Xyloporosis
Xylopyranoside
Xyloryctidae
Xylosandrus
Xylose
Xylosidase
Xylosma
Xylosone
Xylosteini
Xylosteus
Xylota
Xyloterinus
Xyloterus
Xylotomy *(Xylotomic)*
Xylotrechus
Xylotrupes
Xyridaceae
Xyris
Xysmalobium
Xysticus
Xystromutilla
Yabea
Yaks *(Yak)*
Yalan *see* Molinate
Yamatocallis
Yambeans
Yamogenin
Yams *(Yam)*
Yarns *(Yarn)*
Yarrow
Yaupon *see* Ilex
Yearbooks *(Yearbook)*
Yearlings *(Yearling)*
Yeast *(Yeastlike, Yeasts)*
Yellowing
Yellows
Yenhusomidine
Yenhusomine
Yersinia

Yews *see* Taxus
Yield *(Yielded, Yielding,*
    Yieldings, Yields)
Yllenus
Yoghurt *(Yoghourt,*
    Yoghourts, Yoghurts,
    Yogourt, Yogurt, Yogurts)
Yolks *(Yolk, Yolked)*
    X-Ref: Intravitelline
Young *(Younger, Youngest)*
Youth *(Youths)*
Yponomeuta
Yponomeutidae
    X-Ref: Argyresthiidae
Yttrium
Yucca
Yuccoside
    X-Ref: Protoyuccoside
Yuma
Zaboba
Zabrus
Zacerata
Zacompsia
Zadiprion
Zagrammosoma
Zalerion
Zaluzianskya
Zamia
Zamora
Zanclognatha
Zanders *(Zander)*
Zannichelliaceae
Zantedeschia
Zanthomonas *see*
    Xanthomonas
Zanthoxylum
Zapota *see* Sapodilla
Zatropis
Zea *see* Maize
Zearalanol *see* Zeranol
Zearalenone
Zeatin
Zeaxanthin
Zeazine
Zebra
Zebrina
Zebus *(Zebu)*
Zectran *see* Mexacarbate
Zedoaria *see* Curcuma
Zeia *see* Agropyron
Zeilites *(Zeolite)*
Zein
Zeiraphera
Zelandopsocus
Zelandopsyche
Zele
Zeleny
Zelkova
Zelleromyces
Zelotes
Zelus
Zen
Zenobiinae
Zeolites *(Zeolitic)*
Zephyrantheae
Zephyranthes

Zeranol *(Zearalenol)*
    X-Ref: Ralgro,
    Zearalanol
Zercon
Zerconidae
Zerlate *see* Ziram
Zerumbone
Zerynthia
Zeta
Zethus
Zetzellia
Zeuzera
Zeuzeridae
Zeyheria *(Zeyhera)*
Zeyherol
Zieria
Zignoella
Zinc
Zincoxen
Zineb *(Mancozeb)*
    X-Ref: Cynkotox,
    Perocin
Zingeria
Zingerone
Zingiber
Zingiberaceae
Zingiberales
Zinnia *(Zinnias)*
Ziram
    X-Ref: Zerlate
Zirconium
Zizania
Ziziphus *(Zizyphus)*
Zlotniki
Zoalene
Zodion
Zoellnerallium
Zolone
Zonabris
Zonaroic
Zones *(Zone)*
Zoning
Zonocerus
Zonosemata
Zoocenoses *(Zoocenological,*
    Zoocenosis)
Zoogeography
    (Zoogeographic,
    Zoogeographical)
Zoohygiene *(Zoohygienic,*
    Zoosanitation)
    X-Ref: Zoosanitary
Zoologists *(Zoologist)*
Zoology *(Zoologic,*
    Zoological)
Zoonoses *(Zoonosis,*
    Zoonotic)
Zooplankton *(Zooplanktons)*
Zooprophylaxis
    (Zooprophylactic)
Zoos *(Zoo)*
Zoosanitary *see* Zoohygiene
Zoospores *(Zoosporangia,*
    Zoosporangium, Zoospore,
    Zoosporic,
    Zoosporogenesis)

Zootechnics *(Zootechnic,*
    Zootechnical,
    Zootechnicians,
    Zootechny)
Zootermopsis
Zooxanthella
Zopfia
Zopfiella
Zophobas
Zophodia
Zoraptera
Zorius
Zorotypus
Zostera
Zosteraceae
Zoxazolamine
Zoysia *(Zoisia)*
Zucchini
Zygaena
Zygaenidae
Zygiella
Zyginidia
Zygnema
Zygnemataceae
    (Zygnemataceous)
Zygnematales
Zygoclistron
Zygogynum
Zygomatic
Zygomorphic
Zygomycetes
Zygomycosis
Zygopetalum
Zygophyllaceae
Zygophyllum
Zygoptera *(Zygopteran)*
Zygorhizidium
Zygorhynchus
Zygosaccaromyces
Zygospermella
Zygospores *(Zygospore)*
Zygosporium
Zygote *(Monozygous,*
    Zygosity, Zygotes,
    Zygotic)
    X-Ref: Monozygotic
Zygowillia
Zymogen *(Zymogens)*
    X-Ref: Proenzyme
Zymography *(Zymogram,*
    Zymograms)
Zymolyase
Zymosan